W0081210

Edible Nanostructures

Edible Nanostructures

Edited by

Alejandro G Marangoni
University of Guelph, Guelph, Ontario, Canada
Email: amarango@uoguelph.ca

David Pink
St Francis Xavier University, Antigonish, Nova Scotia, Canada
Email: dpink@stfx.ca

THE QUEEN'S AWARDS
FOR ENTERPRISE:
INTERNATIONAL TRADE
2013

Print ISBN: 978-1-84973-895-8

A catalogue record for this book is available from the British Library

© The Royal Society of Chemistry 2015

All rights reserved

Apart from fair dealing for the purposes of research for non-commercial purposes or for private study, criticism or review, as permitted under the Copyright, Designs and Patents Act 1988 and the Copyright and Related Rights Regulations 2003, this publication may not be reproduced, stored or transmitted, in any form or by any means, without the prior permission in writing of The Royal Society of Chemistry or the copyright owner, or in the case of reproduction in accordance with the terms of licences issued by the Copyright Licensing Agency in the UK, or in accordance with the terms of the licences issued by the appropriate Reproduction Rights Organization outside the UK. Enquiries concerning reproduction outside the terms stated here should be sent to The Royal Society of Chemistry at the address printed on this page.

The RSC is not responsible for individual opinions expressed in this work.

The authors have sought to locate owners of all reproduced material not in their own possession and trust that no copyrights have been inadvertently infringed.

Published by The Royal Society of Chemistry,
Thomas Graham House, Science Park, Milton Road,
Cambridge CB4 0WF, UK

Registered Charity Number 207890

Visit our website at www.rsc.org/books

Printed and bound by CPI Group (UK) Ltd, Croydon, CR0 4YY

Preface

At first sight, the idea of engaging in a field where the basis of the physical structures involves many components, ranging in size from molecules to macroscale assemblies and leading to extreme complexity, might cause one to simply opt out – until one realizes that biologists do this every day. The key to the enormous success in such fields is the systematic reduction in the number of uncontrollable parameters. One might then draw a correspondence between the complex fields of biology and food science, but one could not be further from the truth. The science of biological systems, if one includes medicine, is barely 2000 years old. Modern biology goes back a few hundred years. The study, and understanding, of food and its preparation likely goes back tens and probably hundreds of *centuries*. But the scientific study of the fundamental components of food goes back less than a century. It is only 20 years since Athene Donald wrote her article about the physics of food [A. M. Donald, *Rep. Prog. Phys.*, 1994, 57, 1081–1135] – long after the study of the physics of biological systems had become respectable – but, already, food science was moving into a new era. In the twentieth century, mass production of processed food came into its own with little regard for the long-term health of the consumers. It is only when human health became of central importance to society as a whole that the components of processed foods became of concern. What is in it and how it can be made healthier without sacrificing any of what is enjoyed became, and remains, a challenge. But this has opened a doorway to research and applications that have

Edible Nanostructures
Edited by Alejandro G Marangoni and David Pink
© The Royal Society of Chemistry 2015
Published by the Royal Society of Chemistry, www.rsc.org

driven the adoption of new techniques and has brought food science to the outskirts of the modern research paradigm: collaborative work with scientists outside the field.

With this development has come another one: the journey from the large-scale to the small-scale. Not the "small-scale" of reducing all phenomena to non-interlocking components, but the appreciation of how the small-scale components interlock to give rise to the larger, more complex static and dynamic structures. The latter can go by many names: cooperative phenomena, spontaneous self-assembly and long-range order are only three aspects of the creation of the large-scale from the small-scale. But, and this is the key point, understanding the small-scale units enables one to manipulate the large-scale structures – it enables one to create new large-scale structures.

These are two aspects of modern food science research: collaborative work with other disparate disciplines and characterizing the small-scale structures to enable the manipulation and creation of larger-scale structures.

These are reasons why this book is timely. An understanding of the small-scale structures and their transition *via* aggregation, or some other process, into large-scale structures using new techniques is the wave of the present. These encompass not-so-new techniques such as atomic-scale microscopy and confocal laser scanning microscopy, and newer approaches *via* synchrotron X-ray scattering and neutron scattering, to relate the nanoscale to the mesoscale. The recognition that food research is part of research into soft matter is a step in the right direction [*Faraday Discussions*, 2012, **158**]. But it is *edible* soft condensed matter. The implications are not trivial: in the end the products must be edible. This puts a severe constraint on what substances can be, ultimately, used. Finally, amongst the new techniques is computer simulation. Although mathematical models, such as the "mean field" Avrami model, have been in use for many years, the use of computer simulation to understand equilibrium and non-equilibrium characteristics of systems is relatively new to food science. Thus, to study nanoscale structures one must likely go beyond "mean field" models and address, for example, the problems of recognizing the importance of "bound" water and oil molecular layers in molecular conformational and spatial reorganization. It is likely that such new directions will expand beyond university research laboratories into industry.

Accordingly, this book should find importance as a textbook for students of this developing field of edible soft condensed matter, both in graduate schools and at advanced undergraduate levels. It should also provide current researchers with jumping-off points to further their own work. It is likely to be the first unifying step towards the new field of edible soft condensed matter.

The editors are grateful for the efforts of Ms Judy A. Campbell, whose substantial and excellent work on copy editing has enable the book to be completed. They also express their thanks to the numerous funding agencies, especially the Natural Sciences and Engineering Research Council, for the support of this important field.

David A. Pink,
Antigonish NS Canada

Contents

Edible Nanostructures
Edited by Alejandro G Marangoni and David Pink
© The Royal Society of Chemistry 2015
Published by the Royal Society of Chemistry, www.rsc.org

CHAPTER 1

Edible Nanostructures: Introduction

ALEJANDRO G. MARANGONI* AND DAVID A. PINK*

Center for Food and Soft Materials Science, Biophysics Interdepartmental
Group, Guelph-Waterloo Physics Institute, Dept. of Food Science,
University of Guelph, Guelph, ON N1G2W1, Canada
*Email: amarango@uoguelph.ca; dpink@stfx.ca

Interest in soft materials has experienced tremendous growth in the
past decade. This stems from the fact that these materials include
many important industrial products, such as plastics, foods and lubricants. This study is particularly relevant today because of the need
to find more green, sustainable and environmentally friendly inputs
for manufacturing. The formation of such materials is contingent on
the ability of relatively simple molecules to self-assemble into highly
complex structures. Many foods are edible soft materials and their
study can make use of the myriad of experimental techniques developed over the last century for characterizing such materials. One
need not constrain oneself by the requirement to study food only in
the "physically relevant" regime. Insight into material structures and
dynamics frequently requires information from outside the temperature or pressure ranges of direct applicability, and the study of
foods need not be bound by constraints. Current efforts geared
towards the improvement in the nutritional quality of food products
and security of food systems requires a materials science level approach to food structuring to minimize unhealthy components and

Edible Nanostructures
Edited by Alejandro G Marangoni and David Pink
© The Royal Society of Chemistry 2015
Published by the Royal Society of Chemistry, www.rsc.org

deliver nutritionally functional compounds, as well as to decrease cost and increase availability.

The process of creating a supramolecular structure does not simply entail finding a thermodynamic global free energy minimum. Kinetic factors play a key role in determining which local free energy minima are chosen along the formation pathway. Environmental effects become even more important (than molecular effects) beyond the microscopic world, where heat and mass transfer effects will strongly influence the formation of nanostructures, microstructures, and eventually a network (Figure 1.1).

Several research groups have now observed that several food soft materials of industrial interest, such as triglyceride (fat) crystal networks, display such structural hierarchy, and that the elasticity of such materials is dependent on the interactions between clusters of nanocrystals (Figure 1.2).[1,2]

Furthermore, the final size and morphology of these nanostructures and microstructures is strongly influenced by heat and mass transfer phenomena. For example, when triglycerides are cooled from the melt to a temperature below their melting point, *i.e.*, when they are supercooled, they undergo a liquid–solid transformation to form crystalline nanoplatelets with characteristic polymorphism and polytypism.[3,4] These primary nanocrystals aggregate, or grow into each other, to form clusters, which further interact, resulting in the formation of a continuous three-dimensional network (Figure 1.3).[5,6]

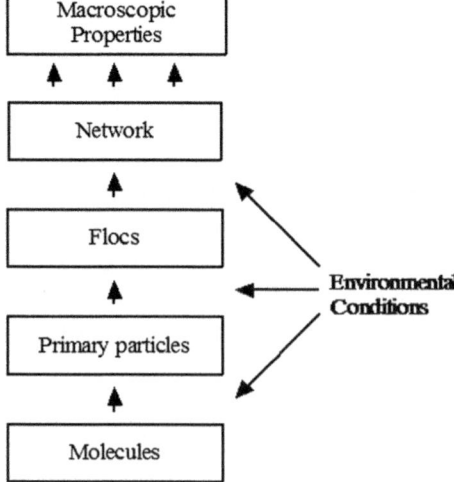

Figure 1.1 Structural hierarchy in many food soft materials.

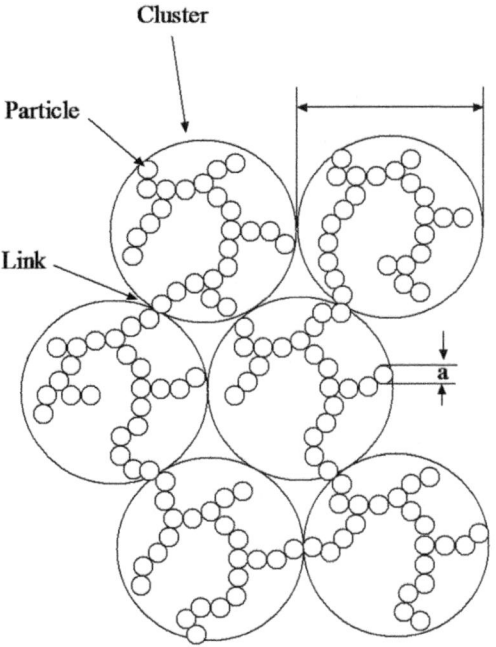

Figure 1.2 Idealized structure of a nanostructured colloidal network.

The macroscopic properties of such systems are somehow affected by all these levels of structure, as well as external fields, which should all be taken into account when predicting and engineering material functionality.

In order to truly understand, and eventually predict, the macroscopic properties of food, it is necessary to characterize and define the different levels of structure present in the material and their respective relationship to a macroscopic property. In our experience, the macroscopic properties of many soft materials cannot always be directly related to molecular structure. Rather than always invoking "molecular interpretations" to explain the macroscopic properties of materials, relationships between the appropriate level of structure and macroscopic properties should be sought instead. Knowledge of the relationships between molecular composition and phase behavior, solid state structure, growth mode, static structure, dynamic structure, and macroscopic properties will eventually allow for the rational design of specific macroscopic properties.

Discovery and knowledge – the unification of data accumulation and understanding, thereby giving rise to a single overarching picture of the world – form one of the basic aspects of the scientific paradigm.

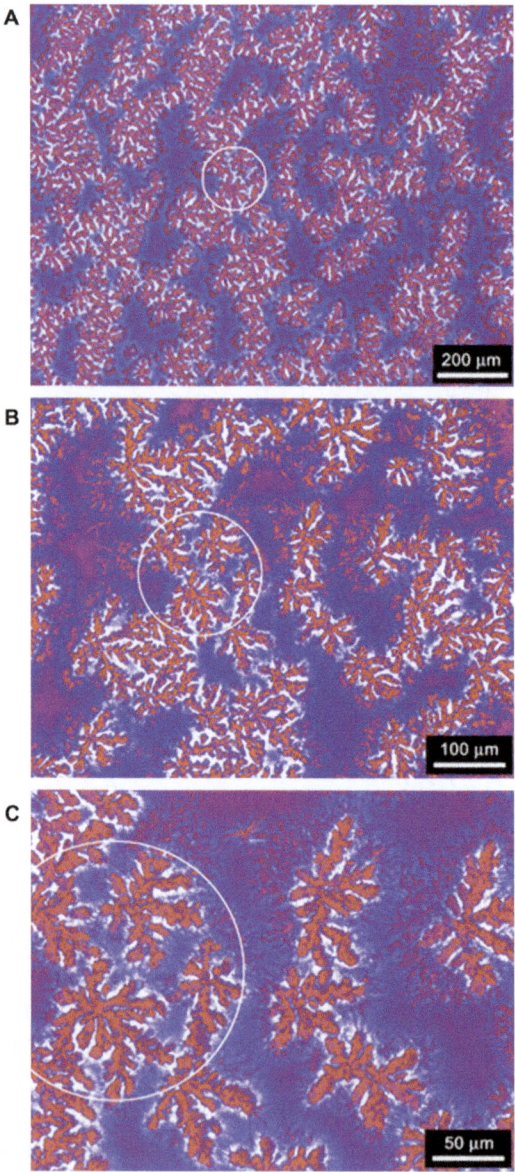

Figure 1.3 Duotone polarized light–phase-contrast micrograph (50% overlay) of a crystal network of the high melting fraction of milk fat in triolein.

But there is also a less-appreciated journey: the journey from the large-scale to the small-scale. Not the "small-scale" of reducing all phenomena to non-interlocking components, but the appreciation of how the small-scale components interlock to give rise to the larger, more complex static and dynamic structures. The latter can go by many names: cooperative phenomena, spontaneous self-assembly and long-range order are only three aspects of the creation of the large-scale from the small-scale. But, and this is the key point, understanding the small-scale units enables one to manipulate the large-scale structures – it enables one to create new large-scale structures.

REFERENCES

1. A. G. Marangoni, N. Acevedo, F. Maleky, E. Co, F. Peyronel, G. Mazzanti, B. Quinn and D. Pink, *Soft Matter*, 2012, **8**, 1275.
2. A. G. Marangoni, S. S. Narine, N. C. Acevedo and D. Tang, *Structure and Properties of Fat Crystal Networks*, ed. A. G. Marangoni and L. H. Wesdorp, CRC Press, Boca Raton, FL, 2nd edn, 2013, pp. 173–232.
3. N. C. Acevedo and A. G. Marangoni, *Cryst. Growth Des.*, 2010, **10**, 3327.
4. N. C. Acevedo and A. G. Marangoni, *Cryst. Growth Des.*, 2010, **10**, 3334.
5. D. A. Pink, B. Quinn, F. Peyronel and A. G. Marangoni, *J. Appl. Phys.*, 2013, **114**, 234901.
6. F. Peyronel, J. Ilavsky, G. Mazzanti, A. G. Marangoni and D. A. Pink, *J. Appl. Phys.*, 2013, **114**, 234902.

CHAPTER 2

Fat Nanostructure

CHLOE O'SULLIVAN,[a] NURIA ACEVEDO,[b] FERNANDA PEYRONEL[a] AND ALEJANDRO G. MARANGONI*[a]

[a] Department of Food Science, University of Guelph, Guelph, ON N1G2W1, Canada; [b] Department of Food Science and Human Nutrition, Iowa State University, Ames, IA 50011, USA
*Email: amarango@uoguelph.ca

2.1 INTRODUCTION

Fats and oils appear in many areas of our day-to-day lives, from the lotions we use, to the vitamins we take, the biodiesel in our car, to the food on our plate. They play an essential role in our diets and affect the health of the global population. Research in the area of edible fats and oils is exciting and constantly evolving, with applications in nutrition, disease prevention, drug delivery, and food product development. An understanding of the chemical and physical nature of the material is fundamental in all these areas, from molecular packing to crystal network formation. Every level of structure influences the functionality and applications of the lipid.

2.1.1 Edible Fats and Oils in Our Diet

Edible fats and oils are made up (over 95%) mostly of mixtures of triacylglycerol (TAG) molecules – one glycerol molecule esterified to three fatty acids – and contain small amounts of diacylglycerols,

Edible Nanostructures
Edited by Alejandro G Marangoni and David Pink
© The Royal Society of Chemistry 2015
Published by the Royal Society of Chemistry, www.rsc.org

monoacylglycerols, free fatty acids, and other fat-soluble compounds.[1] The difference between fats and oils is found in their state at room temperature: fats are solid at 25 °C, while oils are liquid. This is the conventional definition, however these terms are often used equivalently.

Nutritionally, lipids are the main source of energy in our diets, supplying 35 to 40% of our daily calories and containing more Joules per gram than any other macronutrient. In an average Western diet, 50 g to 100 g of fat is consumed per day.[2] Both the type and the quantity of lipid play a role in a healthy diet. The Dietary Guidelines for Americans 2010 recommend that less than 7% of daily calories comes from saturated fats, with the rest of our fat intake being mono- and polyunsaturates. *Trans* fats should be avoided altogether.[3]

Worldwide, we are facing a health crisis due to an increase in fat consumption, particularly of the saturated and *trans* fat varieties. Eating excess lipids can lead to obesity, type II diabetes, cardio-vascular disease and other health complications, causing enormous strain on national healthcare systems and compromising the health and wellbeing of the global population. National dietary guidelines recommending reductions in fat intake are just one example of how governments are dealing with this phenomenon. In more strict measures, the Food and Drug Administration (FDA) of the United States has recently moved towards removing the "generally recognized as safe" status of *trans* fats, a decision that would eliminate them from the market.[4] Denmark has had regulations in place limiting *trans* fats in foods for 10 years now. The food industry has to adapt to make changes like these possible and meet evolving consumer and government demands.

How do we reduce our fat intake? It is not as simple as a quick replacement with another ingredient or a complete elimination; along with their nutritional and caloric value, fats impart texture, mouth feel, consistency, colour, taste, and structure to the foods we enjoy. The underlying challenge faced by researchers and food industry is how to cut down on "bad" fats without compromising the sensory qualities of the edible product.

The means by which fats achieve their functionality is therefore of great importance to industry, and that lies in the chemical and physical nature of these solids. In a solid fat, high melting TAG components crystallize and form a three-dimensional network, trapping the liquid oil component and providing structure to the material. Fats can be characterized chemically using analytical techniques such as chromatography to separate and identify acylglycerol and fatty acid components. Functional properties such as melting behaviour, yield stress, and viscoelastic properties can also be measured with different instruments.[5] The analysis of fat structure, however, is more complex owing to lengthy sample preparation procedures and resolution limits. Details on the nano-, micro-, and mesoscale of this fat crystal network are what give insight into the mechanism of the functionality of the material.

In 2010, Acevedo and Marangoni at the University of Guelph published research outlining the characterization of a nanoplatelet structure in the fat crystal network.[6] This was the first characterization of a fat crystal structure on the sub-micron scale, and served to bridge the gap between fat molecular structure (molecules arranged in lamellae) and crystal network formation. The importance of this nano-sized component in the bulk properties of the fat will be used to better understand structure–functionality relationships in the material.

2.1.2   Fat Structure–Functionality Relationships

The physical properties of a fat are the result of its chemical composition, solid fat content and its underlying crystal network, which is in turn affected by the chemistry of the lipid and its processing conditions.[7] An understanding of how these variables affect fat structure and impact functionality opens the door to rational lipid design. The study of fat structure–functionality relationships is necessary not only for its theoretical value, which is a deeper understanding of material properties and crystal networks, but also for the development of fat products in the food industry. Characterizing the fat crystal network and its relationship to functionality will provide us with the opportunity of engineering healthier fat-containing products without compromising their quality.

The elimination of *trans* fats in high fat products such as shortenings – currently a major push in the food industry – is an example of how this research is applied. *Trans* fats originate in the partial hydrogenation of oils, a process originally created to turn inexpensive vegetable oils into solid fats. During the reaction, some double bonds are hydrogenated and others change from their natural *cis*- to the *trans*- configuration. Shortenings created this way were quickly adopted by industry because of their increased stability and economic value.

One solution to cut out *trans* fat in these products involves replacing partially hydrogenated oils with their fully hydrogenated equivalents. However, this change would result in increased levels of saturated fats and compromise the mouth feel and texture of the product. Saturated fats have higher melting points than the *trans* fat components and can create a waxy consistency. Research into how processing conditions, such as shear or cooling rate, affect the nanoscale of the fat crystal network, the physicochemical properties of the fat and consequently the bulk properties of the lipid in a food system, can lend insight into how to make this replacement possible without changing product quality.[8,9]

With that in mind, the purpose of this chapter is to outline the developments that have been made in characterizing the structure of the fat crystal network and explaining its unique bulk properties. It will start with an introduction to the chemistry of triacylglycerol molecules and discussion of the physical properties of common edible fats and oils. Additionally, nucleation, crystallization and structural hierarchy of the fat crystal network will be covered, with an emphasis on nanoscale structuring agents. Finally, we will explore examples of engineering the fat crystal network, looking at modifying the structural properties of the plastic fat to achieve desirable bulk characteristics.

2.2 EDIBLE LIPID CHEMISTRY

2.2.1 Lipid Components

2.2.1.1 Fatty Acids. Fatty acids (FAs) are long carboxylic acid chains and are the main components of fats; 99% of fatty acids in lipids are found as part of acylglycerol molecules, esterified to a glycerol backbone. Most FAs found in nature are even-numbered, and vary in length from 14 to 24 carbons. Notable exceptions are present in cow's milk, where short chain FAs such as butyric acid (four carbons) are common. As lipid components, fatty acids are an energy source for our body, where their caloric value is dependent on their length and absorption capabilities.[1]

Fatty acids can be named using the International Union of Pure and Applied Chemistry (IUPAC) naming system, identifying the number of carbons in the carboxylic acid chain and adding an "-*oic acid*" ending. However they are regularly referred to by their common names, derived from their principal source. A shorthand numerical system is also used for naming: the first number in the pair corresponds to the number of carbons in the chain, and the second (separated by a colon) represents the number of double bonds in the fatty acid. For

Table 2.1 IUPAC, Common and Shorthand Names of Some Fatty Acids Found in Food.

	IUPAC	Common	Shorthand
Saturated fatty acids	Butanoic	Butyric	4:0
	Hexanoic	Caproic	6:0
	Octanoic	Caprylic	8:0
	Decanoic	Capric	10:0
	Dodecanoic	Lauric	12:0
	Tetradecanoic	Myristic	14:0
	Hexadecanoic	Palmitic	16:0
	Octadecanoic	Stearic	18:0
Unsaturated fatty acids	*cis*-9-Octadecanoic	Oleic	18:1ω9
	cis-9, *cis*-12-Octadecanoic	Linoleic	18:2ω6
	cis-9, *cis*-12, *cis*-15-Octadecanoic	Linolenic	18:3ω3
	cis-5, *cis*-8, *cis*-11, *cis*-14, *cis*-17-Eicosapentaenoic	EPA	20:5ω3
	cis-4, *cis*-7, *cis*-10, *cis*-13, *cis*-16, *cis*-19-Docosahexaenoic	DHA	22:6ω3

example, 16-carbon saturated fatty acid is hexadecanoic acid under IUPAC naming, but is known by its common name palmitic acid and short form 16:0 (Table 2.1).

2.2.1.1.1 Unsaturated Fatty Acids. Unsaturated fatty acids contain one or more double bonds in their carbon chain. In natural sources, these double bonds are present in the *cis*- configuration (hydrogen atoms on the same side of the double bond) and most polyunsaturated FAs have methylene separation between their double bonds.

The presence of double bonds in the hydrocarbon backbone introduces a "kink" into the otherwise straight chain. The melting point of lipids high in unsaturated fatty acids is generally lower, because the kink leads to weaker interactions between these molecules. Unsaturated fatty acids pack loosely and require less energy to change from the solid to the liquid phase (Figure 2.1).

The terms "omega-3" and "omega-6" can be used in unsaturated FA nomenclature. The number refers to the position of the first double bond from the methyl end of the fatty acid. This notation is useful because the biological activity of a fatty acid depends on the interactions of enzymes with the carbonyl end of the fatty acid.[10,11] Oleic acid, an 18-carbon, monounsaturated fatty acid is the most common acid found in food lipids.[12] Certain omega-3 and omega-6

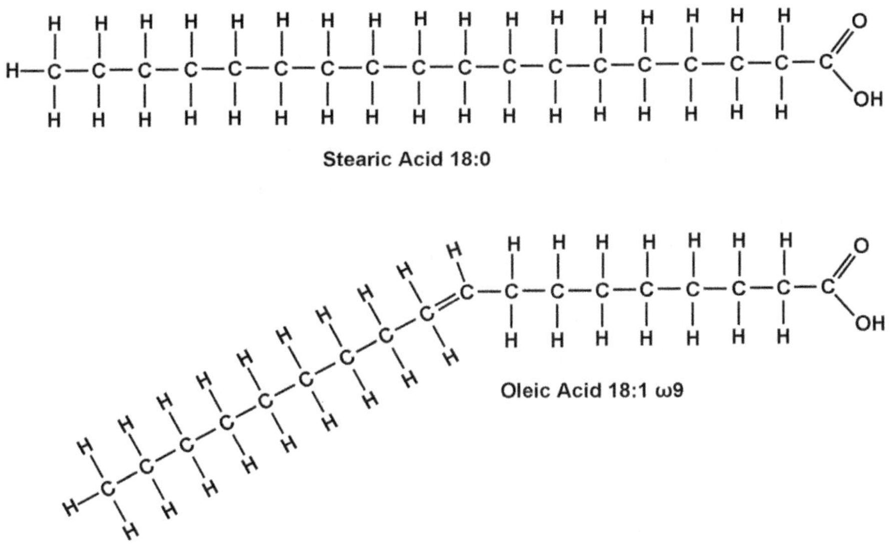

Figure 2.1 Structure of saturated and unsaturated fatty acids. The double bond in unsaturated fatty acids introduces a kink into the hydrocarbon chain.

fatty acids are considered essential fatty acids because they cannot be synthesized in our bodies.

2.2.1.1.2 Saturated Fatty Acids. Saturated fatty acids are fully hydrogenated, containing no double bonds. These saturated carboxylic acids can pack tightly, creating maximum hydrogen-bonding between chains (see Figure 2.1). Fats high in saturated FAs, therefore, have higher melting points.

Once consumed, saturated fatty acids have different effects on cholesterol levels from unsaturated FAs. Most saturated FAs increase low density lipoprotein (LDL)-cholesterol levels while unsaturated FAs decrease their levels in the blood.[13,14] Since elevated LDL-cholesterol is associated with a risk of heart disease, diets low in saturated fatty acids have been encouraged.[3]

The most abundant saturated fatty acid in plant and animal fats is palmitic acid (16:0), followed by stearic acid (18:0).[1]

2.2.1.2 Acylglycerols. Acylglycerols have a glycerol ($HOCH_2$–$CHOH$–CH_2OH) backbone and are esterified to one, two or three fatty acids. Triacylglycerols are the most common among these, with three fatty acid chains, but diacylglycerols (DAGs) and monoacylglycerols (MAGs) are also present as minor components in fats and oils. DAGs and MAGs are commonly used in food industry as emulsifiers; however, for the purposes of this chapter we will concentrate on the chemical and physical properties of the triacylglycerol molecules.

2.2.1.2.1 Triacylglycerols. Three fatty acids are esterified to a glycerol backbone to make up a triglyceride molecule. The Fischer projection of a TAG has the three glycerol-carbons on the vertical, numbered 1 to 3 from top to bottom (Figure 2.2). Stereospecific numbering (*sn*) is used to specify the carbons on the glycerol backbone, to differentiate between structural isomers *e.g.* *sn*-1,2-dipalmito-3-stearin and *sn*-1,3-dipalmito-2-stearin. Physical and chemical properties of individual TAGs vary in their composition and positioning of fatty acids. The TAG exhibits chirality when the fatty acids attached to C1 and C3 are different. The two enantiomers react differently with enzymes in the body as a result of their chirality.[1]

The arrangement of fatty acids on the glycerol backbone also affects their bioavailability once consumed. Pancreatic lipase cleaves fatty acids from the *sn*-1 and *sn*-3 positions, making these acids less available for uptake because they form insoluble calcium salts.[15]

The oils and fats we consume are composed of mixtures of TAGs, made up of different fatty acids in different positions on the glycerol backbone. This chemical diversity helps explain their various

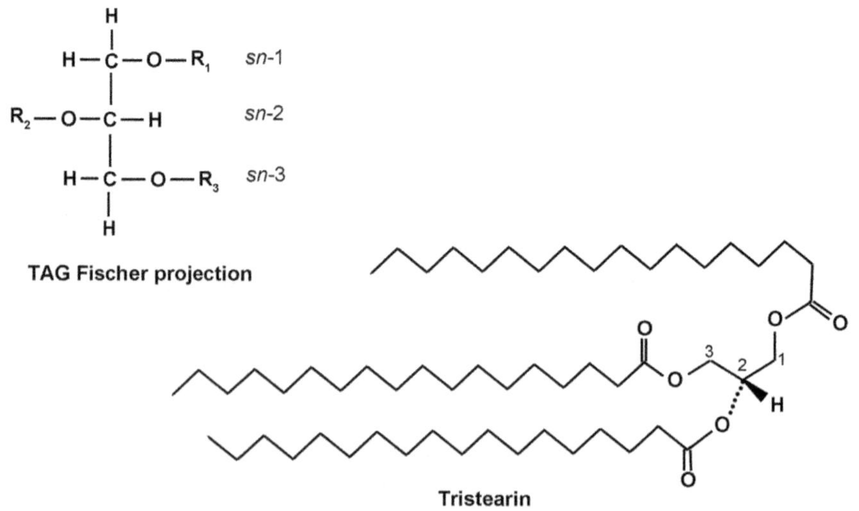

Figure 2.2 Structural representations of triacylglycerol molecules. In the Fischer projection, carbons on the glycerol backbone are numbered 1 to 3 from top to bottom. In the tristearin skeletal structure, carbons on the glycerol backbone are also numbered.

functions in our foods. Within this variety, some oils and fats contain distinct TAG profiles. Cocoa butter, for example, is unique in its high composition of triacylglycerols with oleic acid in the *sn*-2 position; it contains over 80% POP (palmitic-oleic-palmitic), POS (palmitic-oleic-stearic) and SOS (stearic-oleic-stearic).[16] Milk fat contains a wide variety of fatty acids, particularly short chain and branched acids, which are rare in many other naturally occurring fats.[10] Meat from ruminants such as cows, sheep and goats is high in stearic acid (18:0), and animal fats generally contain higher content of saturated fatty acid in the *sn*-2 position than plant fats.[10] Common vegetable oils such as olive, canola, or soybean oil are high in long chain un-saturated TAGs while common fats such as milk fat, cocoa butter and palm fat are high in saturated fat content.

The TAG composition of fats can be analysed by chromatographic or spectroscopic methods to give the molecular view of the material: its TAG composition and characterization, and the fatty acid profile.[17] The chemical profile of a fat does contribute to the overall physical properties of the bulk material, some of which are described in the following section.

2.2.2 Physicochemical Properties of Fats

2.2.2.1 Rheology. Fats are solid materials made up of a crystalline triglyceride network entrapping liquid oil molecules. Because of this composition, they can be described as colloidally structured gels or crystalline solids, exhibiting behaviour similar to both types of material.[6]

Plastic materials, the classification used to describe matter ex-hibiting this dual behaviour, behave as solids below the yield stress, σ^*. In this solid range, small deformations in the material occur but no bonds in the crystal network are broken. When subject to forces exceeding the yield stress, bonds within the material break perman-ently and it flows like a liquid.

Fats exhibit non-Newtonian behaviour, their response to stress deviating from ideal behaviour. For example, a fat could flow when subject to a force below the yield stress, or break over a range of forces, as opposed to at a precise point.[10] Materials that show this non-Newtonian behaviour are called viscoelastic. The viscoelasticity of a fat can be explained by bonds within the fat crystal network stretching, breaking and reforming under different stresses.[18]

The rheological properties of the fat can be measured in the la-boratory using small deformation rheology, which is used to characterize the relationship between stress and strain in the material. Experiments are carried out in the linear viscoelastic region, where strain in the fat is proportional to stress. The results give values for G' (shear storage modulus) and G'' (shear loss modulus), indications of the elastic and the viscous nature of the material respectively.[5]

Many studies have looked at the relationship between fat rheological properties and the solid fat content of the material, as well as the structure of the crystal network including particle shape, size, and network interactions.[8,18,19] Recently the link between structure and rheology was extended to the nano- length scale,[9] where crystal unit dimensions were found to be inversely proportional to G' and σ^*. The elucidation of the relationship between fat rheological properties and the crystal network is most useful when optimizing processing conditions for high fat-containing products - the yield value of a fat, which can be modified with shear processing, can in turn affect the spreadability and overall quality of the product.[20]

2.2.2.2 Solid Fat Content (SFC). Since fat is a material structured by solid lipid particles, its total solid fat content (SFC) has great impact on its structure and rheological properties. The state of the lipid can be expressed as an SFC value (fraction from 0 to 1, or percentage), where a fat with a higher content of high melting TAG will have a higher SFC and this will lead to a harder material. For example, butter has a higher SFC than a spreadable margarine at the same temperature.

Solid fat content is also temperature dependent – a property that explains the functional behaviour of some high-fat products. For example, margarine must be solid at refrigeration temperature but spreadable and not completely liquid at room temperature; these qualities are associated with changes in SFC. Complex TAG blends (like most natural fats) show a change in SFC from 0 to 1 over a range of temperatures, while pure TAG systems transition from 0 to 1 at exactly the melting point of the molecule.

SFC is measured in the laboratory by pulsed proton nuclear magnetic resonance spectroscopy (pNMR), which excites the TAG protons and records a signal based on their relaxation times. Protons in the solid phase are in closer contact with other atoms and therefore relax more quickly than protons in the liquid phase. This allows one to quantify the composition of solid TAG in the fat with a numerical value of solid fat content.[5]

Solid fat content not only imparts structure to the fat, but also can modify the structure of the fat crystal network. It is used as a variable representing supersaturation when comparing the average dimensions of fat crystals in different complex TAG mixtures.[21] Other characteristics such as the thermal properties of the fat can be used along with SFC to give more information on fat behaviour in complex food systems.

2.2.2.3 Thermal Properties. The melting point and enthalpy of phase transition of a fat are used to explain, in part, the varying functionalities of different fats. For example, in chocolate, the hardness of the cocoa butter crystal network at room temperature gives the confection its desirable "snap", and its thermal profile explains its pleasurable mouth feel, when it turns to liquid at body temperature.[22]

Usually a natural fat will exhibit a wide melting peak, as all the TAGs present in the material have different characteristic melting temperatures. This can be observed by looking at the thermal profile of a fat using differential scanning calorimetry (DSC), a technique that measures heat flow and temperature changes in a material against a standard. In natural fat samples, phase transitions appear as wide peaks instead of fine spikes in heat flow *vs.* temperature graphs (Figure 2.3).[5]

Shear, cooling rate and other external fields can also affect the thermal profile of the material by manipulating crystal size, shape, aggregation, and/or orientation.[9,19,23–25]

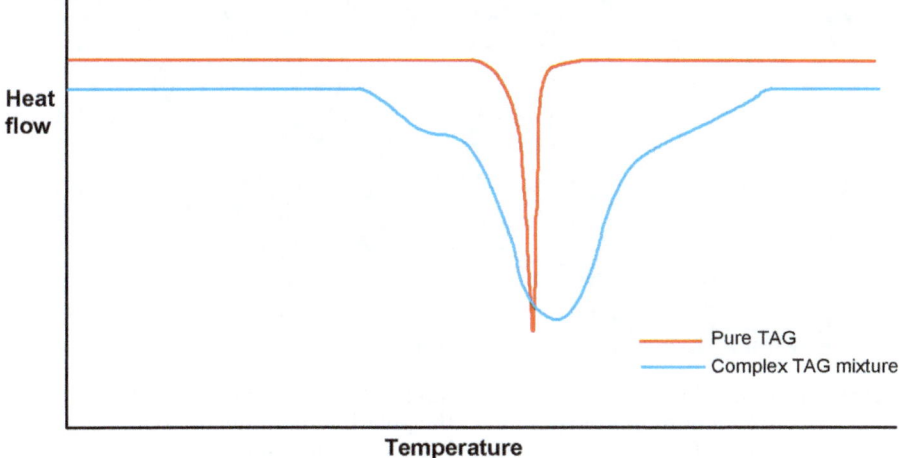

Figure 2.3 Schematic DSC melting curves of pure TAG and complex TAG mixtures.

2.3 CRYSTALLIZATION BEHAVIOUR

The process of fat crystallization is governed by the principles of thermodynamics: the system will change state when it results in an overall lowering of free energy. Formation of crystals happens in a step-by-step process where supersaturated conditions are established, nucleation occurs, and crystal growth follows. However not all crystals are created equally; depending on the TAG structure and crystallization conditions, different unit cells and subcells will be formed that will impact the physicochemical properties of the fat. Along with these variations in packing arrangements, larger crystals can exhibit different morphologies, forming needle-like structures, microplatelets, fractal-like structures, spherulites and many others (Figure 2.4).

The following will be a brief introduction to the crystallization behaviour of fats, giving background on the formation of crystal network from melt, and an understanding of the length scales involved in the greater structuring of the fat. For more information, an in-depth review on the crystallization behaviour of fats is available.[26]

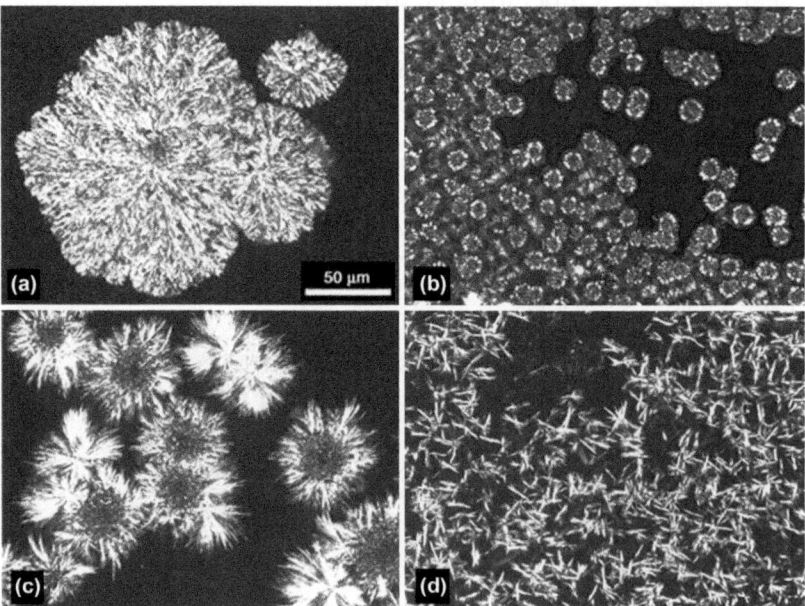

Figure 2.4 Cocoa butter (CB) samples crystallized under different conditions show different morphologies under the polarized light microscope. The image shows CB samples crystallized at 15 °C after 28 days (a), crystallized at 22 °C after 28 days (b), crystallized at 24 °C after 21 days (c), and crystallized at 26 °C after 28 days (d).

2.3.1   Nucleation

The first step in crystal formation requires supersaturation of the melt. As the melt cools to below the crystallization point of the highest melting TAG (supercooling), those molecules become supersaturated in the liquid melt. The degree of supercooling $(T_m - T)$ increases the chemical potential difference $(\Delta\mu)$ between solid and liquid TAG and drives nucleation [see equations (2.1) and (2.2)].

$$\ln\beta = \frac{\Delta H_m}{RTT_m}\,(T_m - T) \tag{2.1}$$

$$\Delta\mu = RT\ln\beta \tag{2.2}$$

Where β is a measure of supersaturation, R is the gas constant, T is the temperature of the crystalizing melt, T_m is the melting temperature of the neat solid, and ΔH_m is the melting enthalpy of the neat solid.

As the solubility of the TAG decreases, there is a stronger drive to crystallize. In other words, as the difference between actual temperature and the melting temperature increases, there is a greater difference in chemical potential and a push towards crystallization.

Crystallization can be expressed with the Gibbs–Thompson model [see equation (2.3)], which illustrates the change in Gibbs free energy associated with the formation of a solid nuclei (ΔG_n).

$$\Delta G_n = A_n\gamma - V_n(\Delta\mu / V_m^s) \tag{2.3}$$

Nucleation is driven by the second term of this equation, illustrating a lowering of free energy of the system by the creation of a lower energy species (crystal). This energy loss is proportional to the change in chemical potential between solid and liquid $(\Delta\mu)$, the volume of the nucleus (V_n), and the molar volume of the solid (V_m^s). The first term describes an energy penalty due to the formation of a solid–liquid interface when a crystal is formed. This positive contribution to the Gibbs free energy is proportional to the surface area of the crystal (A_n) and the surface free energy per unit area (γ). Because of this "give and take" relationship, there exists a critical radius above which the Gibbs free energy of the system lowers (Figure 2.5). Nuclei that are smaller than this critical size will dissociate, while larger ones will grow.

Nucleation in fats usually occurs heterogeneously, where the crystal nucleus is formed on some impurity within the material. The

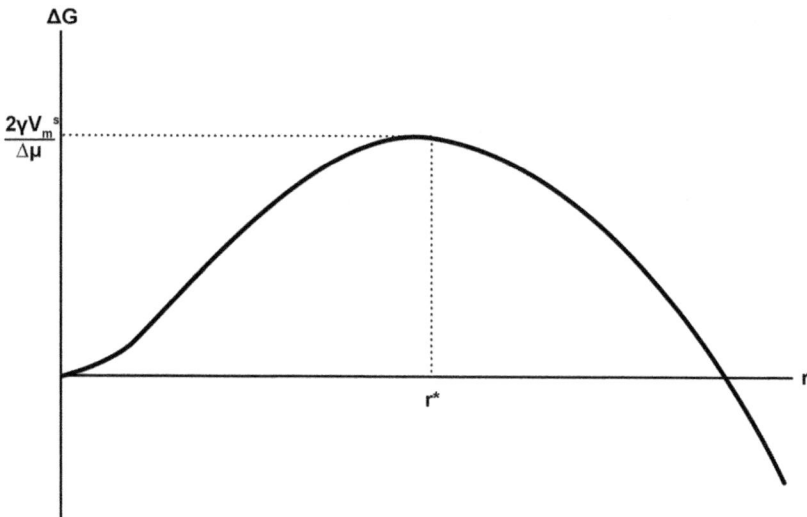

Figure 2.5   The critical radius (r^*) of a nucleus required for crystal growth, according
to the change in Gibbs free energy (ΔG) of the system.
(Adapted from ref. 10.)

resulting nuclei can be composed of molecules in different packing
arrangements, referred to as *polymorphs*, which will be discussed in
the following section. The rate of nucleation affects the number of
nuclei formed, the size of the crystals and the density of the fat crystal
network.[18,27,28] Factors that affect nucleation therefore influence the
mechanical and functional properties of the fat.

2.3.2   Crystal Growth Kinetics

After nucleation, crystal growth occurs, as more TAG molecules from
the melt are incorporated into the solid. The solid builds on different
surfaces of the crystal nucleus, leading to a wide variety of possible
crystal morphologies: needles, fractals, spherulites, and many more.
The development of the crystal is dependent on heat and mass
transfer. The TAG molecules move within the melt and when they
crystallize the energy is dissipated in the surrounding liquid.

Processing conditions such as shear and cooling rate can affect
heat and mass transfer and influence the crystal growing process. For
example, shear rates can increase molecular movement in the melt
and promote nucleation, leading to smaller crystal sizes.[21,29] High
shear can also break crystal structure and prevent aggregation, resulting in smaller clusters.[29] Since crystal size and network density

affect the mechanical properties of the fat, these trends in nucleation and crystal growth kinetics are important to consider for fat functionality.

The effect of supercooling, influencing both nucleation and crystal growth, is seen in the size and number of crystal nuclei formed. Crystal growth and nucleation rates increase to different degrees with increasing supercooling, until a maximum is reached. At higher levels of supersaturation, molecular immobility limits both crystal growth and nucleation. The end result is that at high and low supercooling crystal growth dominates while at intermediate levels nucleation is favoured.[10]

As crystallization continues, smaller crystallites aggregate and crystal structure on the mesoscale (1 μm to 200 μm) becomes visible. This can be observed by polarized light microscopy, which is used to visualize crystalline TAGs against liquid TAG background, based on their birefringence. Crystalline material in the fat glows while the liquid matrix remains dark.[6,30] This imaging technique allows for the visualization and characterization of crystal structures as small as 1 μm to 2 μm, where the many distinct morphologies can be seen. This characterization is important since measurements of crystal morphology, such as fractal dimension, greatly affect the rheological properties of the material.[19] The breakdown of the crystal structure on different length scales will be discussed further in Section 2.4.

2.3.3 Polymorphism

Upon nucleation, TAG molecules arrange in lamellae, adopting distinct conformations depending on the chemistry of the molecules. When fatty acids in the *sn*-1 and *sn*-3 positions are in the same orientation relative to the glycerol backbone, the TAG is said to be in the "tuning fork" configuration. When hydrocarbon chains in the *sn*-1 and *sn*-2 positions are opposite that in the *sn*-3 position, the TAG is in the "chair" configuration (Figure 2.6). These fully extended molecules then stack, forming lamellae that are two FAs, three FAs, or four FAs thick, depending on their configurations. The triacylglycerol crystal unit, the repeating structure that forms the fat crystal, has a long *c*-axis the height of the lamella, and short *a*- and *b*- axes (see Figure 2.6).

Polymorphism is the ability of one material to have different crystalline structures. For triacylglycerols, this variation occurs in the packing of the ethylene units of the fatty acid chains, a unit referred to

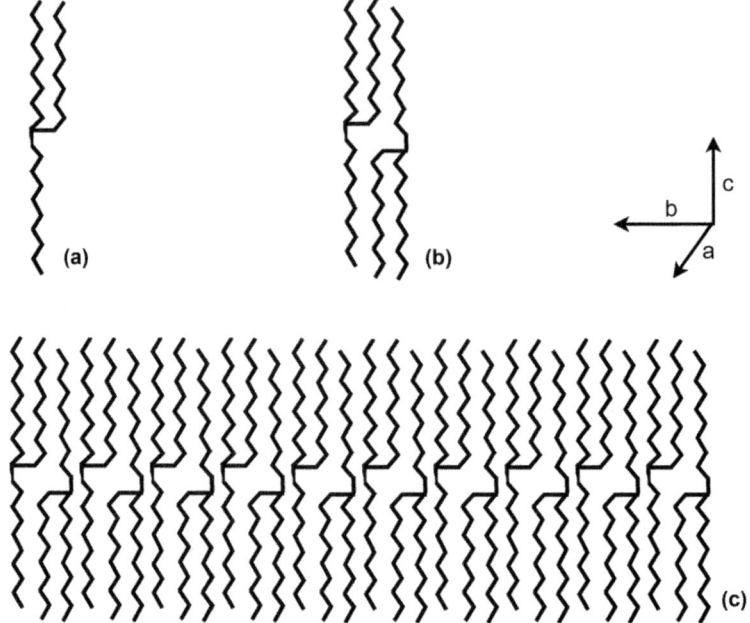

Figure 2.6   TAG in the tuning fork configuration (a), a crystal unit cell (b), and a single lamellar layer (c).
(Adapted from ref. 10.)

as the crystal *subcell*. This subcell is the smaller repeating unit that makes up the unit cell.

There are three common packing arrangements for TAG molecules: α, β', and β, which are hexagonal, orthorhombic perpendicular, and triclinic parallel arrangements respectively. Although all three polymorphic forms are possible to achieve directly from the melt, they vary in their activation energies and thermodynamic stabilities. In terms of crystallization activation energies (E_A), $\alpha < \beta' < \beta$, making the α-form the kinetically favoured product (Figure 2.7). However, because of their different packing, the β-polymorph is lowest in energy, followed by β', then α (see Figure 2.7). This means that transitions from α to β' to β will occur spontaneously over time as the system progresses to the most stable state.

The melting profile of the fat is affected by the polymorphic form of its crystals: the β-polymorph has the highest degree of crystallinity and the highest melting point, followed by the β'-form and the α-polymorph, which is the most loosely packed. Changes in crystalline packing can thus be observed when looking at the thermal profile of the fat using DSC, where multiple peaks representing the melting and

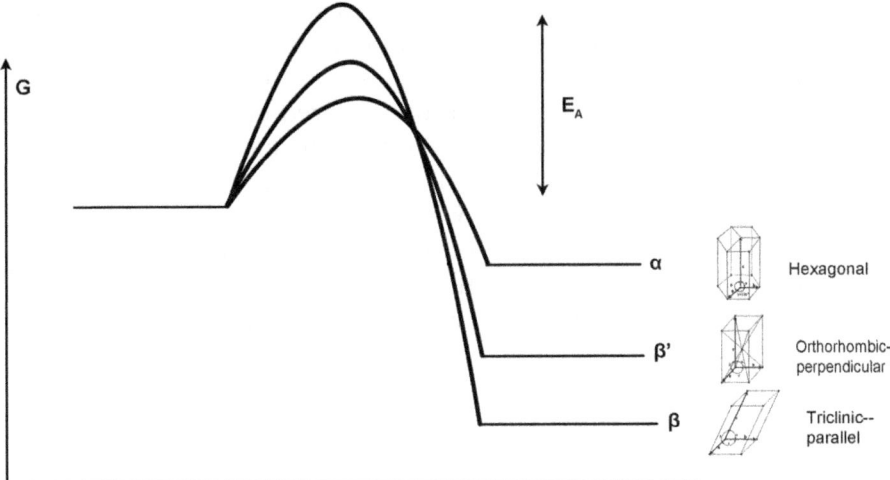

Figure 2.7 Activation energies and thermodynamic stability of different TAG polymorphic forms. The polymorphic forms are characterized by different
hydrocarbon packing arrangements: hexagonal, orthorhombic and triclinic.
(Adapted from ref. 10.)

reordering of the system from the least to the most stable polymorph
are visible.[5]

Because polymorphic form also affects fat bulk properties, certain
food products require specific forms for their desired characteristics.
Margarines and spreads, for example, rely on the β'-polymorph for
their glossy appearance and water incorporation abilities, while
shortenings favour the β-polymorph for their flakiness.[10]

Successful chocolate manufacturing is based on the formation of
the 'Form V' polymorph. Out of six identified polymorphic structures
(subsets of α, β', and β), only Form V packing leads to a chocolate with
desirable melting point, mouth feel, and appearance.[31,32] In the
tempering process, the chocolate melt must be cooled to around
27 °C, reheated to 30 °C, and cooled again to foster the formation of
Form V crystals and discourage other polymorphs from forming.[26]
When crystals of Form V transform to Form VI (due to long-term or
high-temperature storage), large fat crystals can form on the surface
of the chocolate, creating a white, flaky covering referred to as fat
bloom.[33] This is what one might see when opening an old bag of
chocolate chips.

Similar factors that affect fat crystal structure also affect the polymorphic form of the crystal. Shearing, for example, has been shown to

accelerate transition from a less stable to a more stable arrangement.[23,33] Shear forces either input enough energy to overcome the energy barrier of transformation, or melt less stable crystal nuclei, allowing for crystal growth on the more ordered crystals.

Looking at the effects of cooling rate and solid fat content on polymorphism in blends of fully hydrogenated canola oil (FHCO) and canola oil (CO), Maleky, Acevedo and Marangoni found that fast cooling rates of 100% FHCO favoured the formation of the α-polymorph (the least organized system) while systems diluted with up to 50% CO cooled at the same rate were in the β-form.[27] In this case, the decrease in viscosity associated with the higher fraction of canola oil allowed for rearrangement of the molecules into the more stable form.

Along with analysing the melting profile of the fat for signs of multiple melting and rearrangement peaks, the polymorphic state of a system can be measured directly using powder X-ray diffraction analysis (XRD). Distinct diffraction patterns at wide angles (16–35°) give an indication of the subcell packing of the triacylglycerol network. An in-depth explanation of the technique is given by Peyronel and Campos.[5]

Ultimately, the polymorphic form of the lipid is relevant in the larger crystal structure because it lays the foundations for the formation of the crystal nanoscale. As crystallization progresses, molecules arrange into lamellae ($\sim$30–60 Å) which stack to form nanoplatelets, (the nanoscale structuring unit of the fat crystal network).[6] The height of a single lamellar layer in the stack corresponds to the *c*-axis of the crystal unit cell (see Figure 2.6).

2.4 CRYSTAL STRUCTURE HIERARCHY

The system of the fat crystal network has been characterized on many levels, detailing the transition from liquid triacylglycerol melt to solid fat upon cooling. We have discussed briefly the arrangement and packing of TAGs into lamellae, but where do they fit into the formation of a larger crystal network? In a big picture model, the lamellae make up the molecular structure of the "primary crystals", which come together to form polycrystals, which build crystal aggregates in turn (Figure 2.8). These final aggregates are visible crystals on the micrometre scale.

The nature of the "primary crystal" in the network has been the subject of recent research. With the development and refinement of new techniques in sample preparation and observation, Acevedo and Marangoni at the University of Guelph were able to successfully

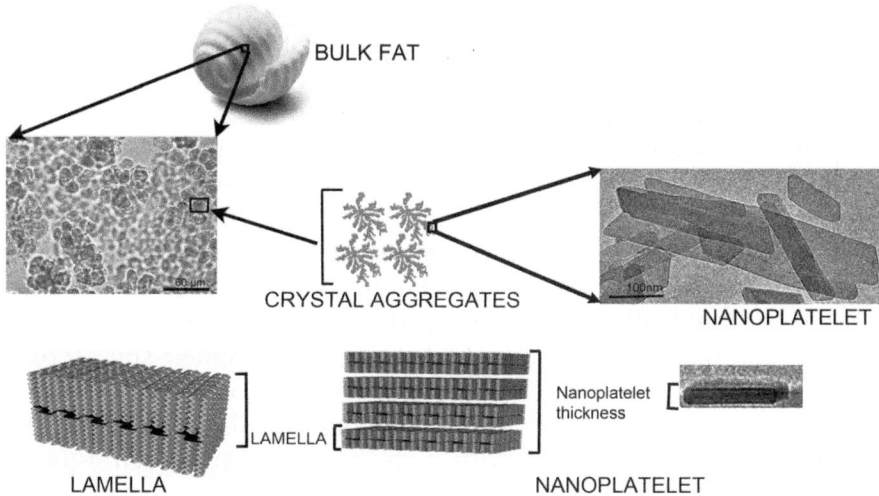

Figure 2.8 Representation of the structural levels present in a triacylglycerol crystal network. The crystalline unit is a platelet with sizes within the range of several nanometres; in turn, nanoplatelets are composed of stacks of TAG lamellae. On the microscale, crystal clusters and larger aggregates can be observed, which then self-assemble to constitute a tridimensional network.
(Adapted from ref. 7.)

isolate and characterize the fat crystal nanoplatelet, a unit on the sub-micron scale that structures the fat crystal network.[6]

In this section, the isolation and characterization of the nanoplatelet will be discussed, as well as the nature of the larger structures within the crystal network, as studied using ultra small angle X-ray scattering (USAXS) and computer simulations. This will serve to place the nanoscale within the structural hierarchy of the greater fat crystal network.

2.4.1 Nanoplatelets

Although the presence of a smaller platelet-like structuring unit was hypothesized in previous work,[23,34,35] the isolation, characterization and individual imaging of these platelets required the design of a special technique.

Fat crystals have been observed successfully at ambient temperatures using polarized light microscopy; however, this technique has a limit of resolution of 1–2 μm and cannot be used to visualize smaller structuring elements. Various other methods such as cryogenic scanning electron microscopy (cryo-SEM),[30,36] confocal scanning light microscopy, and multiple photon microscopy have been used to visualize fat crystals with varying degrees of success.[24] All these techniques have in common a limit of resolution on the micrometre scale when imaging native lipid crystal structure.

Cryogenic transmission electron microscopy (cryo-TEM) offers a low limit of resolution, but provides another obstacle to visualizing the crystal network. Under cryogenic conditions, the liquid oil component of the fat freezes and the contrast between the matrix and the crystalline TAG structure is poor. This problem is resolved with the removal of liquid oil by treatment with cold organic solvents[36,37] or aqueous detergents,[38] both of which have been applied successfully with microscopy to probe the solid fat network. Acevedo and Marangoni developed their extraction technique using these findings, combining crystal extraction with a rigorous isolation technique to image and characterize individual nanoplatelets.[6]

The cryo-TEM imaging technique Acevedo and Marangoni derived consists of a treatment of the fat with cold isobutanol to remove liquid oil, followed by homogenization with a rotostator to break up the crystal network. This solution is then filtered, resuspended in cold solvent, homogenized once more and finally sonicated to distribute the crystals in the solvent. Electron microscopy imaging is done rapidly after the solution treatment so the nanoplatelets do not aggregate and can be imaged individually.

The resultant TEM images show platelets with dimensions 150×60×30 nm to 370×160×40 nm from samples processed under non-shear conditions and at varying concentrations of FHCO in SO (Figure 2.9). These dimensions have been shown to vary with changes in fat composition and external fields.

The lengths and widths of the nanoplatelets are visible in most instances; however, some platelets are oriented perpendicular to the slide and their height is measurable. On those "side-view" platelets, the layers of TAG lamellae are visible within the structure, giving a striped appearance (the dark lines mark the intersection of two lamellae). The width of a single lamella, measured to be 4.23 ± 0.76 nm, corresponds well to the small angle X-ray diffraction peak at 4.5 nm for pure FHCO. Also, the total height of the nanoplatelet (30–40 nm) as measured from the microscope images is in agreement with the Scherrer analysis of the (001) crystal plane reflection peak. Platelets in this system were 7–10 lamellar layers thick. The agreement between microscopy measurements and XRD data is significant since it confirms that the crystals being imaged are indeed small structuring units of the larger crystal.

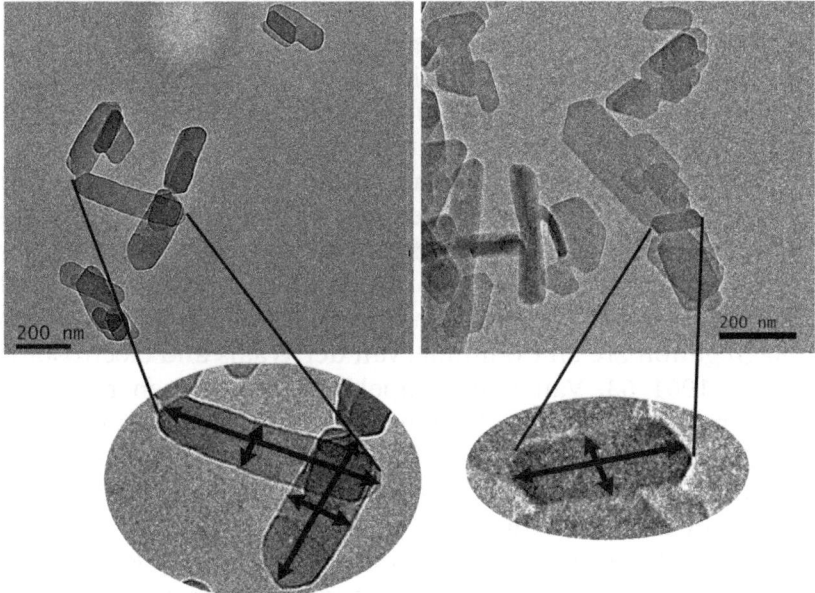

Figure 2.9 Examples of cryo-TEM micrograph sections with selected nanoplatelets in which lengths and widths were measured by image analysis using ImageJ software. Platelet lengths and widths are marked with arrows.

So far, this nanoplatelet imaging technique has been applied to six systems: fully hydrogenated canola oil (FHCO) in high oleic sunflower oil (HOSO) blended[6] and interesterified,[21] followed by cocoa butter (CB),[25,39,40] fully hydrogenated soybean oil (FHSO) in soybean oil (SO) blends,[8,9,29] FHCO in canola oil (CO) blends,[27] tristearin in triolein,[41] and cocoa butter substitute with SMS.[42]

2.4.2 Crystal Aggregates

The observation of crystalline nanoplatelets (CNPs) in different lipid systems led to the question: how do these CNPs aggregate to form the structures observed under light microscope? This question has been addressed by combining modelling and computer simulation with experiments using the technique of ultra small angle X-ray scattering (USAXS). USAXS enables one to study lipid systems *in situ*, in a single non-destructive experiment. This technique covers a length scale from 100 Å to 15 μm and yields information about the structures of the minority phase present. For example, when working with a binary system in which a pure solid lipid is mixed with a pure liquid lipid, USAXS gives information about the crystalline component as long as the solid content does not exceed that of the liquid. If one has a pure solid material then there will be observable results only if the solid contains voids which become the minority phase.

The physical interactions that take place in a system crystallizing from the melt fall under two categories: those responsible for assembling molecules to form the fundamental crystalline unit and those that are responsible for the formation of aggregates once the fundamental unit particle has been formed. The forces that take part in the aggregation are all Columbic: van der Waals and electrostatics. As early as 1961–63, Van den Tempel and Nedeerveen referred to triacylglycerols as forming flocculated solid particles that were clumped together *via* Van der Waals forces.[43,44] Van der Waals forces have been studied in colloidal systems for over a century, both experimentally and theoretically.[45,46]

Among the three van der Waals forces – Debye, Keesom and London – the last is believed to be responsible for creating the fat crystal networks.[47] London–van der Waals dispersion forces arise through fluctuations in the electromagnetic field. These fluctuations induce changes in atomic charge distributions, resulting in the formation of fluctuating atomic dipoles. When the internal states of the electromagnetic field are summed, an effective direct force between the fluctuating atomic dipoles is the net result.[48]

It has been argued that the van der Waals forces can be computed using the Lifshitz theory[49] or the Lennard-Jones 6-12 potential.[47] The Lifshitz approach is adequate for bulk measurement but not for making predictions at the molecular or nanoscale level; it includes the three cases of van der Waals forces and focuses on the dipole–dipole interaction. Although the Lennard-Jones 6-12 potential uses only the C_6 coefficient in the London–dispersion interaction applied between pairs of particles, its application and predictions have proven to be successful for atomic scale molecular dynamic simulations.[47]

Knowledge of the parameters obtained from fitting the USAXS data is necessary to understand the results from the fat crystal scattering experiments. At the Advanced Photon Source (APS), Argonne National Laboratories,[50,51] the scattering cross section as a function of the scattering vector (q) is detected by a photodiode detector and is reported as absolute intensity ($I(q)$), in units of cm^{-1}. Two relevant parameters obtained from $I(q)$ are the size of the individual scatterer and the fractal dimensions. The scatterer is the basic unit at a particular length scale (L) that causes scattering, where L is related to q by equation (2.4):

$$L = 2\pi/q \qquad (2.4)$$

Smaller features in the material are therefore observed for large q values, while larger features are studied at very small q values.

If, for a certain range of q, $\ln(I(q))$ is linearly proportional to $\ln(q)$ then the system is considered fractal for the spatial range corresponding to that range of q.[52,53] From the slopes of these linear regions, either the mass fractal dimension (D_m) or the surface fractal dimension (D_s) can be computed. It has been shown that when $I(q) \sim q^P$ then P is related to the fractal dimension D as follows:[54]

If $1 \le |P| < 3$, then $D_m = |P|$
If $3 \le |P| < 4$, then $D_s = 6 - |P|$ and $2 < D_s \le 3$

However, if $|P| > 4$, all that is known is that the surfaces of the aggregating particles are diffuse.

USAXS data obtained for lipid systems have been analysed using two models: (1) The Unified Fit model, introduced by Beaucage,[55,56] which considers the sizes of the scatterers and the morphology of the system as a function of q, and (2) a Guinier–Porod model that gives information regarding the shape of the scatterers.[57] The first model yields the correct analytical dependence upon q as $q \to 0$, while the Guinier–Porod expression is valid for $I(q)$ values as $q \to \infty$. Both models assume that the total scattering intensity can be written as a linear combination of scattering from different regions of q-space, referred to as "levels".

The parameters associated with each level obtained from the Unified Fit model are the radius of gyration (R_g) and the exponent P described above, while for the Guinier–Porod model, the parameter s describes the shape of the scatterer. If $s = 0$, the scatterers are spheres; if $s = 1$, they are cylinders; and if $s = 2$, they are platelets. In both models, a level is described by a Guinier[58] or a "knee" and an associated power-law regime, or the Porod's region,[59] is illustrated in Figure 2.10.

Figure 2.10 shows the interpretation given to the USAXS results obtained for mixtures of 5, 10, 15 and 20% tristearin (SSS) mixed in triolein (OOO).[41] The results in the largest q-region $(4 \times 10^{-3} < q < 2 \times 10^{-2} \text{ Å}^{-1})$ show that $P_1 = 4$, indicating the presence of crystalline nanoplatelets. The size of the CNPs, obtained as the

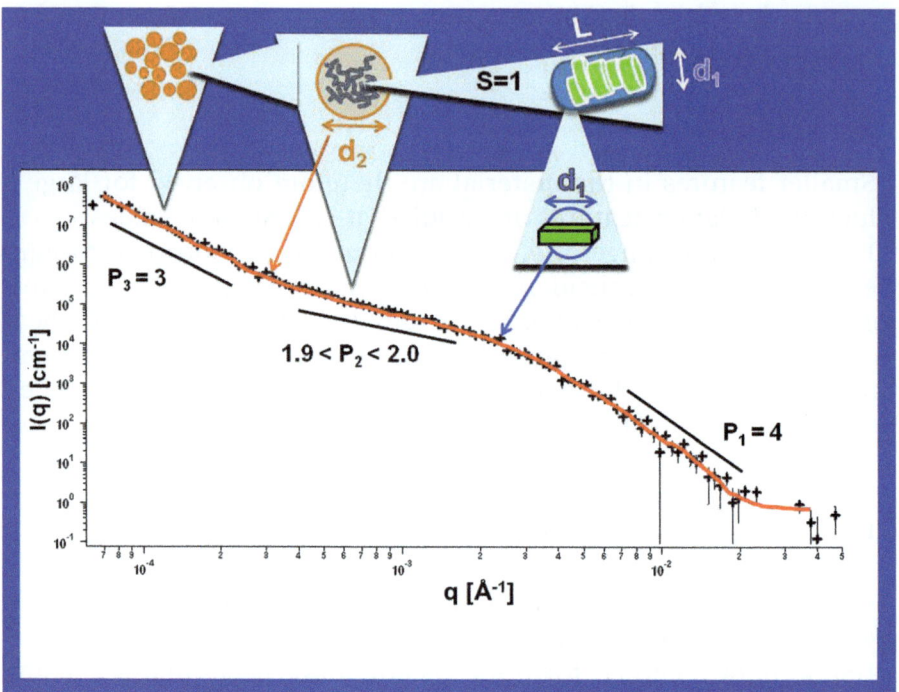

Figure 2.10 USAXS intensity as a function of scattering vector for a sample containing 20% tristearin in triolein. The dark line shows the fitting using the Unified Fit model.

spherical diameter from the radius of gyration (R_g), ranges between $\sim$110 nm and $\sim$430 nm.

These CNPs then form cylindrical aggregates with dimensions L and d_1 (see Figure 2.10), which are especially noticeable in samples crystallized at fast cooling rates. These structures appear to be TAG-woods – cylindrical stacks of CNPs – predicted by theoretical models (see Chapter 10, Computer Simulation of Food Nanostructures).[60] The presence of these TAGwoods was confirmed by fitting the USAXS data with the Guinier–Porod model, yielding a value of $s = 1$.

These cylinders then aggregate to form fractal structures that exhibit mass fractal dimension (D_m) values between 1.9 and 2.0 (see Figure 2.10, middle region). Both the cylindrical aggregates (TAG-woods) and the fractal structures shown in USAXS analysis are in agreement with the modelling and computer simulation predictions.[41,60] It was also shown that these aggregates were uniformly distributed in space, yielding a fractal dimension of $D = 3$ (Figure 2.10, smallest q region) (see Chapter 10, Computer Simulation of Food Nanostructures). However, Peyronel *et al.* showed that if the 20% SSS in OOO system was not at equilibrium, then a smaller *D* of 2.2 in the smallest *q* region was obtained.[41]

Quinn *et al.* carried out simulations in complex lipid systems (containing more than two distinct TAGs), again using only London–van der Waals interactions between CNPs.[61] Some of those predictions were confirmed by USAXS experiments carried out by Peyronel *et al.*,[62] demonstrating the benefits of combining modelling with *in situ* experiments to interpret results.

2.5 MODIFYING CRYSTAL STRUCTURE

Armed with an understanding of the microstructure and nano-structure of the fat crystal network, we can now explore how structures on both length scales are affected by changes in system composition and external fields. Because the network nanoscale was not considered prior to the development of new isolation and imaging techniques, the relationship between nano- and mesoscale crystal components has only recently begun to be characterized.

Considering the application of this research, an understanding of how these changes in crystal structure affect the physicochemical properties of the fats and functionalities is of great importance. This is the ultimate goal of crystal network characterization on all length scales – understanding how fat bulk properties such as oil binding capacity, strength, or melting profile are affected by the composition and processing conditions of the material.

The following section will show how composition and external field manipulations affect crystal network structure, specifically on the nanoscale. In all studies mentioned, fat systems were imaged using the cold solvent/cryo-TEM treatment and imaging method developed by Acevedo and Marangoni to examine nanoplatelet size along with other crystal network characteristics. Factors that will be examined are blend supersaturation (solid fat content), crystallization temperature and cooling rate, shear processing under turbulent and laminar fields, interesterification, and emulsifier addition.

2.5.1 Supersaturation and Solid Fat Content

When looking at blends of fully hydrogenated canola oil (FHCO) in high oleic sunflower oil (HOSO), Acevedo and Marangoni found an inverse relationship between system supersaturation and nanoplate-let size. In a sample of 100% FHCO, the mean length of nanoplatelets was 148 nm and the mean width was 63 nm. In a diluted sample of 30% FHCO in HOSO, the mean length was 369 nm and the mean width was 157 nm.[6] It is important to note that, although the sizes of the nanoplatelets changed with composition, the general morphology remained the same. Further studies of FHCO in CO systems at different concentrations showed the same trend: an increase in nanoplatelet length and width with a decrease in solid fat composition.[27]

An increase in supercooling of the melt increases the chemical potential difference between solid and liquid TAG and promotes nucleation. This result is seen in experiments as a decrease in mean nanoplatelet dimensions.

In more complex fat systems, supersaturation is difficult to compute. These systems contain non-uniform TAG mixtures with multiple distinct high melting point TAGs, which all contribute to the solid fat network. In these cases, solid fat content (SFC) is used as a measure of supersaturation.

Looking again at FHCO in HOSO mixtures and their interesterified blends, Acevedo and Marangoni found that the effect of supersaturation in these systems is dominant over other external field effects.[21] This is due to the elevated kinetic barrier which is already in place under high supersaturation conditions: the high viscosity of the melt in high SFC blends discourages the growth of crystals and any effect of increased cooling rate or other fields on nucleation is less pronounced.

It is interesting to note that decreases in platelet size at the nanoscale were accompanied by increases in mesocrystal dimensions when increasing SFC under static conditions.[6,27] This contrasting behaviour can be explained by the fact that, while nanoplatelet dimensions are dependent upon nucleation and crystal growth, mesocrystals are further affected by aggregation mechanisms. An increase in high melting TAG increases the total solid content in the system and lends more material for the formation of larger crystals.

Increasing solid fat content has a direct relationship with oil binding capacity. A higher portion of crystalline material lends more surface area to bind liquid oil and decreases oil loss.[9] Under high shear, this trend is not observed because the network is destroyed and oil binding capacity is compromised.

The elastic modulus of a fat is a function of the total solid content in the material.[63] However, in Acevedo and Marangoni's studies on sheared blends there is evidence that G' is not solely dependent on SFC. When FHCO in CO blends were sheared, SFC increased compared to their statically cooled equivalent mixtures, but this change was not accompanied by higher G' values. This indicates that the structural changes of the network caused by the processing also played a role in the rheological properties of the material.[9]

2.5.2 Crystallization Temperature and Cooling Rate

Crystallization temperature and cooling rate are two factors easily manipulated in fat processing systems which can impact crystal growth and structure. Previous studies have shown a decrease in mesocrystal size with increase in cooling rate,[24,28,64,65] as this quickly establishes high supersaturation in the melt causing nucleation to dominate over crystal growth.

In blends of FHCO in HOSO, Acevedo and Marangoni noted that an increase in cooling rate from 1 °C min^{-1} to 10 °C min^{-1} also resulted in a decrease in nanoplatelet size.[21] This reduction in size was from 10% to 50% in length and width, and from 5% to 10% in thickness depending on the FHCO portion in the blend. The larger size differences were seen in lower solid fat content blends (50% FHCO compared to 100% FHCO), where crystal growth was not already affected by molecular mobility.[21] Further studies by Maleky, Acevedo and Marangoni on FHCO in CO blended systems showed similar results with changes in cooling rates.[27]

The rheological properties of fat–oil blends processed under varying cooling rates were found to be dependent not only on the total fat content of the blend, but also on their crystal structures and solid particle distributions. In blends cooled at 10 °C min^{-1}, the network was structured by smaller nanoparticles and mesocrystals, forming disperse networks and contributing to a higher shear storage modulus (G') compared to those cooled at 0.7 °C min^{-1}.[27]

Using a model relating mechanical properties of the fat to the arrangement, size, volume, and interactions between crystals [see equation (2.5)],[63,66] Maleky, Acevedo and Marangoni were able to analyse this storage modulus behaviour.

$$G' \approx \lambda \Phi^{1/(3-D)} \tag{2.5}$$

where λ is a constant dependent on crystal size and interactions between crystals, Φ is the crystal volume fraction, and D is the fractal dimension, a measure of the spatial distribution of solids in space.

Interestingly, the ratio of $G'_{fast} : G'_{slow}$ was approximately equal to $\lambda_{fast} : \lambda_{slow}$. This suggests that the rheology of the fat is dependent on crystal size as well as the spatial distribution of those solids in the network.[27] Moreover, the ratios of mesoscale and nanoscale crystal sizes at fast and slow cooling rates were much larger than $\lambda_{fast} : \lambda_{slow}$, indicating that interactions between the crystals must play an important role in the physical properties of the fat.

Using a scraped surface heat exchanger, Acevedo, Block and Marangoni studied the functionalization of FHSO in SO blends by processing the mixtures at two different shear rates and three different wall temperatures: $-10\,°C$, $0\,°C$ and $20\,°C$.[8] Larger nanoplatelets were seen in all blends when wall temperature was higher, regardless of the shear rate used. At high temperatures, the melt experiences a slower cooling rate and has a lower degree of supercoiling, hence the promotion of crystal growth over nucleation.

2.5.3   Shear Processing

Shear processing of fat blends can be done in a turbulent field, where there is no set direction, or in a laminar field, where force is applied in a single direction. Turbulent shear processing is conducted industrially using a scraped surface heat exchanger (SSHE) or paddle mixer, and is used to speed up crystallization or achieve desired polymorphic forms.[33] The effect of laminar shear on blends is similar to that of turbulent shear fields, with the added factor of particle orientation. Because shear occurs during many steps of the food manufacturing process, its effect on the physical properties of all ingredients, including fat, must be considered. In the same way as a mixing step is crucial when combining ingredients to bake a cake, shear is used strategically on the industrial scale to achieve the desired consistency and final product characterization.

Looking at the functionalization of FHSO in SO blends, Acevedo and Marangoni used low shear processing ($30\ s^{-1}$) in a SSHE and laminar shear at two different rates to process the fat mixtures.[9] Their results showed an increase in nanoplatelet length by almost 200% with shear processing compared to statically cooled samples. This increase in size was seen in all sheared samples, but was most pronounced in the SSHE samples.

Mesoscale crystals reacted differently under shear, showing a slight decrease in size at low shear rates and no significant size difference under high shear processing.[9] The rheological behaviour of the blends, however, was closely linked with the nano-length scale, with G' and $\sigma*$ inversely proportional to nanocrystalline dimensions.

Blends processed in the SSHE showed the lowest G' values: a decrease of 75% to 85% compared to statically cooled samples. Yield stress also decreased after shear processing, to a greater degree than G'. The lower elastic modulus and yield stress were explained by damage to the network caused by the blades of the SSHE. This damage also affected the oil binding capacity of the network, resulting in a 5- to 46-fold increase in oil loss for sheared samples.

Shear fields are especially useful when processing cocoa butter, where the target polymorphic form is not kinetically favoured. The presence of shear during crystallization speeds up the transformation to the desirable β_V form, and affects crystal structure. In research done by Maleky, Smith and Marangoni, high shear (340 s^{-1}) in a laminar shear cell or paddle mixer was used to process cocoa butter to examine its influence on the fat crystal network.[39] Along with smaller nanoplatelet dimensions in sheared samples, more uniform distributions of sizes of the nanocrystals were seen, due to the even heat and mass distribution in these samples. Statically processed fats may have different microenvironments for crystallization due to uneven fields or poor heat and mass transfer in the melt, causing variation in nucleation and crystal growth rates in different regions.[67,68] This would lead to a wide distribution of nanoplatelet sizes in the fat. Looking at further changes, the average domain size of the sheared nanoplatelets, as measured by XRD and Scherrer equation calculations, was smaller than that of statically cooled samples by about one lamellar layer.[39]

The functional consequences of the high shear-treated cocoa butter were in agreement with the previous results on lipid blends: G' and σ^* were inversely related to nanoplatelet size, so the smaller nanoplatelets in sheared samples corresponded to higher elastic moduli and yield stresses. The melting profiles of the sheared blends were also affected by their crystal network changes. Analysis of their melting profiles showed that sheared samples exhibited lower and narrower melting peaks than their statically cooled equivalents. The lower melting points in sheared samples could be attributed to a smaller domain size (less one lamellar layer) of the nanoplatelets, while the narrow melting peak observed could be due to the uniform size distribution of platelets in the processed samples.[25]

Like other shear fields, laminar shear has been shown to enhance the rate of heat and mass transfer in TAG melt, promote formation of stable polymorphic forms, and affect crystal aggregation.[23,69] In addition, previous work has demonstrated that, in low solid fat systems, unidirectional shear can overpower interparticle interactions and induce particle orientation and growth parallel to the direction of the applied force.[70] The calculation of the rotational Peclet number (Pe_r) for the system [see equation (2.6)] indicates whether the system structuring is dominated by Brownian motion or shear.[70]

$$Pe_r = \frac{\dot{\gamma}}{D_r} \qquad (2.6)$$

where $\dot{\gamma}$ is the shear rate in s^{-1} and D_r is the rotational diffusivity of the crystalline particle; 30 s^{-1} was identified as the critical shear rate above which fat nanoplatelets were oriented by the laminar field.[29]

Looking at the effect of laminar shear on blends of FHSO in SO, Acevedo, Block and Marangoni observed not only orientation of crystalline particles with the external field, but also nanoplatelet growth in the direction of shear. Nanocrystals isolated from sheared samples at 30 s^{-1} and 240 s^{-1} showed 83% and 45% increases in aspect ratio, respectively.[29] Orientation of the nanoparticles was detected by examining the scattering intensities in the small angle XRD pattern. An uneven intensity distribution around the Debye ring (the characteristic circular pattern of reflected X-rays from one crystal plane on the detector) indicates an uneven scattering of the X-rays and an ordering of the crystals along an axis.[39]

These same experiments indicated the presence of a critical shear rate, below which shear forces induced nanoplatelet growth and above which crystal fragmentation and secondary nucleation occurred. This finding helped explain discrepancies in nanoplatelet size trends in differently sheared samples. Smaller nanoplatelets were isolated in high shear-processed blends (300 s^{-1} and 340 s^{-1}),[21,25,39] while larger nanoplatelets were identified in samples after lower shear treatments (30 s^{-1} and 240 s^{-1}).[8,9,29]

Nanoplatelet sizing and orientation caused by laminar shear affect fat functionality in different ways. In one example, Maleky and Marangoni measured the rate of oil migration in laminar-sheared cocoa butter samples.[40] The fat was processed under high shear (340 s^{-1}), causing a decrease in nanoplatelet size; oil migration through the sheared cocoa butter network was measured by placing a dyed fat (model chocolate filling) beside the cocoa butter. As the dye leaked from the "filling", its migration through the cocoa butter was monitored.

Liquid oil moves through the cocoa butter by diffusion[71] and this is partly dependent on particle size and shape. The porous nature of the fat crystal structure explains the ability of oil to migrate through the material. Tortuosity, a variable that takes into account pore size and connectivity, can be computed for the different systems.[72] Directionally sheared samples consistently showed a lower rate of oil loss and higher tortuosity compared to statically and turbulent-sheared samples.

Shear treatment causing changes in fat strength, melting profile and oil migration properties is able to increase the functional possibilities of different fats and open up new and improved applications. For example, a harder, lower melting point cocoa butter could provide more structure to a chocolate product, while maintaining its desirable mouth feel. Alternatively, a chocolate coating made under laminar shear could prevent oil leakage in a truffle, increasing its shelf life and quality.

2.5.4 Interesterification

Interesterification is a process used to chemically alter blends of TAG molecules to obtain a mixture of TAGs with randomized fatty acid distributions. It is commonly used in the food industry to modify blends of saturated and unsaturated fats in order to get desirable physical properties.

When studying the effect of processing on blends of FHCO in HOSO, Acevedo and Marangoni compared the crystal structure of non-interesterified and chemically interesterified mixtures crystallized under different conditions.[21] Change in solid fat content is an expected result of interesterification; in this case, when FHCO (mostly SSS) and HOSO (mostly OOO) blends are interesterified, there is a decrease in tristearin (and an increase in TAG species containing one or two stearic acids), resulting in a decrease in overall SFC.

This chemical modification leads to alterations on the nanoscale crystal structure as well as decreases in bulk crystal volume. When analysing nanoplatelet sizes, comparisons were made between blends and interesterified mixtures with the same solid fat content [*i.e.* a blend of 40% FHCO and 60% HOSO (SFC = 0.29) was compared against an interesterified blend of 60% FHCO and 40% HOSO (SFC = 0.28)]. This was done to look at the effect of interesterification alone on the crystal structure, as changes in SFC would also have a strong influence on the nature of the network.

Results showed that chemical interesterification led to a decrease in average nanoplatelet size compared to the unmodified blend. This result follows the increase in number of individual TAGs containing saturated fatty acids, which nucleate at lower temperatures. Interesterification is therefore another processing technique that can be used to achieve useful properties for otherwise less functional fat blends.

2.5.5   Emulsifiers

Emulsifiers are often added to fat-rich food products to improve functionality and affect fat crystal network formation by altering nucleation, crystal growth and polymorphism.[73]

The influence of unsaturated monoglyceride (UM) on two different blends of FHSO in SO, crystalized statically and under low shear, was studied by Acevedo, Block, and Marangoni and was seen to have a complex relationship with blend functionality.[8] Notably, monoglyceride addition was shown to affect the morphology of the larger crystals in the system, imaged with polarized light microscopy. In blends containing emulsifier, mesocrystals were described as granular particles while pure fat systems showed spherical aggregates. Similar to trends observed with changes in supersaturation, crystal mesoscale and nanoscale were affected differently by the presence of emulsifier.

While crystal size decreased on the mesoscale in the presence of emulsifier, nanoplatelets appeared significantly larger (1.6 times longer and 1.4 times wider at 3% UM). An increase in concentration of monoglyceride from 3% to 5% led to a further increase in nanocrystal dimensions, particularly in the length (2.2 times longer and 1.6 times wider than non-emulsifier blends at 5% UM). The growth of nanoplatelets in the presence of emulsifier was caused by inhibiting nucleation in the melt. Conversely, the UM led to the creation of smaller mesoscale crystals by impeding aggregation of crystallites. The interference of the monoglyceride in the system was due to the effect of cocrystallization, with the UM causing disorder in the TAG crystal structure.[8]

Disorder in the crystal structure also influenced fat functionality: blends with added emulsifier had significantly lower G' and σ^*. The effect of UM on oil binding capacity was composition dependent. Addition of emulsifier to the blend resulted in significantly lower oil loss for 40 : 60 blends and significantly higher oil loss in the higher solid fraction sample.[8]

The effect of unsaturated monoglyceride on fat functionality is shown to be system-specific and complex. The advantage of such a relationship is the existence of an array of emulsifiers that can be added to blends to achieve different changes in functionality.

2.6   CONCLUSIONS

This chapter has discussed fat structuring from the bottom up, with an emphasis on the newly elucidated fat crystal nanostructure. All length scales of the fat crystal network play a role in its functionality, and composition and external fields can be varied to adjust the physical properties of fat. Nanostructure and microstructure in the crystal network react differently to changes in crystallization conditions, but both length scales are reflected in the bulk physical properties of the material.

Possibilities for future research in this area are vast. New applications for underutilized fat blends can be proposed by altering their functionality with changes in processing and composition. With a further understanding of the effects of external fields on nanostructure, one could identify different processing methods (*i.e.*, high shear or fast cooling rate) that induce the same crystal structure formation and functional properties in the fats. In-depth structural analysis of specialty fats can also be performed, for example identifying the aspects of roll-in-shortening crystal structure that explain its unique strength and elasticity. And finally, an understanding of fat crystal networks lends the opportunity to identify and design alternative oil-structuring agents that mimic solid TAG structure. This last area has applications that go beyond food, from environmentally friendly lubricants to green cosmetics.

REFERENCES

1. F. D. Gunstone, *The Chemistry of Oils and Fats*, CRC Press, Boca Raton, FL, 2004.
2. B. A. Griffin and S. C. Cunnane, *Introduction to Human Nutrition*, JohnWiley & Sons, Ltd, Chichester, UK, 2nd edn, 2009, pp. 86–121.
3. Tentative determination regarding partially hydrogenated oils; request for comments and for scientific data and information, https://www.federalregister.gov/articles/2013/11/08/2013-26854/tentative-determination-regarding-partially-hydrogenated-oils-request-for-comments-and-for (accessed May 2014).
4. Dietary guidelines for Americans 2010, http://www.health.gov/dietaryguidelines/dga2010/DietaryGuidelines2010.pdf, (accessed May 2014).
5. F. Peyronel and R. Campos, *Structure–Function Analysis of Edible Fats*, ed. A. G. Marangoni, AOCS Press, Urbana, IL, 2012, pp. 231–294.
6. N. C. Acevedo and A. G. Marangoni, *Cryst. Growth Des.*, 2010, **10**, 3327.
7. A. G. Marangoni, N. Acevedo, F. Maleky, E. Co, F. Peyronel, G. Mazzanti, B. Quinn and D. Pink, *Soft Matter*, 2012, **8**, 1275.

8. N. C. Acevedo, J. M. Block and A. G. Marangoni, *Langmuir*, 2012, **28**, 16207.
9. N. C. Acevedo and A. G. Marangoni, *Food Bioprocess Technol.*, 2013, **2**, 575.
10. D. J. McClements and E. A. Decker, *Fennema's Food Chemistry*, ed. S. Damodaran, K. L. Parkin and O. R. Fennema, CRC Press, Boca Raton, FL, 4th edn, 2008, pp. 155–216.
11. A. P. Simopoulos, *Biomed. Pharmacother.*, 2002, **56**, 365.
12. S. F. O'Keefe, *Food Lipids Chemistry, Nutrition, and Biotechnology*, ed. C. C. Akoh and D. B. Min, CRC Press, Boca Raton, FL, 3rd edn, 2008, pp. 3–37.
13. R. P. Mensink and M. B. Katan, *N. Engl. J. Med.*, 1990, **323**, 439.
14. R. P. Mensink, P. L. Zock, A. D. M. Kester and M. B. Katan, *Am. J. Clin. Nutr.*, 2003, 77, 1146.
15. E. A. Decker, *Nutr. Rev.*, 1996, **54**, 108.
16. M. Lipp and E. Anklam, *Food Chem.*, 1998, **62**, 99.
17. W. W. Christie, *Lipid Analysis: Isolation, Separation, Identification, and Structural Analysis of Lipids*, The Oily Press, Bridgwater, UK, 3rd edn, 2003.
18. D. Tang and A. G. Marangoni, *J. Am. Oil Chem. Soc.*, 2006, **83**, 377.
19. S. S. Narine and A. G. Marangoni, *Food Res. Int.*, 1999, **32**, 227.
20. A. J. Haighton, *J. Am. Oil Chem. Soc.*, 1976, **53**, 397.
21. N. C. Acevedo and A. G. Marangoni, *Cryst. Growth Des.*, 2010, **10**, 3334.
22. M. Lipp and E. Anklam, *Food Chem.*, 1998, **62**, 73.
23. G. Mazzanti, S. E. Guthrie, E. B. Sirota, A. G. Marangoni and S. H. J. Idziak, *Cryst. Growth Des.*, 2003, **3**, 721.
24. M. L. Herrera and R. W. Hartel, *J. Am. Oil Chem. Soc.*, 2000, 77, 1177.
25. F. Maleky and A. Marangoni, *Cryst. Growth Des.*, 2011, **11**, 2429.
26. K. Sato, *Chem. Eng. Sci.*, 2001, **56**, 2255.
27. F. Maleky, N. C. Acevedo and A. G. Marangoni, *Eur. J. Lipid Sci. Technol.*, 2012, **114**, 748.
28. S. Martini, M. L. Herrera and R. W. Hartel, *J. Am. Oil Chem. Soc.*, 2002, **79**, 1055.
29. N. C. Acevedo, J. M. Block and A. G. Marangoni, *Faraday Discuss.*, 2012, **158**, 171.
30. D. Rousseau, A. R. Hill and A. G. Marangoni, *J. Am. Oil Chem. Soc.*, 1996, **73**, 973.
31. R. L. Wille and E. S. Lutton, *J. Am. Oil Chem. Soc*, 1966, **43**, 491.
32. G. M. Chapman, E. E. Akehurst and W. B. Wright, *J. Am. Oil Chem. Soc.*, 1971, **48**, 824.

33. R. O. Feuge, W. Landmann, D. Mitcham and N. V. Lovegren, *J. Am. Oil Chem. Soc.*, 1945, **39**, 310.
34. G. Mazzanti, S. E. Guthrie, E. B. Sirota, A. G. Marangoni and S. H. J. Idziak, *Cryst. Growth Des.*, 2004, **4**, 1303.
35. I. Heertje and M. Leunis, *LWT – Food Sci. Technol.*, 1997, **30**, 141.
36. I. Heertje, M. Leunis, W. Vanzeyl and E. Berends, *Food Microstruct.*, 1987, **6**, 1.
37. G. G. Jewell and M. I. Meara, *J. Am. Oil Chem. Soc.*, 1970, **47**, 535.
38. C. Poot, W. Dijkshoorn, A. J. Haighton and C. C. Verburg, *J. Am. Oil Chem. Soc.*, 1975, **52**, 69.
39. F. Maleky, A. K. Smith and A. Marangoni, *Cryst. Growth Des.*, 2011, **11**, 2335.
40. F. Maleky and A. G. Marangoni, *Soft Matter*, 2011, 7, 6012.
41. F. Peyronel, J. Ilavsky, G. Mazzanti, A. G. Marangoni and D. A. Pink, *J. Appl. Phys.*, 2013, **114**, 234902.
42. F. Peyronel and A. G. Marangoni, *Food Res. Int.*, 2014, **55**, 93.
43. M. Van den Tempel, *J. Colloid Sci.*, 1961, **16**, 284.
44. C. J. Nederveen, *J. Colloid Sci.*, 1963, **18**, 276.
45. P. C. Hiemenz and R. Rajagopalan, *Principles of Colloid and Surface Chemistry*, Marcel Dekker, New York, 3rd edn, 1997.
46. J. N. Israelachvili, *Intermolecular and Surface Forces*, Academic Press Inc., London, 1992.
47. D. A. Pink, *Structure–Function Analysis of Edible Fats*, ed. A. G. Marangoni, *AOCS*, Urbana, IL, 2012.
48. I. E. Dzyaloshinskii, E. M. Lifshitz and L. P. Pitaevskii, *Sov. Phys. Uspekhi*, 1961, **73**, 153.
49. V. A. Parsegian, *Van der Waals Forces: A Handbook for Biologists, Chemists, Engineers, and Physicists*, Cambridge University Press, Cambridge, New York, 2006.
50. J. Ilavsky, P. R. Jemian, A. J. Allen, F. Zhang, L. E. Levine and G. G. Long, *J. Appl. Cryst.*, 2009, **42**, 469.
51. J. Ilavsky, A. J. Allen, L. E. Levine, F. Zhang, P. R. Jemian and G. G. Long, *J. Appl. Cryst.*, 2012, **45**, 1318.
52. R. Jullien, *J. Phys. I Fr.*, 1992, **2**, 759.
53. T. Vicsek, *Fractal Growth Phenomena*, World Scientific, Singapore, 1989, vol. 2.
54. D. W. Schaefer, *MRS Bull.*, 1988, **13**, 22.
55. G. Beaucage, *J. Appl. Crystallogr.*, 1995, **28**, 717.
56. G. Beaucage, *J. Appl. Crystallogr.*, 1996, **29**, 134.
57. B. Hammouda, *J. Appl. Crystallogr.*, 2010, **43**, 716.
58. A. Guinier and G. Fournet, *Small-Angle Scattering of X-Rays*, Wiley, New York, 1955.

59. O. Glatter and O. Kratky (ed.), *Small Angle X-Ray Scattering*, Academic Press Inc., London, 1982.

60. D. A. Pink, B. Quinn, F. Peyronel and A. G. Marangoni, *J. Appl. Phys.*, 2013, **114**, 234901.

61. B. Quinn, F. Peyronel, T. Gordon, A. G. Marangoni, C. B. Hanna and D. A. Pink, *J. Phys. Condens. Matter*, 2014, submitted.

62. F. Peyronel, B. Quinn, A. G. Marangoni and D. A. Pink, *J. Phys. Condens. Matter*, 2014, submitted.

63. S. S. Narine and A. G. Marangoni, *Phys. Rev. E. Stat. Phys. Plasmas. Fluids. Relat. Interdiscip. Topic*, 1999, **60**, 6991.

64. R. Campos, S. S. Narine and A. G. Marangoni, *Food Res. Int.*, 2002, **35**, 971.

65. J. W. Litwinenko, A. M. Rojas, L. N. Gerschenson and A. G. Marangoni, *J. Am. Oil Chem. Soc.*, 2002, **79**, 647.

66. A. G. Marangoni, *Phys. Rev. B*, 2000, **62**, 951.

67. G. Mazzanti, S. E. Guthrie, E. B. Sirota, A. G. Marangoni and S. H. J. Idziak, *Cryst. Growth Des.*, 2004, **4**, 4.

68. G. Mazzanti, S. E. Guthrie, A. G. Marangoni and S. H. J. Idziak, *Cryst. Growth Des.*, 2007, 7, 1230.

69. S. D. Macmillan and K. J. Roberts, *Cryst. Growth Des.*, 2002, **2**, 22.

70. R. G. Larson, *The Structure and Rheology of Complex Fluids*, Oxford University Press, Oxford, UK, 1999.

71. J. R. Galdámez, K. Szlachetka, J. L. Duda and G. R. Ziegler, *J. Food Eng.*, 2009, **92**, 261.

72. C. E. Schaefer, R. R. Arands, H. A. van der Sloot and D. S. Kosson, *J. Contam. Hydrol.*, 1995, **20**, 145.

73. N. Garti, *Phys. Prop. Lipids*, 2002, 265.

CHAPTER 3

Polysaccharide Nanostructures

VASSILIS KONTOGIORGOS

Department of Biological Sciences, University of Huddersfield, HD1 3DH, UK
Email: v.kontogiorgos@hud.ac.uk

3.1 INTRODUCTION

Polysaccharides are carbohydrate polymers in which sugar units are linked together through glycosidic linkages. In living organisms, polysaccharides are the structural polymers that provide support (*e.g.* cellulose in plants or chitin in arthropods) or the sources of energy for plant development (*e.g.* starch). Polysaccharides are routinely used in the food industry, most frequently as thickeners, stabilizers of dispersions (emulsions, foams) or structuring agents of water and air. Thickening solutions and stabilizing dispersions against creaming are two of the most common industrial applications of polysaccharides. These functional properties are used to create formulations with reproducible flow properties, not only during processing but also during the specified shelf-life of the product. The viscosity of a polysaccharide solution exhibits a remarkable increase above the critical polymer concentration (c^*). Polysaccharides normally show Newtonian or pseudoplastic flow behavior at concentrations below or above c^*, respectively. As is evident, concentration, along with other factors, is critical and can be used to control the functionality of polysaccharides. Common polysaccharides that are used to enhance viscosity include xanthan, galactomannans, starches and cellulose

Edible Nanostructures
Edited by Alejandro G Marangoni and David Pink
© The Royal Society of Chemistry 2015
Published by the Royal Society of Chemistry, www.rsc.org

derivatives. Apart from thickening solutions and conferring desirable textural properties, polysaccharides can also be used in more technologically demanding applications that require structuring of water or air, or emulsifying a hydrophobic compound. They can be used for partial or total replacement of fat in reduced-fat formulations by structuring water in the form of a gel. The textural and functional characteristics of the gelled structure should be comparable to those of fat. This is a particularly difficult task, considering the extensive dissimilarities in the chemical structure and physical properties of fats and hydrocolloids. Of particular importance in fat replacement is the melting behavior of the gel, which should resemble that of fat [*i.e.*, a melting point that is close to the body temperature ($\sim 37\ °C$)] with a sharp melting transition so as to impart mouth-melting characteristics to the structure. Another important feature is the structural stability of the gel, required in order to provide a desirable shelf-life for the product.

Quality losses are usually manifested by the presence of a thin layer of water that is expelled out of the structure. This is known as syneresis and is due to rearrangements of the microstructure with time. Syneresis not only results in loss of visual qualities,but, in most cases, is accompanied by losses in texture of the product. Some polysaccharides that are used as fat replacers are polydextrose, microcrystalline cellulose, maltodextrins, and modified starches. As discussed later, mixed polysaccharide systems or mixtures of a polysaccharide with a protein solution may provide, in some cases, superior structuring of water. Air structuring using polysaccharides is another functionality that is employed in the baking industry, and specifically in gluten-free formulations. Hydrophobically functionalized cellulose derivatives [*e.g.* methylcellulose (MC) or hydroxypropylmethylcellulose (HPMC)] are used in applications where thermoreversibility of the gel is required. These polysaccharides self-assemble on heating by means of weak reversible hydrophobic interactions, which lead to gel formation. In gluten-free formulations, the leavening agent (*e.g.* bicarbonate) creates CO_2 bubbles in the dough, which makes it rise. The polysaccharide network that forms on heating during baking (see Figure 3.3, below, for mechanism) not only entraps CO_2, but also provides structural rigidity to the newly formed microstructure. On cooling, the gel reverses to the sol state and the polysaccharide now acts as a water management agent. Formulations of deep fried products (*e.g.* chicken nuggets, fish fingers) may also require similar functionality to prevent oil migration and structure disintegration during frying at high temperatures.[1] Finally, stabilization of flavor oils (*e.g.* limonene) is also possible with the use of appropriate polysaccharides. In this case, the polysaccharide should be able to create fine emulsions without enhancing the viscosity of the solution. This can be achieved by polysaccharides that have been properly functionalized so as to be arranged at the oil–water interface. A typical polysaccharide with this functionality is gum Arabic, which is able to create fine emulsions with minimum increase in viscosity even at concentrations as high as 20%.[2]

Understanding the mechanisms of structure formation demands departure from the traditional approach of analytical and chemical descriptions of polysaccharides and utilization of concepts from materials science. Such an approach is imperative, as research in the last two decades shows that many aspects of food ingredient functionality can be controlled by the interaction of distinct structural elements at various length scales rather than simply by their chemical characteristics.[3] The mesoscopic scale plays a central role in engineering food structure and, for all practical purposes, ranges from 1 nm to 1 μm, although the exact boundaries are not well defined.[4] At this scale, the properties of the material cannot be described adequately by continuum mechanics because interactions among discrete particles come into play.[4,5] The interplay between attractive *vs.* repulsive forces and molecular mobility dictates the stability of the material. Usually, the system is considered stable when the energy barrier between the particles is larger than the thermal fluctuations.[6] Such stability may refer, for instance, to stability against flocculation in emulsions, phase separation in mixed biopolymer systems, gelation in single biopolymer solutions, or stabilization of biopolymer matrices below their glass transition temperature. As foods are metastable materials (out of equilibrium) they are susceptible to structural reorganization through various relaxation mechanisms.[7] Consequently, stability refers to "kinetic stability", emphasizing that the system is arrested at a temporarily stable molecular arrangement that usually matches the technological requirements of the material. Typical examples of such behavior are the α-relaxation in biopolymer glasses in the vicinity of glass transition temperature or the enhancement of inter-chain interactions in gels. The loss of stability in the former example is manifested by the loss of the structural integrity of the material as it enters the rubbery state. In the latter case, syneresis occurs with expulsion of water from the structure, accompanied by significant changes in the mechanical properties of the gel.

The physicochemical responses that influence the functionality and industrial performance of polysaccharides are controlled by the fine structure of the chains at the molecular level. The objective of this chapter is to outline how structure is created and controlled in a wide range of polysaccharide-based systems that are utilized in food applications.

3.2 POLYSACCHARIDE SOURCES AND COMPOSITION

Polysaccharides can be obtained from plants with minimal process-ing (*e.g.* rice or potato starch) or as a result of processing of agri-cultural wastes (*e.g.* pectin). Other sources include extraction from algae (*e.g.* alginates, carrageenan), processing of by-products of the shellfish industry (*e.g.* chitin), or from microbial fermentation (*e.g.* xanthan, gellan). It should be noted that extraction from natural sources or culture media results inevitably in the presence of proteins that, depending on their content, may affect to a various degree the properties of the polysaccharide extract. Irrespective of the protein content in the extract, the fine structure of the isolate depends heavily on the isolation protocol that was followed. For example, the choice of pH, salt concentration, temperature, solvent for precipitation or dry-ing technique (*e.g.* freeze drying *vs.* spray drying) can modify the molecular characteristics of polysaccharides. Modifications may in-clude changes in molecular weight and its distribution, the presence and extent of branching and the extent of functionalization (*e.g.* methyl, acetyl). In many cases the isolated polysaccharide has totally different chemical and physical properties from its source. A typical example is pectin, where although extraction procedures are optimized to tailor isolates having various highly specific functional properties, the structure within the plant cell wall is still largely unknown.[8]

Although in nature there are numerous monosaccharides, the number comprising the polysaccharides is relatively small (Table 3.1). Common sugar units include glucose and mannose, which form the backbone of some of the most important commercial polysacchar-ides. Other sugars or sugar acids, such as galactose, xylose, arabinose or galacturonic, guluronic and mannuronic acids, are commonly found in industrially relevant polysaccharides (Table 3.1). However, the type of linkages, isomeric forms and functionalization of sugars, as well as the branching and periodicity of the monomers in the backbone, result in great structural diversity. Slight structural modi-fications usually change the functionality of the polysaccharide. These modifications include, for example, methylation or acetylation at various positions, the presence of sulfate or other functional

Table 3.1 Predominant monomers and functional groups in polysaccharides, with examples of polysaccharide diversity obtained by varying the linkage type.

Predominant monomers	Predominant functional groups	Linkage variation in glucose homopolysaccharides
Glucose	Carboxyl	β-1→3 (Laminaran)
Mannose	Sulfate	α-1→4 (Amylose)
Galactose	Methyl	β-1→4 (Cellulose)
Arabinose	Acetyl	β-(1→4, 1→3) (β-glucan)
Xylose	Propyl	α-(1→4, 1→6) (Amylopectin)
Rhamnose	Hydroxypropyl	α-(1→2, 1→3, 1→4, 1→6) (Dextran)
Fructose	Amide	
Galacturonic acid		
Mannuronic acid		
Guluronic acid		
Glucuronic acid		
Glucosamine		

groups, or differences in the anomeric type of monosaccharides that make up the polysaccharide (Table 3.1). A notable example is that of amylose and cellulose, which both consist of glucose. Glycosidic linkages between glucose units in amylose are α-D-(1→4), whereas in cellulose they are β-D-(1→4), resulting in totally different functional properties, not only within the plant (structural *vs.* source of energy), but also when they are used as food ingredients. This difference at the molecular level also has implications in the higher level of structure. For example, cellulose chains are able to assemble and form fibrous semi-crystalline structures with unique mechanical properties, whereas amylose is a flexible chain that has the ability to form crystals under certain conditions (*e.g.* in bread staling). By further varying the linkage type and anomeric form between glucose units, a range of different polysaccharides can be obtained with various functional properties (Table 3.1).

3.3 POLYSACCHARIDE CONFORMATIONS

Polysaccharide structuring starts at the molecular level, where they are generally encountered with either ordered or disordered conformations.[9] A polysaccharide forms when several monosaccharide units, usually more than 20, are connected together *via* glycosidic linkages. Polysaccharides are commonly divided into homopolysaccharides or heteropolysaccharides on the basis of the number of different sugars in the structure (Figure 3.1a).

Homopolysaccharides contain a single sugar unit on the backbone (*e.g.* amylose), whereas heteropolysaccharides have more than one (*e.g.* pectin). The sequence of sugar residues in the chain forms the *primary structure* of the polysaccharide. For example, in homopolysaccharides that contain only one sugar residue, the primary structure consists of a sequence of the same sugar unit (Figure 3.1a). In hetero-polysaccharides the repeating motif may be a disaccharide or longer segment (*e.g.* in carrageenan or gellan), resulting in more complex primary structures (Figure 3.1a and Table 3.2). Further classifications are possible, for example, according to the source, type of sequence, charge, *etc.* Table 3.2 shows examples of repeating patterns of common polysaccharides that are used in industrial applications.

The sugar units have the ability to rotate around the glycosidic linkage with two torsion angles (φ, ψ) (Figure 3.1b). Although the pyranose ring also shows flexibility, its effect on the conformation of polysaccharides is negligible when compared to the effect of the rotations around the glycosidic bonds.[10] Therefore, the conformations that affect the interactions of polysaccharides at the molecular level can be understood by studying the conformations of disaccharides. Figure 3.1b shows the different possible torsion angles in a polysaccharide. Angle φ is located between the anomeric carbon and the oxygen of the glycosidic linkage of the first monomer, and ψ between the oxygen of the glycosidic linkage and the non-anomeric carbon of the second monomer. Introduction of branching at C6 gives one more possible angle of rotation (ω) about the C-5 and C-6 bond (Figure 3.1b). The conformation of a polysaccharide chain can be specified by the relationship between the φ and ψ torsion angles. Because of the ability of sugar monomers to rotate about the linkages, polysaccharides may adopt *secondary structures*. When $\varphi_1 = \varphi_2$ and $\psi_1 = \psi_2$ (and all the subsequent φ, ψ sets in the chain are the same) the chains adopt a helical conformation in the solid state. However, in solutions $\varphi_1 \neq \varphi_2$ and $\psi_1 \neq \psi_2$ and the chains generally tend to adopt random coil conformations.[10] The most stable conformation is usually the one that results in the lowest energy, because some are not allowed owing to steric hindrances. These steric hindrances are short range, between neighboring residues, or long range, with sugar units that are remote in the chain but near in space (Figure 3.1c). The long-range interactions result in excluded volume effects that depend on the qualities of the solvent (ionic strength, pH). As a result of these interactions, polysaccharides may adopt one or more of the three idealized conformations (*secondary structures*): random coil (*e.g.* pullulan), ribbons (*e.g.* cellulose) or helices (*e.g.* κ-carrageenan).[9,11,12]

Table 3.2 Classification of polysaccharides according to charge, with examples of repeating units for some common polysaccharides and their sources.

Class	Examples of repeating units	Examples of Polysaccharides	Sources
Neutral		Starch, cellulose, β-glucan, glycogen, curdlan	Cell walls, tubers or roots, microbes, mammals
		Guar, locust bean gum, fenugreek gum	Seeds
		Agarose	Algae
		Arabinoxylans	Cell walls

Table 2 (*Continued*)

Class	Examples of repeating units	Examples of Polysaccharides	Sources
Anionic		Pectins	Cell walls
		Alginates	Algae
		κ-Carrageenan	Algae
Cationic		Chitin, chitosan	Arthropods

Interactions of polysaccharide chains at the molecular level depend most commonly on the quality of the solvent. In good solvents, interactions between solvent and chain segments are favorable, resulting in extended conformations and high solubility. In poor solvents, interactions of chain segments with themselves are favored, resulting in aggregation. At a specific temperature called the θ-temperature, the long-range interactions no longer influence the conformations of the chains and the short-range interactions become predominant (Figure 3.1c). The interplay between the interactions of polysaccharide and solvent molecules determines whether the biopolymer will be able to form stable structures at greater length scale, most commonly gels. When polysaccharides are charged the situation becomes more complex because charges also affect chain conformation. To control these interactions it is possible to manipulate a range of factors such as concentration, temperature,

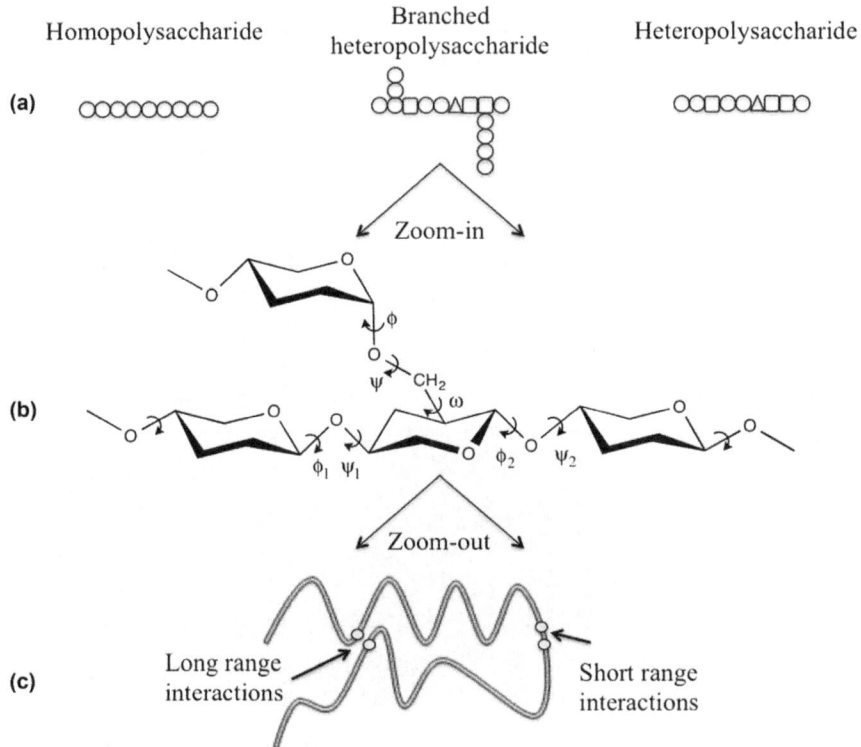

Figure 3.1 (a) Illustration of different types of polysaccharides. (b) Torsion angles between adjacent monosaccharides in a polysaccharide chain. (c) Interactions within a polysaccharide chain.

polydispersity, ionic strength and pH, or to add crosslinkers such as calcium cations, as in the case of low methoxylated pectins or alginates. It is very important to understand how the various factors depend on each other because deviations from optimum conditions usually influence the ability of the polysaccharide chains to associate into a three-dimensional network.

It is evident from the above discussion that polysaccharide structures do not fit into a simple description, owing to the multitude of factors that need to be controlled simultaneously. Various experimental techniques are available to study conformations at various length scales, such as X-ray diffraction, light scattering, small angle X-ray scattering, nuclear magnetic resonance (NMR) or atomic force microscopy (AFM). AFM is one of the few techniques that allows visual observation of a single polysaccharide chain. It generates images by sensing the surface of the molecule with the aid of a sharp probe. Because this technique minimizes sample preparation it is possible to image polysaccharides in a "near native" state of the macromolecule.[13] Images obtained using AFM under various experimental conditions (Figure 3.2) illustrate the great diversity in chain

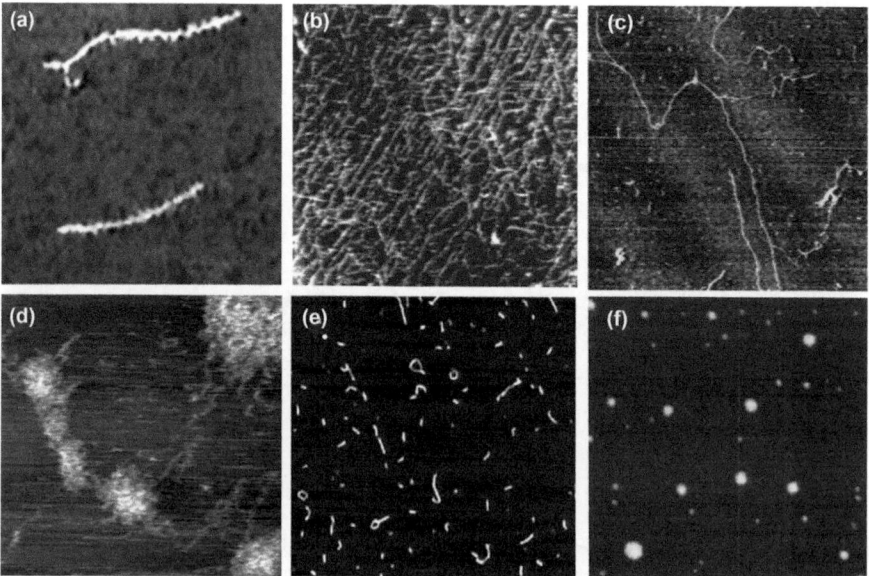

Figure 3.2   Atomic force microscopy images of nanostructure of various polysaccharides: (a) xanthan, (b) κ-carrageenan, (c) pectin, (d) gellan, (e) β-glucan and (f) arabic gum. Scan size is 2.5×2.5 μm for a, c and d and 2×2 μm for b, e and f.
(Reprinted with permission from refs. 14–19.)

conformations of polysaccharides. Xanthan,[14] κ-carrageenan[15] and pectins[16] (Figure 3.2a, b and c) form elongated structures, whereas gellan (Figure 3.2d) forms short rods,[17] each corresponding to polysaccharide-specific conformations. On the other hand, intra-chain aggregation in β-glucan[18] dispersions is evidenced by the presence of large aggregates with linear chains protruding away from the structure (Figure 3.2e). This is a typical behavior when intra-chain interactions are strong. Finally, arabic gum[19] shows globular structures as a result of the presence of protein moieties on the polysaccharide backbone (Figure 3.2f). It should be stressed that these images represent the conformations of polysaccharides under the specific conditions that were used to capture them, and they tend to change depending on the composition of aqueous medium. However, they demonstrate the complexity that is involved in polysaccharide structuring at nanometer length scales.

3.4 STRUCTURING USING POLYSACCHARIDES: HIGH MOISTURE REGIME

A bottom-up approach to structuring requires the biopolymer chains to assemble and form well-defined "building blocks" at nanoscale level, which may interact and further develop to a macroscopic structure at higher length scales. The macroscopic structure is usually "soft" owing to the characteristic mechanical properties of the resulting material (*e.g.* low yield point, viscoelasticity). Such structuring occurs *via* weak, reversible, non-covalent interactions, *i.e.*, hydrogen bonding, hydrophobic, ionic, and van der Waals interactions, steric and excluded volume effects. The aggregated system represents a minimum energy structure or equilibrium phase and exhibits short-range, localized ordering in contrast to the long-range atomic order of crystals.[20] The previously mentioned forces that are responsible for the ordering between molecules are both attractive and repulsive and the balance between them determines the stability of the structure. Repulsive interactions in polysaccharides in aqueous solvent are mostly due to steric and excluded volume effects.[21] Excluded volume is the volume that one part of a long chain cannot occupy when it is already occupied by another part of the same chain. Furthermore, when atoms in a chain are too close to each other their electron clouds overlap, resulting in steric repulsion. Both events influence the polysaccharide conformation and its ability to form macrostructures. Attractive forces are the result of the van der Waals interactions and hydrogen bonding that stem from dipole–dipole interactions.

These forces are important in gel formation by polysaccharides, particularly if we consider the multitude of hydroxyl groups in polysaccharide chains that are available to interact with water or with each other. Ionic forces predominate when polysaccharides are charged. This occurs very frequently when monomers have reactive groups available such as carboxyl or sulfate (*e.g.* carrageenan, pectin or alginate). Bridging of adjacent chains and subsequent gel formation is frequently mediated by the presence of cations (*e.g.* Ca^{++}, K^+). Finally, the hydrophobic effect is important when polysaccharides are functionalized with hydrophobic groups such as methyl, acetyl, propyl, *etc.* (*e.g.* cellulose derivatives or pectin). This confers new properties on polysaccharides, such as gel formation on heating or the ability to arrange at interfaces and act as emulsifiers.

At this juncture, we should stress there is an important difference between polysaccharides and other biological molecules that have the propensity to self-assemble at nanoscale. Self-assembly of polysaccharides is not as easy as for small amphiphilic molecules (*e.g.* mono- or diglycerides, surfactants) or proteins, because dispersions of polysaccharides in aqueous solutions exhibit very low interfacial tension, conferring water solubility on the molecule. In contrast, with casein micelles, the most characteristic self-assembled food nanostructure, the specific balance of the hydrophobic to hydrophilic amino acids not only allows formation of the nanostructure, but also helps in retaining the individual character of the micelles. In polysaccharides, self-assembly requires the chemistry of the monomers to be modified by the introduction of appropriate functional groups. For biomedical applications and drug delivery, self-assembled polysaccharide nanostructures are currently being used in a wide range of applications. These are mostly based on chitosan or dextran derivatives and various glycosaminoglycans.[22,23] In such applications, the polysaccharide nanoparticle is usually required to deliver a specific functionality to cells or tissues but is not required to build macroscopic superstructures. In these cases the individuality of the nanoparticles should be retained and aggregation phenomena must be avoided. In contrast, food structuring with polysaccharides requires creation of structures up to the macroscopic length scale with specific mechanical properties and technological functionality. Therefore, the individual character of the nanoparticle is rarely important in food structuring applications, and association at atomic or mesoscale requires further aggregation to create a three-dimensional macrostructure, namely, a gel. Gelation involves attractive interactions among polysaccharide chains, which convert the solution into a three-dimensional metastable viscoelastic "soft" solid occupying the same volume as the solution.

As discussed above, polysaccharide chains in water will interact with each other (inter-chain interactions), with themselves (intra-chain interactions), and with water molecules (chain–solvent inter-actions). Inter-chain interactions usually lead to gel formation, whereas intra-chain interactions result in aggregation of the poly-saccharide and precipitation. In gels formed by neutral polysacchar-ides, the length scale is controlled to some extent by the mesh size of the network. Similarly to semi-dilute polymer solutions, the mesh size can be adjusted by the polysaccharide concentration, affecting their mechanical properties directly. In gels formed by charged poly-saccharides, mesh size can be also adjusted by carefully tuning pH and ionic strength or by addition of crosslinking ions.[3,24] The pH influences, in most cases, the degree of dissociation of the carboxyl group of uronic acid residues, whereas in chitin and chitosan, the pH influences the dissociation of the amino group. When the pH is above the pK of the charged group, repulsive interactions maintain the chains in extended conformations. Ionic strength can also be used to tailor the interactions and conformations in polyelectrolytes. Charged polysaccharides at low salt concentrations (low ionic strength) tend to adopt extended conformations as electrostatic repulsion keeps charged groups apart. Electrostatic screening provided by counterions at higher concentrations (usually 0.1 M NaCl) contracts chains to form more compact conformations, affecting solubility and the ability to gel.[25,26]

Gels are classified according to the nature of their interactions into covalently crosslinked, entanglement or physical networks.[27] In food systems, the most predominant gels are those that are formed *via* physical interactions. Interactions at the molecular level involve the creation of structures with short-range order, such as helices, "egg boxes", ion-assisted bridging or junction zones. Depending on the strength of these interactions, gelation may be reversible or ir-reversible. Figure 3.3 illustrates three different mechanisms of gel formation using representative examples for κ-carrageenan, methyl-cellulose and mixed linkage $(1 \rightarrow 3)(1 \rightarrow 4)$-β-D-glucan. The κ-carra-geenan gelation mechanism initiates with helix formation and ion-assisted crosslinking of the helices.[28] In κ-carrageenan solution above ~ 60 °C, the chains are in random coil conformation. Cooling below ~ 60 °C induces a coil-to-helix transition and κ-carrageenan coils are able to form double helices. Aggregation proceeds with for-mation of hydrogen bonding between helices, which in turn enables

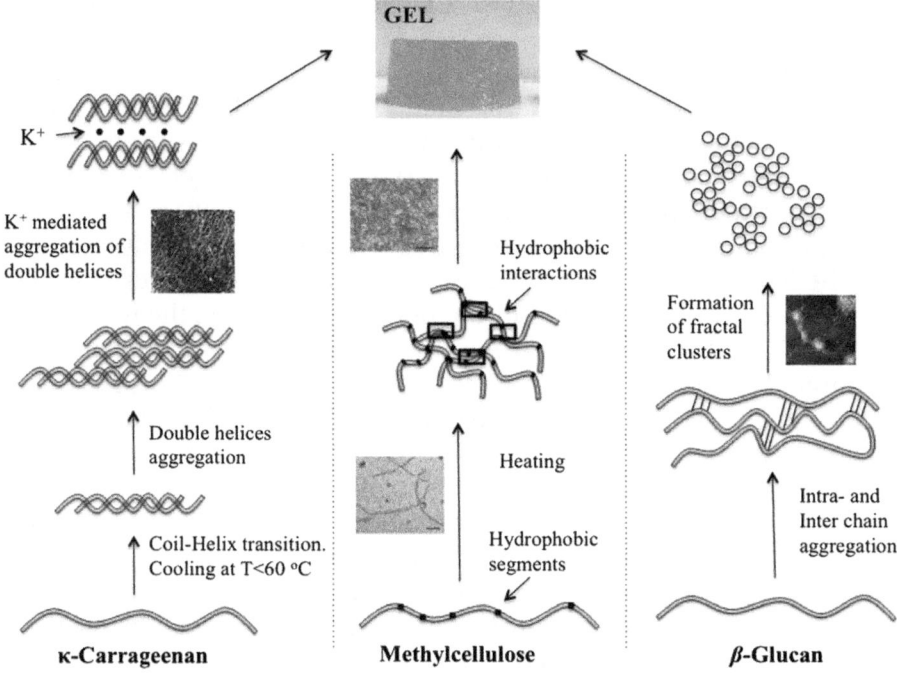

Figure 3.3   Examples of gelation mechanisms and AFM images of structure for representative polysaccharides. Association of chains involves various forces depending on the polysaccharide type, monomer composition and functionalization. (AFM images reprinted with permission from refs. 15, 29 and 18 for κ-carrageenan, methylcellulose and β-glucan respectively.)

formation of a weak three-dimensional network. The introduction of potassium cations to the solution allows crosslinking of helices, owing to the presence of sulfate groups. The mechanical properties of the final gel depend not only on the molecular properties of the κ-carrageenan that is used to create the gels (*e.g.* sulfate content, molecular weight, polydisperisty), but also on the concentration of K^+, ionic strength and the pH of the solution. As is evident in this case, there are several parameters available that can be used to fine-tune the structure and the properties of the gel. Other polysaccharides that gel by means of coil–helix transition include gellan, agar and curdlan.

Hydrophobic interactions among polysaccharides can also be exploited to create gels for food applications, as in the case of hydrophobically modified celluloses [*e.g.* methylcellulose (MC), hydroxylpropylmethylcellulose (HPMC)]. Polymer chains of MC solutions are in disordered conformation at room temperature. On heating, MC chains are capable of interacting with each other to form thermally reversible gels.[29] The mechanism of gel formation is based on the extent of hydrophobic interaction among the MC chains that associate to form a fibrillar gel.[30] As temperature increases, hydrophobic interactions strengthen and chains are able to assemble and form the gel. This gel is thermally reversible and the sol form is recovered as temperature drops below the critical temperature for association. In consequence, the degrees of freedom to control characteristics of the network are the molecular weight and its distribution and the degree of substitution with hydrophobic groups.

A third mechanism of gelation that is commonly encountered in proteins is displayed by mixed linkage $(1 \rightarrow 3)(1 \rightarrow 4)$-$\beta$-D-glucan. This polysaccharide exhibits random coil conformation in hot aqueous solutions. In this case, gel formation progresses by interactions of at least three consecutive cellotriosyl residues that result in conformational ordering with inter- and intra-chain associations, at chain segmental level.[31] Such interactions lead to the formation of a fractal network of particular aggregates that has been described using scaling concepts.[32] Particulate aggregates interact mainly *via* hydrogen bonding, creating fractal clusters that result in the gelled structure. The particulate nature of β-glucan gels has been recently reinforced by AFM imaging[18] (see Figure 3.2f) and particle tracking microrheology,[33] revealing the microheterogeneities that occur during microstructural evolution of the network. Controlling gelation for this type of gel usually requires tailoring of the molecular properties of β-glucan chains to a specific molecular weight and cellotriosyl-to-cellotetraosyl ratio. Particulate gels are most commonly encountered in proteins, where denaturation under specific conditions allows aggregation of the particles to produce colloidal-type, usually irreversible, networks. It is evident that, in all cases, manipulations are directed towards influencing the interactions at the molecular level and affecting the conformational properties of the chains.

Microstructure engineering in polysaccharide systems can also be achieved by varying the processing conditions during gel formation. Application of shear is a way to create new microstructures and should be applied during the conformational ordering process, resulting in *fluid gels*.[34] In this case, the polysaccharide solution is sheared while it undergoes conformational transition, resulting in the production of gel particles *via* a nucleation and growth mechanism. The gel particles grow to a specific droplet size, and stability is obtained if the particles are kept below the gel melting temperature so that re-aggregation is prevented.[35] A range of tools is available to control the microstructures formed, such as the cooling rate, the strength of the shear field, and the concentration and type of polysaccharide. These factors control the droplet size distribution, their shape, and the interactions among the droplets that in turn affect the stability and mechanical properties of the fluid gels.[34] Microstructures can also be fabricated starting from mixtures of phase-separated biopolymers when at least one component is able to gel.[36,37] In this case, a shear field with simultaneous cooling can be used to fabricate the droplet. Shear forces deform the droplet, and cooling induces gelation that kinetically arrests the formed droplets. An example of the effect of the shear field on the morphology of fluid gels can be seen with gellan–κ-carrageenan mixtures (Figure 3.4). From Figure 3.4(a) to (f), the strength of the shear field increases, with concomitant changes in the particle morphology. For instance, at low shear, droplet coalescence takes place before gelation and the particles are bigger (Figure 3.4b) than in the fluid gel created under quiescent conditions (Figure 3.4a). At higher shear rates, the particles become elongated (Figure 3.4c–e) and beyond a specific value the particles obtain non-specific morphology (f).[36] Fluid gels can be used to improve the rheological properties of various products in the food and personal care industries, and to control the release of nutrients in the gut to improve satiety.

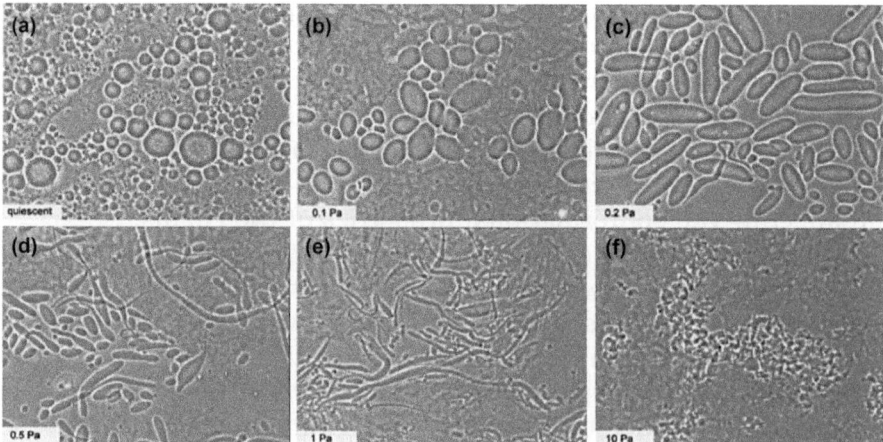

Figure 3.4 Micrographs showing the effect of shear stress on phase-separated gellan–κ-carrageenan mixtures. Stress increases from (a) to (f). Scale bar is 100 μm.
(Reproduced with permission from ref. 36.)

Mixing two different biopolymer species can also achieve micro-structure manipulation and tuning of gel properties. In most cases, mixtures contain two different polysaccharides or a polysaccharide and a protein. Mixing two biopolymers brings about new physico-chemical responses to the system. The mixtures are broadly classified into two groups depending on the nature of interactions between the biopolymer species. Interactions are either *segregative* or *associative* and lead to phase separation or creation of complexes, respectively. Phase separation creates phases that are enriched in one of the two biopolymers, whereas complexation creates complexes that are either soluble or insoluble. The demixing of the biopolymer species depends on the interplay of the interactions between the biopolymer species in the mixture, as described previously. Phase separation primarily depends on the concentration of the biopolymers in the mixture and on the structural characteristics of the chains (*e.g.* molecular weight, charge). Below a concentration threshold the two biopolymers co-exist, whereas beyond the threshold value they phase-separate. The phase behavior is better understood with the use of isothermal phase diagrams of biopolymer mixtures.[38] Figure 3.5(a) illustrates the phase diagram of mixtures of sodium caseinate with β-glucan varying in molecular weight. The solid line represents the binodal, which sets the boundaries of the compatible (below the curve) and the incompatible (above the curve) regions. Compatibility generally increases as the molecular weight of the polysaccharide and nominal concentration of biopolymers in the mixtures decrease. This is a general behavior that is observed in protein–polysaccharide mixtures and influences the stability and rheology of the system.[37–39] These system properties can be adjusted by modifying the concentration and molecular characteristics of the constituent biopolymers, solvent quality, or temperature. The phase behavior also plays a dramatic role in the microstructure of phase-separated mixtures (Figure 3.5b). It is evident that, as the β-glucan molecular weight decreases, a remarkable change in the morphology of the mixtures occurs. The coarse β-glucan-enriched microphases, in the high molecular weight samples, are gradually transformed into fine droplets as the size of the chains decreases, a situation that influences the rheology and textural properties of the mixture.[40] Furthermore, a remarkable change in the continuity of the mixtures occurs as polysaccharide concentration increases (from left to right).[40] Mixtures in which sodium caseinate is the continuous phase change progressively to continuous β-glucan systems, passing from the bi-continuous counterpart. When such a mixture is gelled under the appropriate conditions, as a result of microstructure manipulation,

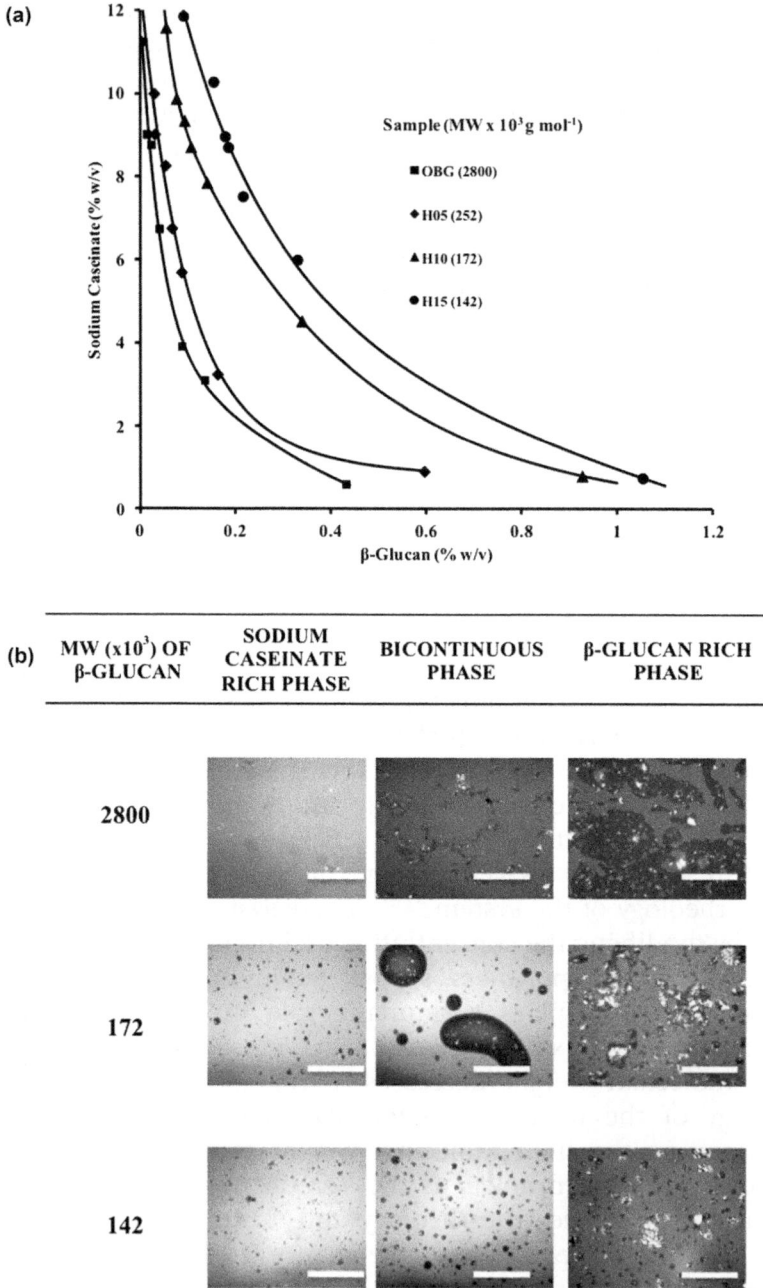

Figure 3.5   (a) Isothermal phase diagrams at 5 °C and (b) microstructural evolution
of β-glucan–sodium caseinate mixtures varying in polysaccharide molecular weight and concentration. Scale bar: 100 μm.
(Reproduced with permission from ref. 40.)

the thermal and mechanical properties vary greatly.[41] At this stage the gels will have distinct mechanical properties, depending on the continuity of the system. We can distinguish three classes, where the gel is: a) biopolymer-A continuous, b) biopolymer-B continuous, or c) bicontinuous. The three types of gel will have completely different rheological, thermal and microstructural properties.[42] Gels that involve synergistic interactions between polysaccharides can be created in a similar manner. Interaction creates gels with properties distinct from those that are created in the absence of the second polysaccharide. For example, mixtures of galactomannans with carrageenans create firmer gels compared to those without galactomannans. Furthermore, interactions between xanthan and galactomannans lead to gelation, although neither of the solutions is able to gel alone.[26] Mixed polysaccharide systems have been explored extensively in the literature for various applications such as reduction of fat and calories, control of texture and mouth feel of various food formulations, or simply reduction of the cost of existing formulations.[42]

3.5 STRUCTURING USING POLYSACCHARIDES: LOW MOISTURE REGIME

The previous discussion focused on the behavior of polysaccharides in solution under conditions that promote gelation. We saw that microstructural elements of polysaccharides may form disordered or short-range ordered structures. The typical level of solids in such a gelling system depends on the chemical properties of the polysaccharide, but in most cases is in the range of 0.5–2%. However, in low-moisture systems that contain biopolymers, water fails to hydrate them adequately. This restricts molecular mobility and conformational rearrangements, and the structure of the material is distinct from its high-moisture counterparts. We can normally distinguish two solid states in polysaccharide systems: the *crystalline* and the *amorphous*. In most cases, branching and chemical heterogeneity restrict crystallization. However, some polysaccharides, either in their native state or under appropriate sample preparation conditions, may give distinct X-ray diffraction patterns revealing the formation of structures with long-range order. On the other side of the spectrum, amorphous solid state lacks long-range order and polysaccharide chains are in a completely disordered state. In that case, glass transitions dominate the physicochemical and mechanical responses of the system. At this point we should mention that this solid state is not encountered in lipid systems.

Studying long-range order of polysaccharides is a difficult task because large crystals cannot be provided for X-ray diffraction studies. Furthermore, powder diffraction X-ray patterns are difficult to interpret because of the molecular complexity and polycrystalline nature of the structures. Polycrystalline materials are composed of aggregated small crystals of different size and orientation. In polysaccharides and some synthetic polymer systems, these materials also include amorphous regions in their structure. Typical polysaccharides that acquire a polycrystalline character during their biosynthesis are starch,[43] cellulose[44,45] and chitin.[46] In cellulose and chitin, for instance, acid hydrolysis of the amorphous regions results in fabrication of new materials that consist of aggregates of cellulose or chitin crystals at various length scales. These materials find applications in the food and pharmaceutical industries as fat substitutes, texture modifiers, tablet binders or additives that reinforce polymer composites. Starch granules present another example of the ability of sugar polymers to form complex crystalline structures controlled by the molecular composition of the material. Maize starch powder X-ray diffraction patterns, for example, show the crystalline and non-crystalline regions of the structure (Figure 3.6a).[47] Furthermore, increasing amylose concentration in the granule decreases the crystallinity of starch, which is attributed mostly to the formation of double helices of amylopectin.[48] To overcome some of the difficulties that are posed by the absence of well-defined single polysaccharide crystals, fiber X-ray diffraction may be used to study the molecular orientation of polysaccharides. In this case, a fiber is prepared that consists of oriented microcrystalline and amorphous regions,[10,49] the extent of which depends on the particular architecture of the polysaccharide (Figure 3.6b).[50]

When crystalline solids melt and form liquids, with subsequent temperature reduction the liquid may crystallize again. Crystallization can be frequently delayed or inhibited, depending on the cooling rate of the liquid solution. When such a liquid solution is cooled below its melting point, it enters a *supercooled* state. With further reduction of temperature in the absence of crystallization, the viscosity of the liquid increases significantly and it eventually undergoes a *glass transition*. The formed amorphous solid-state structure is called "glass". Biopolymer solutions rarely crystallize on cooling (*e.g.* amylose recrystallization), but glass formation often plays an important role in the physical stability and textural properties of the food matrix. Glasses in food systems may be obtained either by removal of water (*e.g.* dehydration or extrusion processes) or by cooling of high-solids

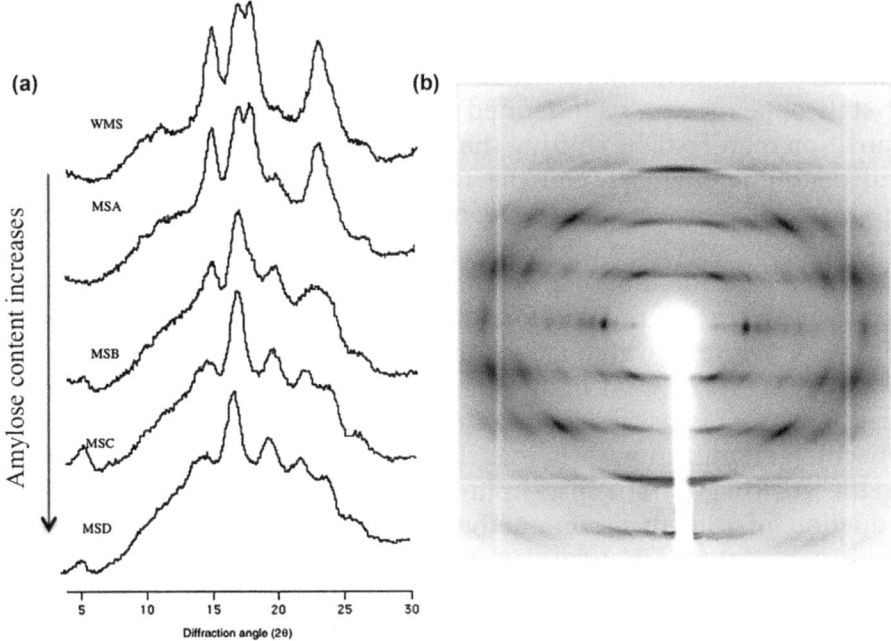

Figure 3.6 (a) Powder X-ray diffraction patterns of maize starch varying in amylose content and (b) fiber X-ray diffraction patterns of ι-carrageenan. (Reproduced with permission from refs. 47 and 50.)

biopolymer solutions below a specific temperature. What happens microscopically at the glass transition is that, on the time scale of observation, the translational and rotational motions of the atoms or the molecules that give rise to the viscous flow have ceased. Below glass transition temperature (T_g), during the measurement period, the atoms are vibrating only about their equilibrium positions. The resulting glassy system is expected to be stable below T_g, whereas, above T_g, the difference between T_g and the storage temperature T $(T - T_g)$ controls the rate of physical and chemical changes.[51] It was stated earlier that, below the glass transition, molecular motions (albeit restricted) persist. This mobility is mainly local and restricted to atom or bond vibrations, or reorientation of small groups.[51–53] Sub-T_g relaxations are named according to their position relative to the main α-relaxation (glass transition), which is due to cooperative motions of the molecules or polymer chains. At lower temperatures, β- and γ-relaxations take place and are linked to rotation of lateral groups (such as –OH or –NH) or to changes in the conformation of the main chain in the case of biopolymers.

Melting and glass transition events can be followed by differential scanning calorimetry, which distinguishes between first (melting) and second (α-relaxation) order transitions. Typically, melting of crystals appears as a well-defined endothermic peak, whereas glass transition manifests by shifting the heat capacity baseline. Identifying and distinguishing between the two transitions pinpoints the processing and storage requirements of polysaccharide-based structures. Thermal properties are ultimately controlled by the fine structure of the polysaccharide, but also by the water content of the system. For instance, starch gelatinization and glass transition temperatures vary with water content, which affects the functional characteristics of the material (Figure 3.7). At high water contents the major endothermic peak (~ 70 °C) is assigned to melting of crystalline regions of amylopectin (gelatinization), whereas the peak at about 110 °C is assigned to the melting of amylose–lipid complexes (Figure 3.7a).[54] With reduction of moisture content below 30% the gelatinization peak disappears, as starch granules cannot absorb water and hydrate. As water content is lowered further (<18%), glass transition events of the amorphous regions of starch granule appear on heating (Figure 3.7b).[55,56] These move to higher temperatures as the plasticization effect of water (see below) is diminished with the decrease in water content. Other relevant events that can be followed using calorimetry include gel "melting" and changes in the protein denaturation temperature. In the case of gels, melting is not a typical first-order transition because there is no actual crystalline structure present. Rather, it refers to the "detachment" of the contact points (*e.g.* junction zones) with increase in temperature.

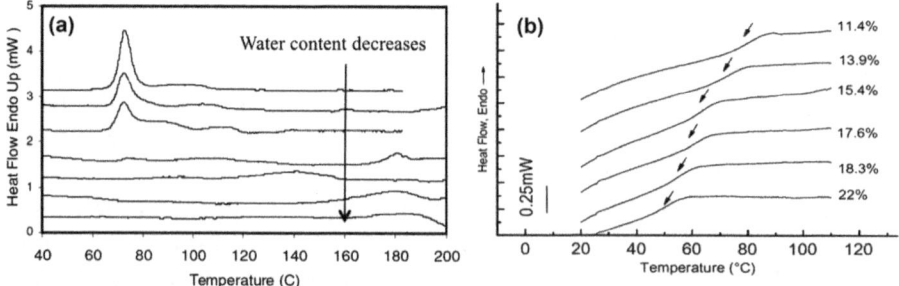

Figure 3.7   Differential scanning calorimetric thermograms of starch at a wide range of moisture content (a) 74–9% and (b) 22–11%.
(Reproduced with permission from refs 54 and 56.)

The microstructure of glasses depends on the kinetics of glass formation or, in other words, on the rate that the system arrives at its pseudoequilibrium (the rate of cooling or water removal). High cooling rates (or fast water removal) arrest the system at a more disordered (more "open") state than slower cooling rates (or slow water removal). Such a process results in structures that are not in thermal equilibrium with their surroundings. Owing to the low temperature motions, or with storage near the α-relaxation temperature (usually between Tg and the temperature where β-relaxations occur), the thermodynamic properties such as enthalpy, entropy and volume will tend to evolve towards their equilibrium values, a process that is called *physical aging*. Aging significantly affects the properties of glassy materials and preparation of the glassy phase; storage should be carefully controlled because the microstructure and glass transition of biomaterials are interdependent.[57] Because of physical aging, the material is subject also to microstructural rearrangements that may have implications to the stability of the system.[52] To account for the variations in the dynamics of the material undergoing a glass transition, the fragility parameter m is introduced[58] to distinguish systems in which relaxation mechanisms (*e.g.* viscosity) are highly dependent on temperature above T_g (m between 100 and 200, "fragile") from those that are less dependent (m between 16 and 100, "strong").[59,60] Such a classification has important implications in various technological processes that may allow tailoring the technological performance of polysaccharide matrices. Variations in the parameter m between two glassy polysaccharide structures may result in significant changes in stability because the rate of relaxation mechanisms (*i.e.* the speed by which the system approaches equilibrium) is influenced significantly in the vicinity of T_g. Furthermore, various processes that involve fast removal of water (*e.g.* extrusion, flaking) or rapid cooling (*e.g.* in the confectionary industry) may benefit from the understanding of relaxation mechanisms of the materials that are utilized in the formulations. Several polysaccharide systems are reported as "strong", indicating moderate dependence of relaxation mechanisms on temperature. For example, in pullulan,[61] chitosan and chitosan blends[62] or pullulan–starch blends,[63] the fragility parameter m varies between 30 and 96, depending on the molecular weight and the moisture content of the materials.

The glass transition temperature depends on the molecular weight of the polysaccharide and the presence of low molecular weight compounds called *plasticizers*.[59] The most common plasticizer for polysaccharide matrices is water, but other small molecules can also show plasticization effects (*e.g.* glucose, sorbitol or glycerol). Plasticizers increase the free volume of the system, thus increasing the molecular mobility of the chains. The result of increased molecular mobility is that the glass transition occurs at lower temperatures than it would in the absence of a plasticizer (see Figure 3.7b). Consequently, by intelligent manipulation of the water content, the microstructural and textural characteristics can be precisely controlled. Furthermore, engineering of novel materials such as edible film coatings[64,65] or encapsulation matrices[66] can be also achieved. Edible films are mostly prepared from polysaccharides, although proteins can also be used as starting materials. Such a film is a low-moisture polysaccharide system that comes into direct contact with the surface of the food. Films provide a barrier to moisture loss or uptake and control gas exchange of food with the environment (*e.g.* O_2 or CO_2). They can be also used to control microbial growth when antimicrobial compounds are introduced. It is easy to realize that the properties of films have a profound dependence on the plasticization effect of any water that may migrate from the food or the atmosphere to the film. This plasticization may reduce the glass transition temperature to the storage temperature of the product, thus altering the effectiveness of the film.

Encapsulation of active ingredients such as flavoring, coloring or nutrients is also accomplished by the use of polysaccharides. This technology allows protection of the encapsulated material from oxidation and losses due to evaporation, light or interactions with food ingredients. Encapsulation usually proceeds with immobilization of the desirable component into a glassy polysaccharide matrix. In the operating environment (*e.g.* the mouth, stomach or intestines), the active component will be released in a controlled manner from the matrix to provide its functionality (*e.g.* flavor or nutrient release). Similarly to edible films, the capacity of the encapsulating matrix to stabilize the ingredients depends on the properties of the glassy polysaccharide matrix and the plasticization effect of water.

3.6 CONCLUSIONS

The evolution of structure formation has been reviewed for a range of polysaccharide systems. Although polysaccharides consist of a relatively small number of monosaccharides, they have the capacity to form a wide range of structures. The interactions among the chains are those that primarily control how the structure will evolve and stabilize. Depending on the water content of the system, it is possible to distinguish two regimes in which polysaccharides can form completely different structures with distinct physical and mechanical properties. In high-moisture systems, polysaccharides are able to form gels, making it possible to structure water or air. At the opposite extreme, where moisture content is low, the glassy state and the related relaxation phenomena control the structural stability of the material, whereas some native materials also show structures with long-range order. The greatest drawback for materials based on polysaccharides is their metastable nature, *i.e.* their sensitivity to structural evolution in time. In food applications, this is usually manifested by a limited shelf-life stability and changes in functional properties during storage. Further work should focus on exploring how to limit the kinetic processes that influence these changes, so as to provide novel polysaccharide materials with improved functional properties.

REFERENCES

1. C. Primo-Martin, T. Sanz, D. W. Steringa, A. Salvador, S. M. Fiszman and T. van Vliet, *Food Hydrocoll.*, 2010, **24**, 702.
2. O. G. Phillips and A. P. Williams, *Handbook of Hydrocolloids*, Woodhead, Oxford, 2009.
3. J. Ubbink, A. Burbidge and R. Mezzenga, *Soft Matter*, 2008, **4**, 1569.
4. D. Frenkel, *Physica A*, 2002, **313**, 1.
5. T. C. Lubensky, *Solid State Commun.*, 1997, **102**, 187.
6. E. Dickinson, *Soft Matter*, 2006, **2**, 642.
7. R. G. M. Van Der Sman, *Adv. Colloid Interface Sci.*, 2012, **176-177**, 18.
8. B. R. Thakur, R. K. Singh, A. K. Handa and M. A. Rao, *C. R. Food Sci. Nutr.*, 1997, **37**, 47.
9. W. C. Steve and W. Qi, *Food Carbohydrates*, CRC Press, New York, 1st edn, 2005.
10. V. S. R. Rao, P. K. Qasba, P. V. Balaji and R. Chandrasekaran, *Conformation of Carbohydrates*, Harwood Academic Publishers, Amsterdam, 1998.
11. S. E. Harding, K. M. Varum, B. T. Stokke and O. Smidsrod, in *Advances in Carbohydrate Analysis*, ed. C. A. White, JAI Press, Greenwich, 1st edn, 1991, vol. 1, 63–144.
12. D. A. Rees, *Polysaccharide Shapes*, Chapman and Hall Ltd, London, 1977.
13. V. J. Morris, in *Modern Biopolymer Science*, ed. S. Kasapis, I. T. Norton and J. B. Ubbink, Academic Press, San Diego, 1st edn, 2009, p. 365.

14. T. A. Camesano and K. J. Wilkinson, *Biomacromolecules*, 2001, **2**, 1184.
15. S. Ikeda, V. J. Morris and K. Nishinari, *Biomacromolecules*, 2001, **2**, 1331.
16. V. J. Morris, A. P. Gunning, A. R. Kirby, A. Round, K. Waldron and A. Ng, *Int. J. Biol. Macromol.*, 1997, **21**, 61.
17. S. Ikeda, Y. Nitta, B. S. Kim, T. Temsiripong, R. Pongsawatmanit and K. Nishinari, *Food Hydrocoll.*, 2004, **18**, 669.
18. J. K. Agbenorhevi, V. Kontogiorgos, A. R. Kirby, V. J. Morris and S. M. Tosh, *Int. J. Biol. Macromol.*, 2011, **49**, 369.
19. S. Ikeda, T. Funami and G. Zhang, *Carbohyd. Polymers*, 2005, **62**, 192.
20. R. Kelsall, W. I. Hamley and M. Geoghegan, *Nanoscale Science and Technology*, John Wiley & Sons, Ltd, Chichester, 2005.
21. M. Rubinstein and R. H. Colby, *Polymer Physics*, Oxford University Press Inc., New York, 2003.
22. S. Boddohi and M. J. Kipper, *Adv. Mater.*, 2010, **22**, 2998.
23. Z. Liu, Y. Jiao, Y. Wang, C. Zhou and Z. Zhang, *Adv. Drug Deliv. Rev.*, 2008, **60**, 1650.
24. P. de Gennes, *Scaling Concepts in Polymer Physics*, Cornell University Press, Ithaca, New York, 1979.
25. O. Smidsrød and A. Haug, *Biopolymers*, 1971, **10**, 1213.
26. M. Djabourov, K. Nishinari and S. B. Ross-Murphy, *Physical Gels from Biological and Synthetic Polymers*, Cambridge University Press, Cambridge, 2013.
27. G. M. Kavanagh and S. B. Ross-Murphy, *Prog. Polym. Sci. (Oxf.)*, 1998, **23**, 533.
28. L. Piculell, *Curr. Opin. Colloid Interface Sci.*, 1998, **3**, 643.
29. R. Bodvik, A. Dedinaite, L. Karlson, M. Bergstrom, P. Baverback, J. S. Pedersen, K. Edwards, G. Karlsson, I. Varga and P. M. Claesson, *Colloids Surf. A: Physicochem. Engin. Asp.*, 2010, **354**, 162.
30. J. R. Lott, J. W. McAllister, S. A. Arvidson, F. S. Bates and T. P. Lodge, *Biomacromolecules*, 2013, **14**, 2484.
31. A. Lazaridou and C. G. Biliaderis, *J. Cereal Sci.*, 2007, **46**, 101.
32. V. Kontogiorgos, H. Vaikousi, A. Lazaridou and C. G. Biliaderis, *Colloids Surf. B*, 2006, 49.
33. T. Moschakis, A. Lazaridou and C. G. Biliaderis, *J. Colloid Interface Sci.*, 2012, **375**, 50.
34. P. W. Cox, F. Spyropoulos and I. T. Norton, in *Modern Biopolymer Science*, ed. S. Kasapis, I. T. Norton and J. B. Ubbink, Academic Press, San Diego, 1st edn, 2009, p. 199.
35. I. T. Norton, D. A. Jarvis and T. J. Foster, *Int. J. Biol. Macromol.*, 1999, **26**, 255.

36. B. Wolf, R. Scirocco, W. J. Frith and I. T. Norton, *Food Hydrocoll.*, 2000, 14, 217.
37. I. T. Norton and W. J. Frith, *Food Hydrocoll.*, 2001, **15**, 543.
38. V. B. Tolstoguzov, *Food Hydrocoll.*, 2003, **17**, 1.
39. V. B. Tolstoguzov, *Nahrung*, 2000, **44**, 299.
40. J. K. Agbenorhevi, V. Kontogiorgos and S. Kasapis, *Food Chem.*, 2013, **138**, 630.
41. V. Kontogiorgos, C. Ritzoulis, C. G. Biliaderis and S. Kasapis, *Food Hydrocoll.*, 2006, **20**, 749.
42. S. Kasapis, *Crit. Rev. Food Sci. Nutr.*, 2008, **48**, 341.
43. S. Perez, M. P. Baldwin and J. D. Gallant, in *Starch: Chemistry and Technology*, ed. J. BeMiller and R. Whistler, Academic Press, London, 3rd edn, 2009, pp. 149–192.
44. A. C. O'Sullivan, *Cellulose*, 1997, **4**, 173.
45. P. Zugenmaier, *Prog. Polym. Sci.*, 2001, **26**, 1341.
46. J.-B. Zeng, Y.-S. He, S.-L. Li and Y.-Z. Wang, *Biomacromolecules*, 2011, **13**, 1.
47. N. W. H. Cheetham and L. Tao, *Carbohydr. Polym.*, 1998, **36**, 277.
48. H. F. Zobel, *Starch – Stärke*, 1988, **40**, 1.
49. T. Yui and K. Ogawa, in *Polysaccharides: Structural Diversity and Functional Versatility*, ed. S. Dumitriu, Marcel Dekker, New York, 1st edn, 2005, p. 99.
50. S. Janaswamy, K. L. Gill, O. H. Campanella and R. Pinal, *Carbohydr. Polym.*, 2013, **94**, 209.
51. D. Champion, M. Le Meste and D. Simatos, *Trends Food Sci. Tech.*, 2000, **11**, 41.
52. G. Roudaut, D. Simatos, D. Champion, E. Conteras-Lopez and M. Le Meste, *Innov. Food Sci. Emerg. Techonol.*, 2004, **5**, 127.
53. G. Roudaut and D. Champion, *Food Biophys.*, 2011, **6**, 313.
54. H. Liu, L. Yu, F. Xie and L. Chen, *Carbohydr. Polym.*, 2006, **65**, 357.
55. K. J. Zeleznak and R. C. Hoseney, *Cereal Chem.*, 1987, **64**, 121.
56. J. Perdomo, A. Cova, A. J. Sandoval, L. Garcia, E. Laredo and A. J. Muller, *Carbohydr. Polym.*, 2009, **76**, 305.
57. S. Kasapis, *Food Hydrocoll.*, 2012, **26**, 464.
58. C. A. Angell, R. D. Bressel, J. L. Green, H. Kanno, M. Oguni and E. J. Sare, *J. Food Eng.*, 1994, **22**, 115.
59. M. Le Meste, D. Champion, G. Roudaut, G. Blond and D. Simatos, *J. Food Sci.*, 2002, **67**, 2444.
60. B. Borde, H. Bizot, G. Vigier and A. Buleon, *Carbohydr. Polym.*, 2002, **48**, 83.
61. A. Lazaridou, C. G. Biliaderis, V. Kontogiorgos, *Carbohydr. Polym.*, 2003, **52**, 151.

62. A. Lazaridou and C. G. Biliaderis, *Carbohydr. Polym.*, 2002, **48**, 179.

63. C. G. Biliaderis, A. Lazaridou and I. Arvanitoyannis, *Carbohydr. Polym.*, 1999, **40**, 29.

64. F. Debeaufort, J. A. Quezada-Gallo and A. Voilley, *C. R. Food Sci. Nutr.*, 1998, **38**, 299.

65. A. Sorrentino, G. Gorrasi and V. Vittoria, *Trends Food Sci. Technol.*, 2007, **18**, 84.

66. L. Lakkis, *Encapsulation and Controlled Release Technologies in Food Systems*, Blackwell, Oxford, 2007.

CHAPTER 4

Protein Nanostructures

OWEN GRIFFITH JONES

Purdue University, 745 Agriculture Mall Dr., West Lafayette, IN 47907, USA
Email: joneso@purdue.edu

4.1   PROTEINS AS MATERIALS

Protein is perhaps the most essential component of biological structures, providing a versatile and adaptable material for the assembly of cells and organs. Enzymes, the workhorses of the biological cell, are comprised largely of protein. Protein is also a major required nutrient for human consumption in foods, being utilized for the building of protein structures in the body, as well as serving as precursors for a variety of essential biological components, such as neurotransmitters. Thus, the quantity and quality of protein in food products has been a concern since humans first began making the link between the food they ate and their resulting health.

Although it has long been known that eating high-protein foods (*e.g.* meat) provided necessary nutrients to the diet, it was not until the eighteenth century that protein as a material was extracted and recognized as an essential component.[1] In the nineteenth century, the chemist Gerardus Mulder discovered that all animal and plant materials contained a common nitrogen-containing chemical of a typical ratio of carbon, nitrogen, and hydrogen. In correspondence with a highly renowned chemist of the time, Jönz Jacob Berzelius, Mulder referred to these molecules as "protein" because the substance

Edible Nanostructures
Edited by Alejandro G Marangoni and David Pink
© The Royal Society of Chemistry 2015
Published by the Royal Society of Chemistry, www.rsc.org

appeared to be a primary, or first (from the Greek *protos*), substance of biological nutrition.[2] Amino acids were being gradually discovered in the same time frame, and eventually it was noted that proteins were assembled from these fascinating molecular monomers.[3] In fact, the functionality and structure of proteins derives from the diversity of properties within the constituent amino acids. These properties include hydrophobicity, ionization, polarity, and the possession of aromatic groups. A summary of the 20 common amino acids and their essential properties is given in Table 4.1.

Table 4.1 Structure and attributes of amino acids that affect the functional properties of proteins.

Amino acid	Side chain (side chain "R")	$pK_a{}^a$	Hydrophobicity[b]
Alanine	—CH₃	n/a	1.0
Arginine	—CH₂—CH₂—CH₂—NH—C(=NH)—NH₂	12.0	1.1
Aspartic acid	—CH₂—C(=O)—OH	3.9–4.0	−0.1
Asparagine	—CH₂—C(=O)—NH₂	n/a	−0.1
Cysteine	—CH₂—SH	9.0–9.5	0.0
Glutamic acid	—CH₂—CH₂—C(=O)—OH	4.3–4.5	0.5
Glutamine	—CH₂—CH₂—C(=O)—NH₂	n/a	0.5
Glycine	—H	n/a	0.0
Histidine	—CH₂—(imidazole ring, NH, N)	6.0–7.0	1.3
Isoleucine	—CH(CH₃)—CH₂—CH₃	n/a	2.7

Table 4.1 (*Continued*)

Amino acid	Side chain (side chain "R")	pK$_a^a$	Hydrophobicity[b]
Leucine	CH$_3$ | ◄CH$_2$—CH—CH$_3$	n/a	2.9
Lysine	◄CH$_2$—CH$_2$—CH$_2$—CH$_2$—NH$_2$	10.4–11.1	1.9
Methionine	◄CH$_2$—CH$_2$—S—CH$_3$	n/a	2.3
Phenylalanine	◄CH$_2$— (benzene ring)	n/a	2.3
Proline	(pyrrolidine ring, HN—)	n/a	1.9
Serine	◄CH$_2$—OH	n/a	0.2
Threonine	H, OH, CH$_3$	n/a	1.1
Tryptophan	CH$_2$ (indole ring, NH)	n/a	2.9
Tyrosine	◄CH$_2$— (benzene ring) —OH	9.7	1.6
Valine	CH$_3$ | ◄CH—CH$_3$	n/a	2.2

[a]From ref. 145.
[b]Estimated energy for solvation, kcal mol^{-1}, from ref. 146.

Proteins are polymers of amino acids, and the quantity and order of the amino acids in the polymer chain defines their behavior in biological systems. Typically, the environment immediately surrounding protein is composed of water with a variety of ions and small-molecule osmolytes, such as sugars. Because water is a fairly polar solvent, hydrophobic amino acids are poorly solvated, while polar or charged amino acids are better solvated. This translates to a reduction in net energy when the hydrophobic amino acids cluster together, effectively decreasing contact with water molecules. The opposite is true for polar and charged amino acids, which are more likely to be exposed to the bulk water. Specific interactions between amino acids are also possible, such as hydrogen bonding between polar groups or formation of disulfide linkages (*i.e.* covalent bonds) between two cysteine residues. Interactivity with water and these specific intra-peptide interactions then contribute to the structural complexity of protein, which can vary substantially between different protein molecules. Protein structure may be further refined within biological cells during their assembly to possess a very specific conformation. These defined conformations allow proteins to be utilized for structural roles (*e.g.* fibers) or as catalyzing materials (*e.g.* enzymes).

Protein structures are built or further refined in the laboratory by chemical or physical means to create desirable biomaterials with nanometer-scale dimensions. Designing such structures from native proteins requires some knowledge of the native protein's attributes, as well as an understanding of the colloidal physics that governs their assembly and post-assembly behavior. In the following section, general properties of typical proteins are described in context of their capacity to form nanometer-scale structures. Current knowledge of protein nanostructures, both natural and designed in laboratories, will also be summarized with the intention of recognizing the common themes and possible future progress in assembling protein nanostructures for value-added applications. Continual advancement in the assembly of protein nanostructures to serve as controlled delivery systems and structural biomaterials has made this an exciting time to study protein physical behavior.

4.1.1 Protein Sources in Food

As mentioned, food materials are biological constructs and therefore contain significant fractions of protein. While from the nutritive viewpoint proteins are only considered for their amino acid content and bioavailability, proteins serve both structural and functional roles in plants or animals that are highly specific to their amino acid composition and conformation. For instance, a large proportion of the proteins in meat tissues from animal muscle serve a physiological function of providing locomotion, which requires significant strength, viscoelasticity, and response to biological triggers. Other proteins serve a general structural role, such as collagen or elastin, where strength and elasticity are even more important. Milk and eggs are animal products intended for delivery of nutrition to developing animals, and so the proteins found in these products are typically highly bioavailable or carry essential vitamins or minerals. Similarly, proteins in seeds or beans that are commonly utilized as agricultural commodities are rich in storage and structural proteins for the development of new seedlings.

Milk is one of the largest sources of protein in the human diet, and in 2013 the commercial consumption of produced milk in the United States was over 201 billion pounds.[4] Protein constitutes approximately 3.5% (w/v) of bovine milk, and this protein is comprised of casein, whey protein, serum albumin, immunoglobulins, and proteose-peptones.[5] Casein, the most prominent assembly of protein in milk, is composed of several protein fractions with excellent nutritive and surface-active qualities. During cheese and yogurt manufacture, the solubility of casein is reduced by acid or enzyme activity and it becomes an insoluble mass known as "curd". Separation of the curd for the production of most cheeses leaves a liquid product known as "whey", which contains sugars and small proteins. Chief of the whey proteins are β-lactoglobulin and α-lactalbumin, both containing isoelectric points near 5 and molecular weights below 20 kDa. Milk proteins are some of the most studied proteins owing to their high production and good water solubility, and in-depth reviews on these proteins can be readily found in the literature.[5,6]

Oil-producing seeds and legumes, such as soybeans, cottonseed, and rapeseed, possess large quantities of storage proteins in the peripheral tissues of the harvested bean or kernel.[7] The seeds are typically defatted, and protein meal is separated from fibers and other components at either acidic or alkaline pH. Isolated proteins have varying structure, nutritional quality, and functionality, but there is ongoing interest in increasing the utilization of these materials to feed the growing populations of our planet. Soybean globular proteins β-conglycinin and glycinin, which are the 7S and 11S fractions (during centrifugation) of extracted storage proteins, respectively, represent a high proportion (~ 50–80%) of the ubiquitous soy flour and soy protein concentrate that is available in the market.[8] These proteins are composed of several subunits held together by hydrophobic interactions and disulfide bonds, making them relatively large structural starting materials. Soybean proteins are popular ingredients in the food industry, as they have desirable emulsifying and gelling properties. Proteins from other oil seeds are also of developing interest to the scientific community, yet the level of their production has limited their potential utility as materials. Grain proteins are an essential component of the human diet and produced in massive quantities worldwide. Grain products include wheat, rye, and barley. Of the diversity of proteins extracted from grains, glutenin and gliadin from wheat are by far the most studied and best known.[9] Glutenin is a polymeric assembly of globular subunits held together by disulfide bonds with good water solubility. The size and hygroscopic network-forming properties make glutenin essential for the structure of many baked goods, such as traditional bread. Gliadin is a relatively hydro-phobic prolamine fraction that is soluble in ~70% ethanol and tends towards self-association or interactions with glutenin protein.

Other important protein sources include meat tissues, egg, algae, and non-oil producing legumes. Gelatin, the highly elastic fibrous protein obtained from animal collagen, is one of the most widely used materials for biomedical and controlled release applications. Apart from gelatin, production of protein-rich products from these sources, such as protein isolates and concentrates, is currently limited. How-ever, there is a growing interest in finding high-value utility in some of the extracted protein products, such as through the development of nanostructures that could be used for controlled release of bioactives or incorporated into gels or films.

4.1.2 Physical Properties of Proteins

Physical interactions between proteins, which determine their cap-acity to form supramolecular assemblies such as nanostructures, are dictated by the amino acid composition and physical structure. Native protein structure is often tightly controlled, and loss of this native structure leads to changes in the capacity of those proteins to interact in solution. In addition, the charge, hydrophobicity, and physical intactness of the protein influence their ability to assemble into nanostructures. Overall, knowledge of the protein structure and its surface characteristics in the chosen environment (*e.g.* pH, ionic strength, and temperature) is essential for developing targeted strat-egies for the desired assembly of nanostructures.

Tertiary protein structures result from interactions between amino acids within the protein and, for practical purposes, are generally divided into three major types: globular, fibrillar, and random. In globular conformations, the polypeptide chain is folded into a compact spheroid with most of the non-polar amino acids located in the interior and the polar amino acids located at the exterior.[10] The major driving force for this configuration is the hydrophobic effect resulting from the presence of hydrophobic amino acids in the chain (discussed below), yet other interactive forces (*e.g.* van der Waals, hydrogen bonding, covalent bonds, and electrostatics) also increase internal interactions. In particular, disulfide linkages, covalent bonds formed between cysteine residues, are important for maintaining the durability of some globular proteins (*e.g.* β-lactoglobulin and soy glycinin) towards stresses, which include heat and pressure.[11,12] Fibrous proteins, the second major grouping of protein structures, partially originate from amino acid sequences that favor twisting of the polypeptide and are reinforced by both secondary and super-helical quaternary structuring, such as found within collagen and keratin. For example, collagen adopts a helical structure because of its narrowly defined amino acid composition, particularly due to the presence of hydroxyl-proline residues. Helical fibrous structures typically have a much greater surface hydrophobicity due to their highly exposed surface areas, which either requires them to be composed of minimal hydrophobic amino acids or limits their stability in aqueous systems. The last category of protein structures discussed possess random structuring, which does not imply a lack of any structure but rather (for the purpose of this review) a structure that cannot be defined as either globular or fibrous.

As shown in Table 4.1, certain amino acids possess significant hydrophobic character that will impart a tendency for proteins containing those amino acids to be excluded from surrounding water. Hydrophobic interactions may be thought of as an exclusion of interstitial water, which induces a *de facto* agglomeration of those components. At the same time, there is evidence that hydrophobic forces have a considerable range (∼20 nm) and are therefore not wholly dependent on the entropically driven exclusion of water molecules but rather involve instantaneous production of surface dipoles at intermediate separation distances.[13] The net effect of hydrophobicity is to drive internalization of those hydrophobic amino acids within globular conformations, which shield the amino acids from bulk water. Thus, protein hydrophobicity influences the conformation and stability of protein structures to temperature and acid. If the hydrophobic amino acids are exposed (either normally or following unfolding the protein), the solubility in water partially diminishes while the solubility in nonpolar solvents or lipid phases increases, resulting in an increased ability to stabilize emulsions and foams.[14] Protein hydrophobicity also contributes to the encapsulation of oil-soluble bioactives, flavors, and drug molecules by creating a localized environment in which those hydrophobic components are compatible.

Charged amino acids significantly contribute to the solubility of proteins in water, and the quantity of charges on the protein can be utilized to increase or decrease interactivity with other charged components in order to direct structural development. As seen in Table 4.1, some amino acids possess dissociation constants and can be either positively or negatively charged as a function of solution pH. Since a protein is composed of numerous amino acids, the net charge of the protein varies from mostly positive at low pH to mostly negative at high pH, and the point of net-neutral charge (*i.e.* where negative and positive charges balance) is termed the isoelectric point. Charges located on the exposed surface of proteins will experience repulsive or attractive forces with other charged surfaces, including solids, lipid membranes, polysaccharides, and other proteins. Repulsive interactions will typically increase solubility in water, while attractive interactions lead to at least some level of agglomeration. Adding salts or other ionized species to solution significantly diminishes the strength of the effective electrostatic interactions between proteins, and inversely scales with the square of the ionic strength. In this way, electrostatic protein interactions are influenced most commonly by pH and added ions. Since the range of electrostatic interactions is dependent on the capacity of the solvent to communicate that charge (*i.e.* dielectric properties), using solvents other than water will diminish the importance of these interactions between suspended proteins.

Another crucial force in the native structure of proteins and the designed assembly of proteins into nanostructures is hydrogen bonding. Hydrogen-bonding and other polar forces are electrostatic interactive forces that occur between functional groups containing hydrogen in which a significantly electronegative atom, such as oxygen or nitrogen, creates a positive–negative dipole moment. Interactions between positive and negative dipoles on separate molecules lead to a modest attractive interaction. However, because hydrogen bonding is ubiquitous between the amine- and carbonyl-functional groups within amino acids, the combined quantity of all of the individual hydrogen-bond interactions often leads to significant cohesive forces between peptides, contributing to the structural stability of α-helices and β-sheets. In fact, amyloid-like protein fibrils, which are discussed in greater detail below, owe their astounding mechanical resilience to highly coordinated hydrogen bonding in the form of serial β-sheets. Hydrogen bonding is diminished by salts and some denaturant chemicals (*e.g.* urea), but can also be reduced at increased temperature.

Native conformations and dispersion of proteins result from a combination of the numerous interactions between amino acids that are, in turn, influenced by changes in the solution composition or external conditions. For instance, at increased temperatures hydrogen bonding is diminished so that many intramolecular protein inter-actions are weakened in relation to hydrophobic interactions. This promotes the unfolding of the native structure of the protein and increases the likelihood of intermolecular protein interactions, par-ticularly if the proteins possess a significant quantity of hydrophobic amino acids. Since electrostatic and covalent bonds also play a role in native structure, proteins may also be unfolded (or "denatured") as a result of changes in pH, ionic strength, solvent type, adsorption to interfaces, high pressure, dehydration, or other chemical treat-ments.[15] Typically, denaturation is irreversible when it is followed by significant aggregation or intramolecular hydrophobic/covalent interactions, which are often observed in most food proteins. Certain proteins, such as β-casein, display little native structure and are therefore less affected by high temperature treatment. Following de-naturation, newly exposed amino acids may alter the potential interactivity of the entire protein, which is why protein aggregation occurs much more quickly in denaturing conditions (*e.g.* high tem-perature or pressure).

Within food systems, factors of pH, ionic strength, and temperature are the most commonly encountered influences on protein structure (Figure 4.1). At pH values near the isoelectric point (or at increased ionic strength) the net charge of the protein is effectively diminished so that amorphous aggregates (*i.e.* large fractal dimension) form during thermal treatment. As the pH of the solution is increased or decreased in relation to the isoelectric point (alternatively, if the ionic strength is reduced), the effective net charge of the protein increases and aggregation becomes relatively less amorphous. Instead, protein aggregation at pH significantly above and below the isoelectric point produces relatively fibrous structures, as the high net charge on the protein surface makes it favorable for proteins to interact only among discrete regions of the protein surface. Two other important factors that dominate protein structure during thermal treatment at different pH are the increase in disulfide exchange at increased pH, which relatively increases intermolecular protein interactivity, and an in-crease in molten globule character in acidic pH conditions; these factors then favor shorter fibrous aggregates at high pH and longer fibrillar aggregates at lower pH. An example of this change in protein structure following heat treatment at different pH is given in Figure 4.2 for β-lactoglobulin. Further discussion is given later in this chapter on microgel structures (Section 4.2.7), formed at pH values

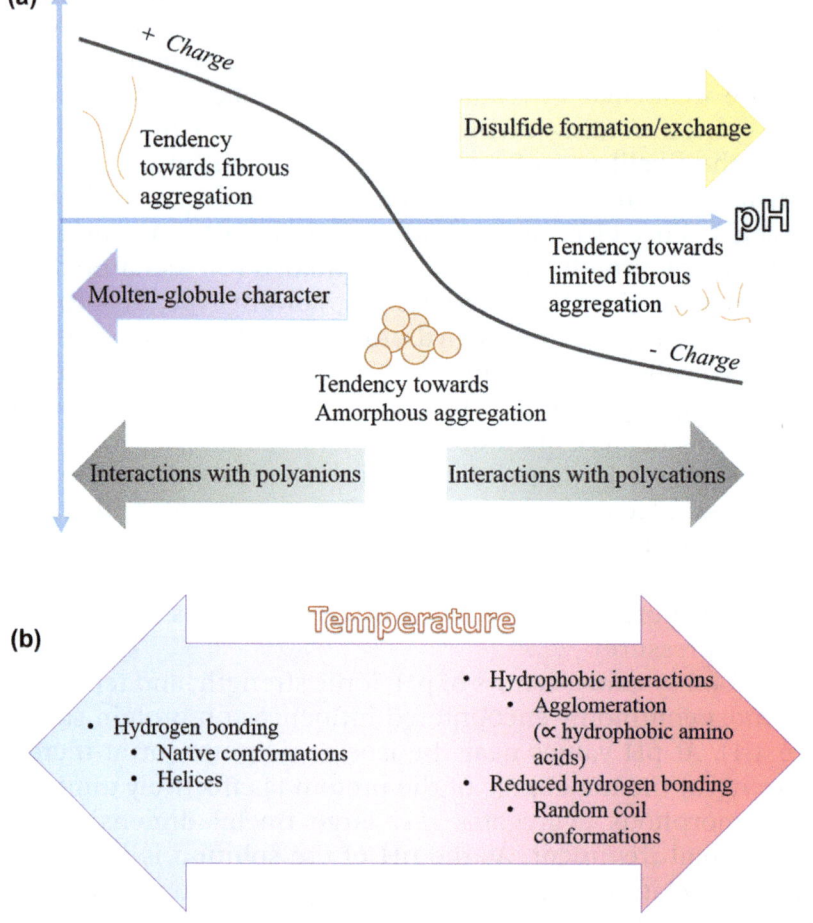

Figure 4.1 Depictions of the influence of (a) pH, ions, and (b) temperature on the physical properties of proteins in solution and their tendency to form particular assemblies.

near the isoelectric point, and amyloid protein fibrils (Section 4.2.8), formed at acidic pH.

Figures 4.1 and 4.2 address common trends in interactivity among soluble globular proteins that could be encountered in most food systems. Poorly soluble proteins, such as prolamins and some egg proteins, possess a significant fraction of hydrophobic amino acids and are therefore insoluble or poorly soluble in water under most conditions. An example of a hydrophobic protein product is zein from corn, which is typically solubilized in aqueous solution with greater than 70% ethanol. For such proteins, solubility in water is only

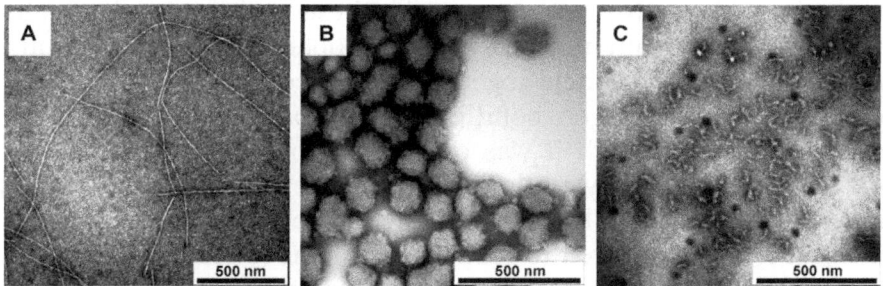

Figure 4.2   Transmission electron microscopy images of β-lactoglobulin heated at (from left to right) pH 2.0, 5.8, and 7.0.
(Reprinted with permission from Jung, Savin, Pouzot, Schmitt, and Mezzenga, 2008, *Biomacromolecules*, **9**: 2477. Copyright 2008 American Chemical Society.)

marginally improved at pH conditions far from their isoelectric point. It is then substantially more difficult to build heat-induced structures from these proteins, such as those shown in Figure 4.2. Instead, poorly soluble proteins are better suited to nanoparticle formation by select precipitation techniques, which are discussed in Section 4.2.5.

4.2   CLASSES OF PROTEIN NANOSTRUCTURE

The assembly of proteins into nanostructures depends on the control of physical intermolecular forces between the amino acids within the exposed polypeptide surfaces in relation to the molecular structure of the protein. Nanostructures are formed naturally, but it is of greater interest to the scientific community to assemble these nanostructures in the laboratory for very specific purposes. This subsection describes what the author perceives as common groups of protein nano-structures and the major principles of how they are formed in order to provide understanding and inspiration for future colloidal scientists. That being said, the following list of nanostructures is not meant to be exhaustive. More detailed information on specific types of protein nanostructures formed in the laboratory and their influence on food systems can be found in recent reviews.[16–18]

4.2.1   An Example from Nature: the Casein Micelle

Casein proteins, including α-, β-, γ-, and κ-caseins, are found naturally bound together as a supramolecular assembly in raw milk with a diameter between 100 and 200 nm.[19] These supramolecular nano-structures, historically referred to by the potentially confusing mis-nomer "micelles", are held together by non-covalent intermolecular bonds, such as hydrogen bonding, ionic cross-linking, and hydro-phobic forces. Although it was originally proposed that the casein proteins possessed truly random coil structure, growing evidence has indicated high levels of local structuring within these protein systems. The reader is encouraged to read the excellent reviews on this intri-cate subject.[20,21] Salt bridges between these casein proteins, particu-larly involving calcium and phosphate, are formed between phosphoseryl residues on the β- and α-caseins and are particularly important for micelle structure, contributing ∼6% to the dry casein weight.[22] These calcium phosphate links are dissolved at lower pH during fermentation or direct addition of acids, which is one of the primary causes of structural changes in cheese curd gel strength among low pH cheese varieties.[23]

Casein micelles self-assemble into approximately spheroidal di-mensions with fluctuations in the density that can be readily resolved in electron microscopy (Figure 4.3a). There exist two major theories concerning the internal casein micelle structure that are

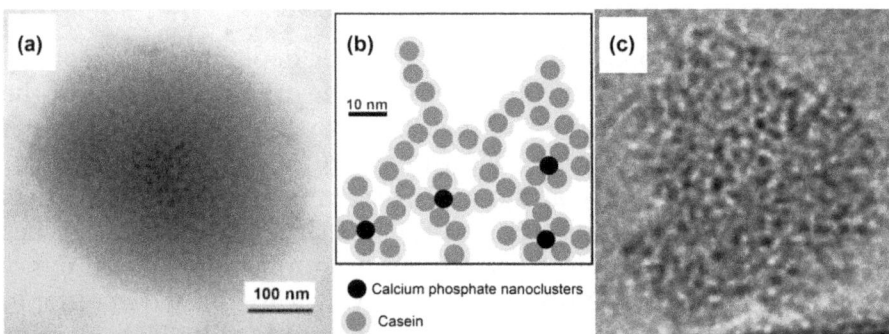

Figure 4.3   Structure of the casein micelle (a) as visualized by transmission electron microscopy after immobilization and freeze-drying,[a] (b) as theorized, based on electron microscopy observations, by assuming an interlocking lattice model,[a] and (c) after reassembly with ions in the presence of vitamin D_2.[b]
([a]Reprinted from *Journal of Dairy Science*, **91**, McMahon and Oommen, "Supramolecular Structure of the Casein Micelle", pp. 1709–1721, Copyright 2008, with permission from Elsevier. [b]Reprinted from *Food Hydrocolloids*, **21**, Semo, Kesselman, Danino, and Livney, "Casein mi-celle as a natural nano-capsular vehicle for nutraceuticals", pp. 936–942, Copyright 2007, with permission from Elsevier.)

differentiated by the pervasiveness of the protein within the micelle. The first theory proposes that micelles are composed of small protein "sub-micelles", or small assemblies of casein proteins. These sub-micelles are attached to one another by calcium phosphate bridges in order to form a cross-linked network of globules. The second theory, based on an extensive review of neutron and X-ray scattering experiments, proposes that the internal structure of the casein micelle may instead be comprised of a dispersed protein matrix with ~4 nm-sized calcium phosphate nanoclusters.[19] Formation of nanoclusters agrees with the model of McMahon and Oommen, where the casein proteins form an interlocking lattice interconnected by hydrophobic bonds and calcium phosphate nanoclusters (Figure 4.3b).[147] It is a comment on the complexity of this micelle system that, after so many years of study, the structure is still not fully understood.

Given that casein is already an assembled protein structure, some research groups have focused on the fixation of this loose casein micelle in order to make a relatively durable and functional nanostructure. As an example, transgutaminase, an enzyme that promotes cross-links between glutamic acid and amines, was utilized to induce internal cross-links within casein micelles and fix the physical structure.[24,25] Similar transglutaminase cross-linking has also been shown to improve gelation properties of casein suspensions[26] and has contributed to the increased gel strength of skim milk gels and yogurt.[27,28] In theory, this casein nanostructure could be utilized as a controlled delivery vehicle by incorporating small molecules prior to cross-linking, although no such applications are known at this time.

Although it is not possible to duplicate the exact assembly of the casein micelle that is found in milk, constituent casein proteins do have a strong tendency to self-assemble into nanostructures. Reassembly of casein proteins in a "micelle"-like structure was experimentally achieved over 30 years ago by adding calcium, phosphate, and citrate ions.[29] The process is essentially a titration of the ions into the bulk protein solution at neutral pH, which allows the gradual formation of a micelle-like structure with a diameter of 50–300 nm. Reassembled casein nanoparticles have been utilized to encapsulate vitamin D_2,[30] as well as polyunsaturated fatty acids.[31] This re-assembled micelle with Vitamin D_2 is shown in Figure 4.3c. The advantage of such nanostructures is the minimal processing necessary for their formation and the exclusive use of natural materials, which is an ideal model for structured nanomaterials intended for food or other commodities.

4.2.2 Electrostatic Complexes

As discussed in Section 4.1.2 and shown in Figure 4.1a, electrostatic interactions between proteins and other colloids/surfaces result from the interplay between charges of opposing sign. Since proteins generally possess a high quantity of ionized amino acids, there is a strong potential for proteins to interact electrostatically with other proteins, charged polysaccharides, or other charged materials. In the presence of a biopolymer (*e.g.* protein or polysaccharide) with similar net charge, there is little or no driving force for intermolecular interactions. However, in the presence of a biopolymer with an opposite net charge, there is a strong driving force for the biopolymers to associate and form electrostatic complexes. The size and charge of the newly formed complexes are then determined by the quantity of biopolymer, the ratio between different biopolymers, their net charge in relation to the pH value, and the ionic strength (influencing effective charge).[32]

Although it is possible to form electrostatic complexes between biopolymers by directly mixing them in solution under conditions where they possess opposing charge, this typically results in a disproportionation in binding between the biopolymers such that aggregates or other phase-separated materials form within microvolumes of the bulk solution. Rather, a more homogeneous distribution of complexes is achieved by mixing the two biopolymers at a pH at which they possess similar charge (*i.e.* where no interactions are expected) and adjusting the pH gradually by titrating with acid or base. This gradual titration of pH slowly induces electrostatic interactions between the biopolymers from weak/minimal interactivity to strong interactivity in order to enhance formation of equilibrated, homogeneous complex structures. For example, to form an electrostatic complex between β-lactoglobulin and chitosan, a positively charged polysaccharide, the protein and polysaccharide would be mixed together at low pH (~2–3) and gradually neutralized until complexes form.[33] To form electrostatic complexes between β-lactoglobulin and pectin, a negatively charged polysaccharide, the protein and polysaccharide would be mixed together at near neutral pH and gradually acidified until complexes form.[34]

Formation of electrostatic complexes among proteins and other proteins or polysaccharides involves three major structures: soluble complexes, complex coacervates, and co-precipitates. During the initial formation of electrostatic complexes, proteins possess an excess of surface charge that is unable to interact effectively with the added biopolymer, so that the protein is very well hydrated and the complex remains soluble. These soluble complexes can then be recognized by a slight, yet significant, increase in scattered light without any noticeable phase separation. Soluble complex formation typically occurs at pH values on the "wrong side" of the isoelectric point (*e.g.* pH > pI for a protein–anionic polysaccharide combination, where the net charges of both protein and polysaccharide are of the same sign), which is due to electrostatic interactions with localized sites on the protein surface.[35] With further pH adjustment, electrostatic interactions increase between the protein and biopolymer until the combined charge is nearly balanced, in which case the complexes separate from bulk solution as a hydrated coacervate phase. It is important to note that coacervates are very well hydrated, and the coacervate phase is a liquid phase (*i.e.* not a precipitated solid), which is discussed in more detail in Section 4.2.4. In some cases, balanced charge interactions between a protein and a biopolymer lead to a poorly hydrated complex, and the resulting complex forms a precipitate phase rather than a coacervate. For example, interactions between globular proteins and polysaccharides of high charge density, such as β-lactoglobulin and λ-carrageenan, will form a precipitate after sufficient association between the two biopolymers.[36]

Other crucial parameters that influence electrostatic complex formation include added salts (*i.e.* increased ionic strength) and the ratio between the protein and the other biopolymer. Since electrostatic complexes form as a consequence of the number of charges on the protein and the other biopolymer, added ions have a very strong influence on their formation by screening the effectiveness of the charge in bulk solution. For instance, when increasing concentrations of sodium chloride are added to mixed solutions of β-lactoglobulin and pectin, the critical pH values for complex formation and complex coacervation both decrease (*i.e.* complex formation becomes less favorable).[34] The ratio between protein and other biopolymers also influences the total number of opposing charges that may interact in bulk solution, and the favorability of forming complexes will decrease as the concentration of one of the components is in excess of the other. Thus, if the total number of charges on the protein and the other biopolymer is approximately balanced at a given pH, adding an excess of protein will unbalance the number of interactive charges and begin to favor soluble complexes rather than a complex coacervate. The same is true when adding an excess of the other biopolymer, so that there exists an optimal protein : biopolymer ratio for complex formation at each given pH value. More information on the phenomenon of electrostatic complex formation and their resultant structures may be found in excellent reviews.[37,38]

One challenge in the utilization of complex coacervates or soluble complexes as functional protein nanostructures is the dynamic nature of electrostatic interactions, which are dependent upon maintaining the solution pH, ionic strength, and biopolymer ratio. This challenge may be resolved by sintering the otherwise dynamic structure using cross-linking or other strong internal interactions. With this aim in mind, thermal treatment above the denaturation temperature of protein was applied to protein–polysaccharide electrostatic complexes in order to induce protein–protein interactions. The resulting structures were found to be similar to protein microgels, which are discussed later in this chapter (Section 4.2.7), as the diameter is typically between 100 and 400 nm and the internal structure resembles a semi-porous gel (Figure 4.4). After thermal treatment, the structures are very stable to further pH changes and added salts.[39] In addition, the size and charge of these "complex microgels" can be readily controlled by fine adjustments in total biopolymer concentration, pH, and ionic strength.[40] Current progress on these emerging complex microgels has been reviewed in the recent literature.[41]

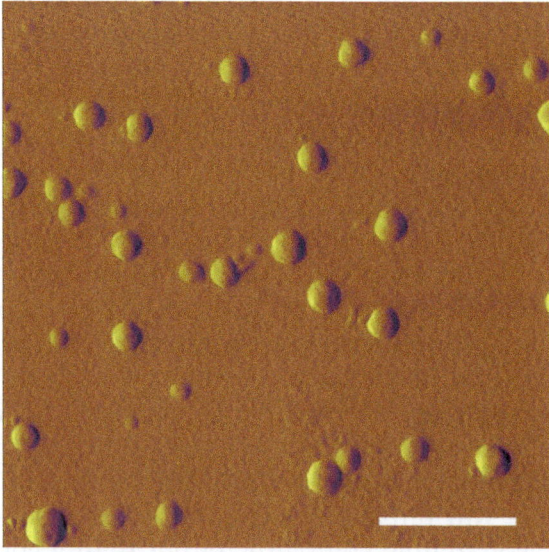

Figure 4.4 Atomic force microscopy (AFM) amplitude image of nanostructure formed by heat treatment of β-lactoglobulin and pectin at pH 4.5. Scale bar = 200 nm.

4.2.3 Self-assembled Conjugates

Phospholipids and other surface-active agents have the unique property of self-assembling into micelles or other complex self-assembled structures (*e.g.* liposomes, cubic phases), as these molecules possess discrete hydrophobic and hydrophilic moieties that favor segregation into water-poor and water-rich environments, respectively. Although proteins may self-assemble during thermal treatment or by desolvation (see Section 4.2.5), most proteins do not possess a significant anisotropic distribution of hydrophobic and hydrophilic functionalities that would facilitate their self-assembly into these sorts of nanostructures. While native casein is incorporated into "micelles" (see Section 4.2.1), casein micelles are actually more of an aggregate species with a predominant distribution of some casein fractions (*e.g.* κ-casein) closer to the "micelle" surface. The only way to induce self-assembly of protein to form discrete nanostructures is then to modify the protein chemically or physically so as to impart hydrophobic and hydrophilic character to limited regions on the protein surface.

One way to increase the anisotropy of the hydrophobic and hydrophilic character of individual proteins is to covalently attach a second polymeric component onto the protein, forming a co-polymer. Ideally, the second polymer should have the opposite affinity for the solvent to the protein so that the co-polymer has a tendency to self-assemble in solution. For instance, a hydrophilic protein should be conjugated with a hydrophobic polysaccharide, lipid, or synthetic polymer so that the protein prefers to exist in the aqueous continuous phase and the hydrophobic polymer forms the internal core of the micelle (Figure 4.5, bottom right). Alternatively, protein can form the "core" of a co-polymer assembly if a hydrophobic or aggregation-prone protein is co-polymerized with a hydrophilic polymer (Figure 4.5, top right). There are a variety of chemical techniques to form conjugates, which are referred to as block-copolymers (conjugation end-to-end) or graft-copolymers (conjugation at various points along the length) in polymer science (Figure 4.5, left). Many of these chemical techniques are unfavorable for food formulations and have a questionable future in food production. There is then a need to develop anisotropic biopolymer systems that promote self-assembly into functional nanostructures while maintaining a clean label that is relatively more attractive to potential consumers by utilizing more "natural" modification techniques.

One modification technique that is performed in nature is the Maillard reaction, which can be utilized to conjugate protein and

Potential Co-polymers
of Protein

Micellar Assemblies of Copolymers
by Type of Added Polymer

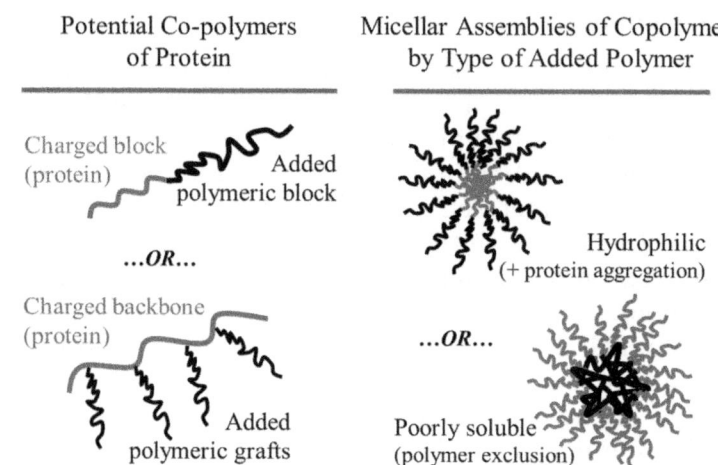

Figure 4.5   Depiction of co-polymers derived from protein (either block- or graft-
co-polymer systems) and their idealized self-assembly into micelles.

polysaccharides at high temperatures. During the Maillard reaction, the reducing end of a carbohydrate (*e.g.* glucose, maltodextrin) forms an imine with a free amine from the protein. The amine of the protein can originate from the N-terminus of the polypeptide, but feasible Maillard conjugate products are more likely to result from reactions with the primary amine of lysine residues. After conjugation, the protein component, which is usually the relatively hydrophobic component, forms the internal phase of a micelle-like structure that can be utilized for encapsulation of hydrophobic compounds. For example, conjugates of caseinate and maltodextrin were formed by dry heat treatment and used to encapsulate vitamin D and epigallocatechin gallate within assembled structures of ∼30 nm diameter, providing improved colloidal stability over a wider pH range and a reduction in oxidation.[42] Similarly, conjugates of bovine serum albumin and dextran were synthesized and formed self-assembled structures for the encapsulation of ibuprofen.[43] In some cases, self-assembly may be assisted by ions, which form ionic bridges within the structures. For instance, self-assembled Maillard conjugates of whey protein and chitosan were internally linked by tripolyphosphate ions, and the resultant structure was utilized for the controlled release of a bioactive catechin molecule (Figure 4.6).[44] Since the components are cross-linked, the internal structure is more likely to resemble a hydrated network that would classify them as microgels (see Section 4.2.7). However, further research on the assembly of co-polymer protein systems and their internal structure is necessary in

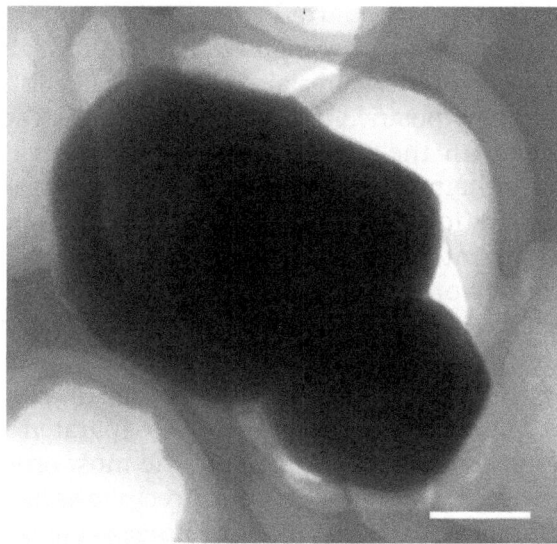

Figure 4.6 Nanometer-scale assembly of whey protein-chitosan maillard conjugates
cross-linked by tripolyphosphate.
(Reprinted from *European Polymer Journal*, **45**, Zhang, Dudhani, Lundin,
and Kosaraju, "Macromolecular conjugate based particulates: Preparation,
characterization and evaluation of controlled release properties", pp. 1960–
1969, Copyright 2009, with permission from Elsevier.)

order to fully understand their potential for development as functional nanostructures.

4.2.4 Simple Coacervate Structures

Coacervation, when loosely translated, means "heap together" and
refers to the induced phase separation of polymers by reducing the
solvent quality. This coacervation process is typically induced by the
addition of unfavorable solvents or co-solutes, although it may also be
induced by changes in temperature. Coacervation is distinct from
other similar processes in that the polymers within the separated
coacervate phase remain partially solvated (*i.e.* not precipitates),
which means that nanoprecipitation or solvent desorption techniques
(discussed in Section 4.2.5) are not synonymous with coacervation. In
practice coacervation can be easily confused with precipitation, and
the best way to verify formation of a coacervate phase is by observation of a macroscopically separated liquid phase.

Simple coacervation is aided by processes that decrease stability in
the bulk aqueous phase yet promote hydration. To the uninitiated

this statement may seem contradictory. Based on the theoretical predictions of Flory, one can recognize that the full dispersion and solubilization of polymeric materials, unlike small molecules, is not dictated by a strongly positive mixing entropy.[45] Instead, mixing entropy diminishes with the size of the polymer so that the full dispersion of large proteins in solution is often energetically disfavored at modest concentrations. At the same time, the polymer may be fully hydrated since the interactions between water and amino acid residues are generally favored. Precipitation will only occur for these large polymers if local water interactions are weak and water is gradually squeezed out of the condensed polymer phase. Still, precipitation can and does occur in polymeric systems, and proof of a true coacervate phase must be obtained by observation of a separated viscous fluid. The conformation of the protein is one of the most crucial factors in determining between coacervation and precipitation, as a random coil conformation is actually less likely to squeeze out water and form a precipitate when compared with a rigid rod conformation. The best example of this difference is gelatin, which will form precipitates at pH values away from the isoelectric pH and coacervates near the isoelectric pH.[46]

Just as with complex coacervates, dispersed coacervate droplets are initially formed when the free energy of forming a second phase becomes favorable. Once formed, these droplets will grow as more of the hydrated polymer separates from bulk solution. After reaching a pseudo-steady state with the dense coacervate phase settled at the bottom of the mixture, trapped solvent gradually leaks out by syneresis and the coacervate droplets coalesce to form a bulk coacervate phase with increased internal entropy.[47] Discrete coacervate droplets are turbid, but after droplets coalesce to form a bulk coacervate phase the system becomes translucent or transparent.

Gelatin is one of the most studied proteins in simple coacervation research. The father of simple coacervation, Bungenberg de Jong, experimentally described the simple coacervation of gelatin near its isoelectric point in aqueous–ethanol solutions as far back as 1934.[48] Gelatin is highly water soluble and coacervation is often induced by adding co-solvents or co-solutes to aqueous solutions in order slightly to perturb the phase stability. For example, sodium sulfate is a co-solute salt that induces simple coacervation.[46] Added solvents that induce gelatin coacervation include concentrated methanol, concentrated ethanol, isopropanol, *tert*-butanol, and dioxane.[49] Use of gelatin has an added advantage in that quenching the temperature of the dispersed coacervate droplets initiates formation of networks, producing very small gelled capsules. While this is certainly a means to create functional protein nanostructures, it should be stressed that fixation of coacervate droplets (whether simple or complex) to form particles with sub-micrometer dimensions is challenging. It is much more common to form particles with dimensions on the order of 10–100 μm.

4.2.5 Desolvated Nanoparticles

Formation of nanometer-sized solid particles can be achieved by very rapid reduction in solvent quality (*i.e.* precipitation) by a technique that may be referred to as desolvation, nanoprecipitation, or solvent desorption. Solid particles formed in this way can be utilized for encapsulation, where the active component of interest is trapped within the solid nanoparticle during the precipitation process. Alternatively, these solid particles can be utilized as textural mimetics (*e.g.* simulation of small fat droplets) or as emulsion stabilizers (*i.e.* Pickering stabilizers). Despite sharing some similarities in preparation procedures with emulsion-templated nanoparticles (see Section 4.2.6) and simple coacervates (see Section 4.2.4), the essential attributes of desolvated nanoparticle formation are: (i) particle formation is spontaneous, (ii) particles are formed within a monophasic solution, and (iii) the resulting particle is a true solid.[50]

Given that nanoparticle formation is influenced by the relative insolubility of the starting material in a given liquid mixture, factors that reduce solubility of specific proteins can be utilized to produce nanoparticles. Most proteins possess endothermic hydration energies, and so reduction in temperature is one method used to precipitate proteins. However, temperature reduction is not a preferred method to create nanoparticles, at least by itself, as it is technically challenging to rapidly lower temperature in the immediate vicinity of individual protein molecules so that the gradual decline in temperature facilitates macro-scale aggregates. Instead, rapid introduction of poor solvents or concentrated salt solutions is ideal for the nearly instantaneous reduction in solvent quality that promotes quick precipitation (*i.e.* spinodal decomposition) and favors formation of numerous nanometer-scale particles.

One of the food proteins most widely utilized to form nanoparticles is the poorly soluble protein zein, which is a prolamine extracted from corn. In the typical method, zein is dissolved in 70–90% ethanol solution (v/v) and added drop-wise to large quantities of aqueous solution. During this process, the ethanol quickly diffuses from the submerged droplet containing the zein, and the sudden increase in zein concentration in the microvolume leads to a rapid precipitation and formation of a nanometer-scale solid particle. Similar particles can also be formed using comparable solvents such as isopropanol.[51] An example of zein nanoparticles formed by nanoprecipitation is shown in Figure 4.7(a). The size of these zein nanoparticles is typically

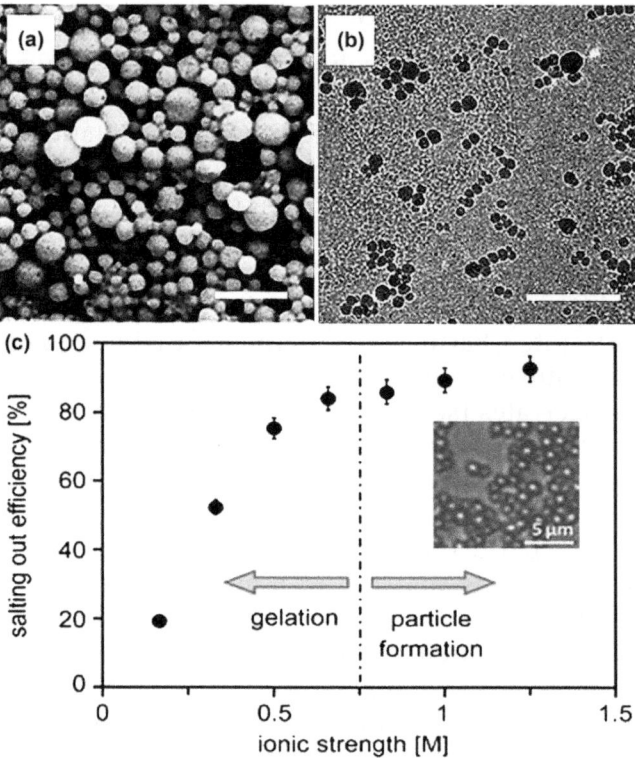

Figure 4.7   Protein nanoparticles prepared by desolvation procedures: (a) scanning electron micrograph of zein nanoparticles,[a] (b) cryo-transmission electron micrograph of whey protein isolate nanoparticles,[b] and (c) relationship between salting-out and ionic strength influencing the eventual formation of silk fibroin nanoparticles.[c]
([a]Reprinted from *Food Hydrocolloids*, **23**, Zhong and Jin, "Zein nanoparticles produced by liquid-liquid dispersion", pp. 2380–2387, Copyright 2009, with permission from Elsevier. [b]Reprinted from *Food Hydrocolloids*, **29**, Guelseren, Fang, and Corredig, "Whey protein nanoparticles prepared with desolvation with ethanol: Characterization, thermal stability and interfacial behavior", pp. 258–264, Copyright 2012, with permission from Elsevier. [c]Reprinted from *Biomaterials*, **31**, Lammel, Hu, Park, Kaplan, and Scheibel, "Controlling silk fibroin particle features for drug delivery", pp. 4583–4591, Copyright 2010, with permission from Elsevier.)

between 100 and 500 nm, depending on the concentration of zein in the original ethanol dispersion, the volume of added droplets, and the ability of the ethanol to diffuse when dispersed in water.[52] Since zein is fairly insoluble in water and possesses an isoelectric point close to neutral conditions (pI ∼ 6), attempts have been made to create a secondary layer of protein (*e.g.* caseinate[53]) or polysaccharide (*e.g.* carboxymethyl-chitosan[54]) on the surface of the zein particle to increase its stability. Zein nanoparticles have shown promise for the stabilization of oil–water interfaces as a Pickering stabilizer,[55] yet significant research challenges remain before these protein nanoparticles find common application.

Although there are numerous poorly soluble storage proteins that could be similarly nanoprecipitated from solutions of organic solvents, the field is relatively new and most such proteins have so far been untested for their capacity to form nanoparticles. An exception to this is the hydrophobic wheat protein gliadin, which is one of the essential components of wheat-based gels and foams. Similar to zein, wheat gliadin forms nanoparticles by rapidly introducing water to ethanolic solutions.[56] Gliadin may also be nanoprecipitated from ethanol-based solutions by factors of pH, temperature, and sodium chloride concentration.[57] Given the potential utility of these precipitated nanoparticles, this technique will likely be expanded in future years to encompass a variety of other hydrophobic plant storage proteins.

Alternatively, nanoprecipitation techniques could be applied to water-soluble proteins by dispersing aqueous droplets of the protein in an unfavorable solvent. For example, whey protein was desolvated in the presence of ethanol to form nanoparticles with a diameter of ∼ 50–100 nm (Figure 4.7b).[58] Nanoparticles have also been crafted from β-lactoglobulin, the major whey protein, by desolvation in acetone.[59] Delivery within non-aqueous solvents greatly limits the applicability of such nanoparticles in food systems. This limitation can be side-stepped by covalently cross-linking the internal structure of the protein nanoparticle within the non-aqueous solvent and then re-dispersing the nanoparticles in aqueous solvent.[60] This was recently shown for a soy protein isolate system, where nanoparticles formed in ethanol were cross-linked by glutaraldehyde, dried to remove ethanol, and re-dispersed in water.[61]

Another frequently encountered natural and water-soluble protein that is used to form nanoparticles is silk fibroin, which is a major protein component obtained from silk that contributes to silk elasticity. While fibroin readily forms natural fibers, such as the silk strands (see Section 4.2.8), many recent studies have instead focused on the formation of nanoparticles or microgels from fibroin. In these studies, fibroin was desolvated in polar aprotic solvents such as dimethylsulfoxide (DMSO),[62] non-polar solvents such as acetone,[63] or strong salt solution (Figure 4.7c).[64] This serves as a good example of how a single protein can be resolved into spherical or fibrous structures depending on the techniques and materials employed.

4.2.6 Emulsion-templated Nanoparticles

One challenge in the formation of nanostructures from coacervation and nanoprecipitation processes is the lack of control of the nanostructure size, which is dependent on the strength of physical forces between the components, the concentration, and inter-particle interactions once the particles have formed. One way to solve this lack of uniformity and control is to create a size template that physically limits the growth of the nanostructures produced. The idea of this is similar to the use of lithographic polymer-based stamps for the creation of micrometer-scale structures. At the nanometer scale, sub-micrometer emulsion droplets serve as excellent templates for the creation of nanostructures[65] and are relatively inexpensive compared to lithographic techniques.

As a quick review, emulsions are biphasic mixtures where the dispersed phase is present as discrete, spherical droplets. The two phases should be sufficiently incompatible with one another that molecules of interest, in this case proteins, partition predominantly in one phase or the other. Formation of emulsions is thermodynamically disfavored, so that increasingly greater amounts of energy are required to reduce the diameter of formed emulsion droplets below 1 μm.

To create nanoparticles from emulsions, protein is partitioned predominantly in the dispersed, spherical droplets. This dispersed phase, which is typically chosen specifically for its volatility, is removed by evaporation. This technique works best for proteins with poor water solubility, as the protein can be solubilized in a volatile organic solvent, such as hexane, that is dispersed within water or a similar polar solvent. After the hexane droplets containing protein are emulsified, the hexane is removed in a rotary evaporator or a similar device that quickly removes the solvent without allowing the droplets to coalesce. Alternatively, both the continuous and dispersed phases can be evaporated simultaneously by lyophilization or freeze-drying to obtain a powder.[66] As the dispersed phase solvent is removed, the protein becomes supersaturated and precipitates as a nanoparticle. Coalescence of droplets and/or flocculation of nanoparticles during this process can lead to nanostructures of undesirably large size, and so parameters should be selected to minimize these mechanisms. While this technique is theoretically powerful for the creation of relatively monodisperse nanoparticles, in practice it is energetically expensive and requires the use of environmentally undesirable volatile organic solvents.

For water-soluble proteins, water emulsions droplets would have to be dispersed in organic solvents or oil, and the water would then be removed by evaporation. However, evaporation of water is energetically expensive and the removal of water would be deleterious to most protein nanostructures or encapsulated materials. Therefore, evaporation is not a preferred method for water-soluble protein nanoparticles. Instead, cross-linking agents are employed in the continuous phase (either organic solvent or oil); these diffuse to the dispersed aqueous phase and form cross-links between the proteins.[67,68] This technique, although similar to nanoprecipitation, creates discrete gels with nanometer-scale dimensions, which would classify them as "microgels" (see Section 4.2.7.), not solid particles.

4.2.7 Microgels

One of the most important functional attributes of proteins is their ability to form percolating networks, which contribute towards strong gels and films. Since the size of most individual proteins is on the order of 1–10 nm, it is possible to form gel-like networks at length scales of several micrometers to ~100 nm by restricting the boundaries of network formation. Thankfully, many polymer systems, including proteins, have self-limited agglomeration processes so that sub-micrometer structures are fairly common. When such discrete polymeric structures are formed and dispersed within a compatible continuous phase, the internal polymeric network is well hydrated. These swollen network particles are then referred to as *microgels*.[69] Microgels have been extensively studied for synthetic polymeric systems as controlled-release vehicles or as interface stabilizers.[70] Research in the past decade has also demonstrated the formation of microgels from protein that offer numerous functional advantages in food and pharmaceutical systems.

While microgels may be assembled from any network-forming protein, whey proteins are perhaps the best studied microgel-forming proteins. Whey proteins, particularly β-lactoglobulin, are small globular proteins with excellent water solubility and denaturation temperatures between 65 and 80 °C that are frequently utilized in the food industry to create cohesive gels.[71] To form microgels from whey proteins, the protein is thermally treated within a small pH window (pH 5.7–6.1) at relatively low concentration (~ 1–3 g l^{-1}).[72] The pH of the solution acts to control the size of the microgels formed by modulating the surface charge (and resultant electrostatic repulsions) between proteins during the thermal aggregation process. This ultimately results in changes to the fractal dimension of the protein aggregates, which go from linear to amorphous as the fractal dimension increases (see Figure 4.2). The influence of the pH in changing the fractal state of aggregates is shown in Figure 4.8; at neutral and acidic pH the aggregates are linear, at the isoelectric point aggregates are relatively amorphous, and in between these pH regions the aggregates, which are microgels, have intermediate fractal dimensions ($1.5 < \alpha < 2.0$).

Microgels, possessing intermediate fractal dimension, have an irregular internal structure and a significant amount of hydration within porous channels.[73] These internal porous channels contain

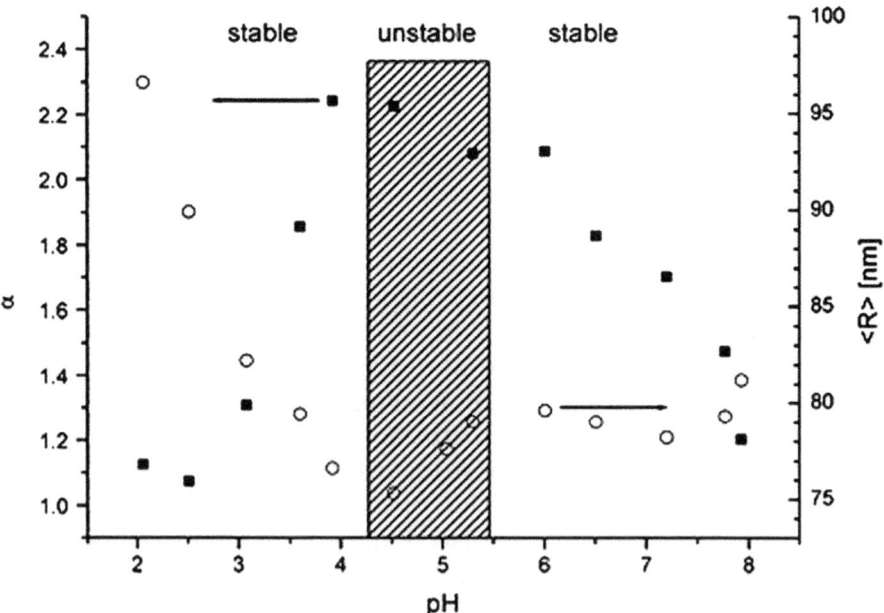

Figure 4.8 Fractal dimension (α) and mean particle radius ($\langle R \rangle$) of whey protein aggregates formed by heat treatment at pH values between 2 and 8. (Reproduced from ref. 76.)

charged and otherwise hydrophilic amino acids, and so the relative internal hydration of the microgel is influenced by the same factors that influence protein solubility. When hydration is increased, the quantity of internal water increases and the microgel correspondingly swells to accommodate the added volume. The opposite is true when hydration of the internal channels is diminished. Thus, microgel swelling can be controlled by relatively minor properties of the solution such as added salts or temperature.[74]

As discussed in Section 4.2.2, microgels from β-lactoglobulin may also be produced by the thermal treatment of β-lactoglobulin–polysaccharide complexes at low pH.[41] Instead of the pH dictating the interactivity between proteins through surface charge directly, the added charged polysaccharide interacts electrostatically with charged segments of the protein and limits the aggregation process during thermal treatment. Microgels formed from β-lactoglobulin–pectin complexes have similar size distributions and structure to microgels formed from β-lactoglobulin alone (Figure 4.9), but the presence of pectin in the microgels improves their physical stability to changes in pH and salt concentration.[75] Because microgel formation is so strongly influenced by electrostatic interactions with pectin, the pectin concentration and added salts have a strong influence on the resultant size of the microgels.[40] Even the type of ion strongly influences the ability of the protein to disentangle from the pectin and form microgels, so that microgel size can be readily tuned by experimental parameters.

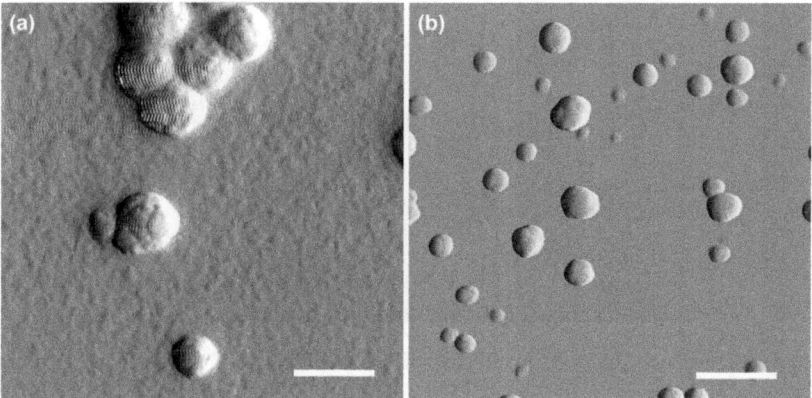

Figure 4.9 AFM amplitude images of microgels assembled after thermal treatment at 85 °C from (a) β-lactoglobulin at pH 5.8 or (b) β-lactoglobulin and pectin at pH 4.75. Scale bars = 200 nm.

Similar to the formation of nanoparticles (see Section 4.2.6), microgel size may also be restricted by utilizing a template. Emulsions serve as excellent templates for mass production of protein nanostructures, where the dispersed, spherical phase contains the majority of the protein. Once the protein is entrapped within the dispersed phase of the emulsion, a treatment may be applied so that the proteins form a gel-like network and a microgel forms within the droplet. For example, a microgel of denatured soy protein isolate and zein was formed within water-in-oil emulsions by the gradual diffusion of glacial acetic acid, which gelled the proteins as the dispersed phase was acidified.[67] Similarly, microgels of gelatin were prepared by thermal set and glutaraldehyde cross-linking.[68] Heat treatment may be utilized to form microgels within emulsions when using proteins that agglomerate at higher temperatures, such as whey proteins contained in water-in-oil-emulsions.[76,77] After the formation of a cross-linked internal gel phase, the microgels are separated by centrifugation to obtain a discrete nanomaterial for advanced applications. An example of such a microgel of whey protein formed from emulsion templating is given in Figure 4.10. The observed

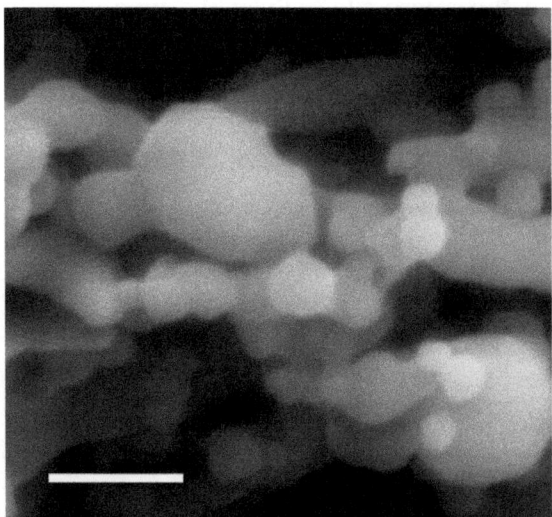

Figure 4.10   Scanning electron microscopy (SEM) image of whey protein microgel cross-linked by transglutaminase for 24 hours followed by heat treatment in a water-in-oil microemulsion. Scale bar = 200 nm.
(Reprinted with permission from Zhang and Zhong, 2009, *Journal of Agricultural and Food Chemistry*, **57**: 9181. Copyright 2009 American Chemical Society.)

polydispersity of the microgels shown in the figure demonstrate that, although the use of emulsion templating is theoretically powerful enough to control microgel size, there are numerous challenges to controlling the stability of the emulsion droplets and the resultant microgels. Thus, this potentially valuable technique to form protein nanostructures requires further investigation.

4.2.8 Fibrillar Protein Structures

In select conditions that favor a gradual alignment of β-sheet segments (such as shown in Figures 4.1 and 4.2), proteins or polypeptides will form linear aggregates held together by non-covalent intermolecular bonds; these are referred to as amyloid fibrils.[78] This continuous alignment and interaction of β-sheets, held primarily by hydrogen bonding, creates an astonishingly strong and rigid structural element, which can be referred to as protofilaments. These protofilaments then assemble laterally to form fibril bundles that are approximately 6–11 Å apart.[79] While hydrogen bonding plays a significant role in the formation of amyloid fibrils, π–π stacking and hydrophobic interactions between amino acid functionality groups are also important for the properties of the assembled fibril.[78] Assembly of proteins and peptides into fibrous nanostructures is quickly emerging as a route to creating new, natural biomaterials for use in food, medicine, and other material applications.[80] Although the study of these protein fibrils has exponentially increased in the past two decades, it is likely that these nanostructures will continue to gain in importance within the coming years.

A small population of protein fibrils assembled under biological conditions (*i.e.* 37 °C, pH 7.4, aqueous solution) have been shown to contribute to the formation of amyloid plaques, and the presence of these plaques has been linked to several harmful disease states. Diseases linked to the formation of amyloid plaques include Alzheimer's disease, Huntington's disease, Parkinson's disease,[81] diabetic-related disorders,[82,83] hemodialysis protein disorders,[84] and prion-related encephalopathic diseases,[85–87] among others.[88] Formation of amyloid fibrils from biological proteins leading to these disease states has been attributed to the malfunction of protein regulatory systems in living cells.[89] The most dangerous characteristic of these amyloid fibrils is their capacity to facilitate amyloid formation in healthy tissues when transmitted, which gives them infectious properties. Only some amyloid fibrils are transmissible in this way, in which case they are referred to as *prions*.[90]

Amyloid-like fibrous protein structures are also found within many other biological structures, as evidenced most significantly by their presence in silk strands. Fibrous assemblies of silk proteins provide the internal resistance to elongation strain that makes natural products such as silk and spider webbing such desirable materials. Silk proteins are a collection of proteins produced by arachnids, *Lepidoptera* spp., and other insects that have a high propensity to form β-sheet secondary structures and are utilized for the creation of cocoons, egg casings, and webs.[91] These proteins are aligned into fibers within tubular-shaped glands that compress and align the proteins, favoring fibril formation by simple molecular confinement. Each of the major types of silk (*e.g.* major ampullate, flagelliform, or aciniform) possess heavy repeating units of several amino acids, including glycine, proline, alanine, and serine, although the identity of the repeating amino acids vary considerably between silk types. These repeat units prefer the formation of strands or tight helices that form β-sheet-like alignments within tubular structures, where coil-like peptides are interspersed to increase dissipative mechanical properties.[92] This proposed internal structure and the morphology of a dragline spider silk strand are given in Figure 4.11. Mechanical properties of silk protein fibrils are comparable to those of Kevlar, which makes them highly desirable for future material development.

Amyloid-like protein fibrils may be assembled from food-grade proteins through extensive thermal treatment at low pH (Figure 4.12). Examples of proteins that form fibrils following thermal treatment at low pH include whey proteins, hen egg white lysozyme, ovalbumin, and soy glycinin. The current theoretical standing on fibril formation indicates that, during thermal treatment, the protein adopts a molten-globule conformational state and is prompted towards β-sheet alignments instead of agglomeration.[93,94] Recent evidence has also indicated that the proteins are also partially hydrolyzed during thermal treatment, and the resulting peptide fragments are then free to align to form cross-β-sheets.[95] While it would be tempting to postulate that even higher temperatures facilitate fibril formation, temperatures near 100 °C and above have been shown to engender greater fibril flexibility and agglomeration of fibrils,[96] which are undesirable for functionality.

Nanotubes are fiber-like protein constructs that are closely related to amyloid-like protein fibrils in their assembly and properties. Just as with amyloid-like protein fibrils, peptide segments align into continuous β-sheet assemblies. While the cross-β-sheet strands among amyloid fibrils will intertwine to form helices or ribbons, the

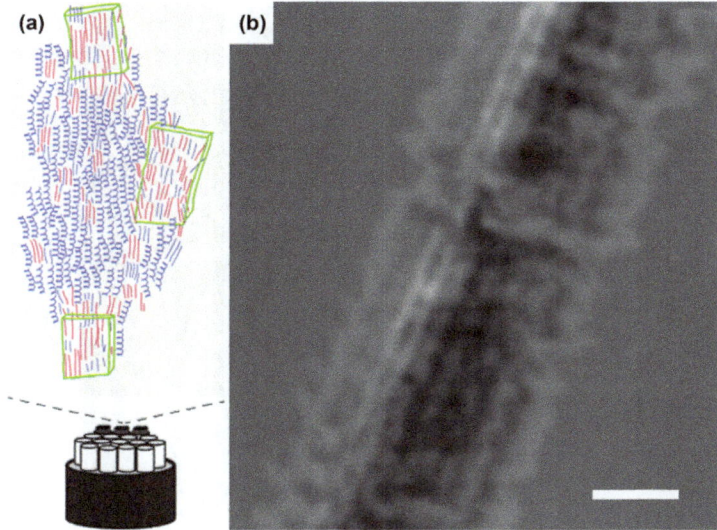

Figure 4.11 Characteristics of dragline spider silk: (a) proposed model of silk fibril structure where β-sheet-rich regions composed of predominantly alanine and glycine are dispersed among helical protein coils to form semi-crystalline regions, all of which are encapsulated within a hardened outer skin,[a] (b) electron micrograph of ultra-thin dragline cross-section.[b]
([a]From van Beek, Hess, Vollrath, and Meier, 2002, *PNAS*, **99**(16): 10266. Copyright 2002 National Academy of Sciences, USA. [b]From Augsten, Muehlig, and Herrmann, 2000, *Scanning*, **22**, pp. 12–15. Copyright John Wiley & Sons.)

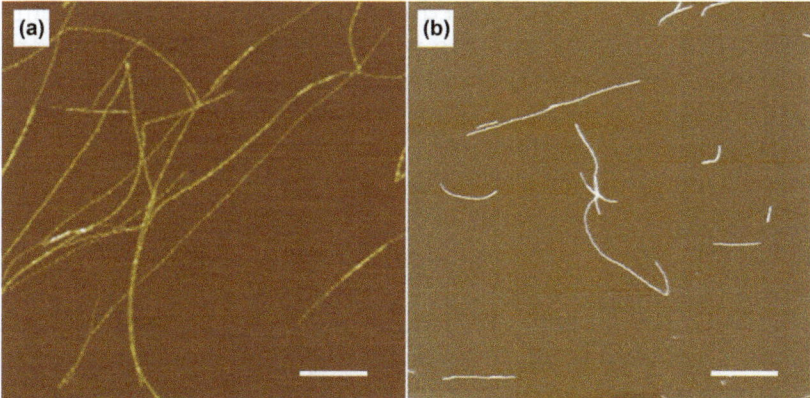

Figure 4.12 AFM images of amyloid-like protein fibrils following thermal treatment at 90 °C for 5 hours at pH 2, composed of (a) β-lactoglobulin and (b) α-lactalbumin. Scale bar = 500 nm.

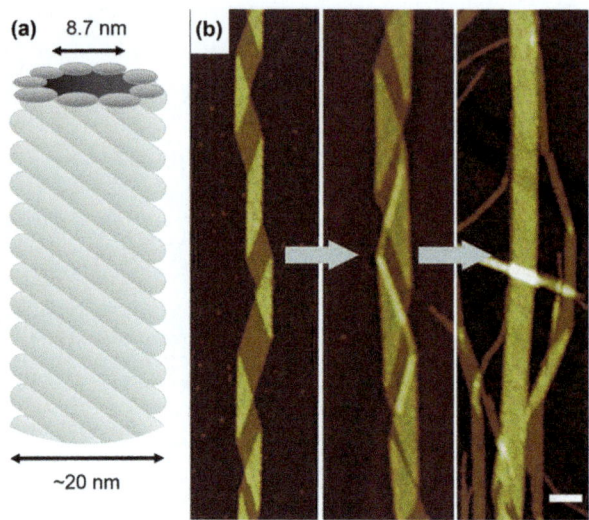

Figure 4.13   Protein nanotube structures: (a) depiction of the structural dimensions
of nanotubes assembled from α-lactalbumin,[a] (b) AFM images of lysozyme fibrils aligned to form (from left to right) a twisted ribbon, a highly
twisted ribbon, and a closed nanotube (scale bar = 100 nm).[b]
([a]Based on Graveland-Bikker, Schaap, Schmidt, & de Kruif, 2006.[106]
[b]From Lara, Handschin, and Mezzenga, 2013, Nanoscale, 5,
pp. 7197–7201.)

cross-β-sheet strands among nanotubes organize in a circular pattern
in a manner that is analogous to the arrangement of surfactants in a
cylindrical vesicle (Figure 4.13a). This cylindrical arrangement leaves
an internal cavity in the center of the nanotube that contains solvent.

Nanotubes have been observed among amyloidogenic proteins or
even synthetic peptides with the propensity to form β-sheets, and it
was proposed that multi-stranded fibrils first flatten into a ribbon and
then curl around a central axis to form these tube structures.[97] This
gradual transition from fibril to ribbon to nanotube has been recently
observed among fibrils assembled from lysozyme (Figure 4.13b).[98]
Protein nanotubes have also been formed from the whey protein
α-lactalbumin, where a microbial protease from *Bacillus licheniformis*
cleaves the protein into oligopeptides that form cross-β-sheets in the
presence of calcium ions at neutral pH.[99] The structure of these
nanotubes has been resolved by microscopy and scattering methods;
they have a cylinder diameter of ∼20 nm, an internal cavity of
∼9 nm, and a length of ∼1 μm.[100] Lastly, nanotubes may be formed
by layer-by-layer deposition of alternately charged polyelectrolytes
within lithographic nanopore arrays, which act as a template to direct

the formation of tubular constructs.[101] Protein nanotubes formed in this manner were recently found to encapsulate the hydrophobic bioactive curcumin within the internal cavity,[102] which indicates that these structures may potentially serve as controlled-release vehicles.

Protein fibril formation may be influenced by other components in the system, and the promotion or inhibition of fibril formation is of intense interest in creating fibril-based biomaterials or inhibiting disease-linked amyloid plaque formation, respectively. Fibril formation is promoted by simple factors such as increasing protein concentration,[103,104] control of ionic strength,[105] moderate shear,[106,107] and reduced pH.[108] In general, conditions that favor formation of fibrils also tend to increase their length and rigidity.

Although fibrils are generally prepared in water, co-solvents are occasionally utilized to facilitate fibril formation. At low to modest concentrations ($\sim$20–50%, depending on the solvent), co-solvents such as trifluoroethanol and ethanol increase interpeptide interactions and promote an increase in β-sheet character.[109,110] These factors contribute to the alignment of peptides into cross-β-sheets that is necessary for protein fibril formation. At higher concentrations, however, co-solvents such as ethanol and acetonitrile have been found to decrease fibril formation.[111,112] Ethanol has even been shown cause deterioration in fibril structure, indicating that the interactions with water are essential for the formation and continued structure of the fibrils.[113] Interestingly, very high concentrations of ethanol have been shown to promote fibril formation from egg white lysozyme,[114] and moderate concentrations of ethanol (following their initial disassembly) were found to promote reassembly of fibrils over prolonged incubation periods (Figure 4.14).[113] These results indicate

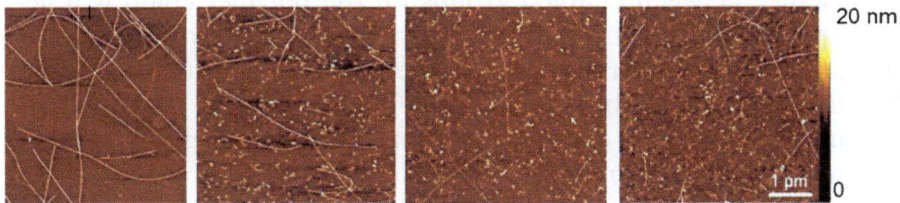

Figure 4.14 AFM images showing the disassembly and reassembly of β-lactoglobulin fibrils in 50% ethanol during incubation at 37 °C for (from left to right) 0, 1, 5, and 8 weeks.
(Reprinted with permission from Jordens, Adamcik, Amar-Yuli, and Mezzenga, 2011, Biomacromolecules, **12**: 187. Copyright 2011 American Chemical Society.)

that assembly of fibrils in the presence of high concentrations of co-solvents may operate under a different mechanism than in predominantly aqueous solvents. In agreement with this hypothesis, fibrils prepared using alcohol co-solvents are far less rigid.[115]

Denaturants such as urea and guanidine hydrochloride may also be utilized to facilitate fibril formation, as they promote the breakdown of the original protein structure to allow the peptides to align into fibril structures. Experiments have typically shown that up to 4–6 mol l^{-1} urea will enhance fibril formation.[108,116] Guanidinium hydrochloride has shown similar enhancement in fibril formation with up to 1–3 mol l^{-1} concentrations.[117,118] Just as with co-solvents, high concentrations of denaturants start to inhibit fibril formation.[119] Denaturants often work by interactions with peptides at the surface of proteins, and high concentrations of these denaturants could actually inhibit cross-β-sheet structures.

Other important components that regulate fibril formation include metal ions, phenolic compounds, and polysaccharides. In particular, metal ions such as aluminium, copper, and iron have been linked to an increased risk of amyloid-related diseases, and there is a hypothesis that this is due to an increase in fibril formation in the presence of these metal ions.[120,121] However, universal promotion of protein fibril formation by metal ions has not been wholly supported by evidence. Instead, there is strong evidence that metal ions will actually inhibit the formation of long, rigid protein fibrils.[122,123] Polyphenolics such as resveratrol and curcumin have also been shown to inhibit fibril formation,[124] which supports the role of polyphenolic nutraceuticals in the prevention of amyloid-related diseases. Charged polysaccharides and other biopolymers also have a slight influence on fibril formation by directly interacting with charged peptide residues. In special circumstances, such as with sulfated glycosaminoglycans, the distance between charged groups on the polymer chain will actually promote β-sheet alignment and fibril formation.[125] In most cases, however, interactions with charged polysaccharides will inhibit the formation of protein fibrils.[125,126] Thus, in general, experimental evidence indicates that components that directly interact with protein tend to inhibit the formation of amyloid fibrils.

Amyloid-like fibrils assembled from proteins have strong potential to serve as functional materials in biological or engineering disciplines because they are readily assembled through simple processing procedures amenable to the food industry and, similar to silk fibers, are highly resilient. Part of this is due to the layered β-sheet structuring of the fibrils, which (reinforced by side-chain interactions) imparts an amazing amount of cohesive strength against elongation strain. Although these β-sheet interactions result from relatively weak hydrogen bonds, the sheer number of coordinated hydrogen bonds generates a remarkable combined strength, just as fabric hook-and-loop fasteners obtain strength through a coordination of multiple weak bonds. After the formation of these layered β-sheets, the structure is not readily reversible by shear or thermal treatment,[127] and the fibrils can be considered a fairly stable structural material.[128] Mechanical measurements of the fibril cross section of various fibers has given Young's moduli (representing potential for composite stiffness) between 0.1 and 20 GPa, depending on the component protein.[129] Nanotubes, however, possess relatively low stiffness, as the Young's modulus of α-lactalbumin nanotubes was on the order of ~0.1 GPa.[130]

Given that protein fibrils possess the desirable attributes of high stiffness and high length-to-width ratio, they are naturally advantageous for the production of strong networks and scaffolds. Fibrils may serve as scaffolds by adhering enzymes, engineered bacteria, or bioactive compounds onto existing fibril networks in order to create an immobilized bioreactor or controlled release system.[131] By building such fibril scaffolds into existing films or gels at the macrostructural level, it would be possible to incorporate fibrils for both structural reinforcement and for functional purposes. For instance, a packaging polymer could be reinforced with a network of the protein fibrils with an immobilized antimicrobial enzyme so that the applied dressing would confer resistance to pathogen growth in the food product. Protein fibril networks can also be coated with metallo-ions to create magnetic or conductive nanowires or to create protein-based liquid-crystalline materials.[132]

Perhaps the most direct utilization of protein fibril network attributes is the formation of gels and films. Because of their strength, protein fibrils have been proposed as an alternative to gelatin-based gel systems, which are disfavored by some consumers for philosophical or health reasons. With fibril lengths on the order of 1–15 μm, the percolation concentration to form gel networks has been typically observed between 3 and 80 g l^{-1}.[133–135] At higher concentration, highly elastic gels are obtained, although the fibrils often aggregate or degrade in the process of gel/film formation.

One significant challenge in utilizing protein fibrils in gel systems for the formation of fibril gels is the loss of fibril integrity and rigidity when they are used in neutral pH conditions. This weakness derives from their assembly in acidic conditions, which promotes the molten globule forms that favor fibril formation. However, when the fibril solutions are neutralized, the protein must pass through the iso-electric point where the protein possesses negligible net charge and has a very strong tendency to aggregate and break into small, non-functional fragments. Formation of these small fragments during neutralization can be partially minimized by increasing the pH rap-idly.[136] An alternative is to utilize basic proteins, such as hen egg white lysozyme, which have isoelectric points that are above neutral pH so that protein fibrils can be formed and used at pH values closer to 7 where the net protein charge is still positive. Another alternative is to utilize fibrils that are nanotubes that are formed enzymatically or with co-solvents, which do not require low pH values to induce the molten globule states. One example of this is the enzymatically formed nanotubes of α-lactalbumin that are formed and fully func-tional near neutral pH conditions. At the same time, fibrils and nanotubes formed near neutral pH have significantly lower stiffness and, therefore, less potential utility as structural materials in gels and films. There is then a current need to develop techniques to improve fibril stability at higher pH while retaining their rigidity and other positive functional attributes.

Recent work has also shown that protein fibrils can serve as structural reinforcement on surfaces, such as oil-in-water emulsions. The motivation for studying fibrils at the interface stems from find-ings that amyloid protein growth may be facilitated at cell lipid bilayers.[137] In the same way, protein fibrils have been found to be surface active, and smaller protein fibrils are particularly well suited to surface stabilization.[138] Longer fibrils, on the other hand, have difficulty in conforming to a spherical emulsion droplet and do not viscoelastically respond to droplet deformations. Thus, long fibrils are better suited for use as secondary interfacial layers. For instance, long, charged protein fibrils may be laminated onto existing sur-factant layers through a multi-layer electrostatic deposition to create a thick and rigid interfacial layer.[139,140]

4.3 PREDICTING FUTURE TRENDS IN PROTEIN NANOSTRUCTURES

Although protein nanostructures exist naturally (*e.g.* the casein micelle; see Section 4.2.1), the development of nanostructures from protein raw materials is still in its infancy. There is growing demand in the marketplace for value-added materials assembled from biopolymers, including protein. Previously, the demand for controlled-delivery vehicles and other biomaterials has been fulfilled by constructs of synthetic polymers derived from petroleum products or other chemically modified forms, yet the public has become distrustful of excessively utilizing synthetic polymers. Proteins offer an excellent alternative to synthetic polymers as they are renewable and biodegradable. Of course, the disadvantage of developing biomaterials and nanostructures from proteins is the potential for allergenicity, as the population of those allergic to specific protein products has increased over the years. Regardless, the benefits of protein-based biomaterials will likely outweigh the risks, and it is likely that protein nanostructure research will continue to grow in the coming years.

All of the protein nanostructures discussed in the previous section show potential for further development, although fibrils and nano-particles will likely dominate future research (just as they have for the past 20 years). Nanoparticles of protein are often prepared with organic solvents or other components that may be undesirable for consumers, yet nanoparticles are relatively robust and/or cost-effective to produce when compared to microgels or coacervate-based systems. In addition, nanoparticles, as solids, have a much greater capacity to serve as controlled-release vehicles for bioactives or drug molecules, as the diffusion of the bioactive or drug molecule is greatly reduced compared to gel- or emulsion-based delivery systems (con-taining significant porosity and liquid-based diffusion, respectively). Fibril nanostructures will perhaps surpass nanoparticle research in coming years because of their excellent mechanical attributes and ease of production. Further development of fibrils, such as by adding metallic ions or attaching biopolymers, can expand the applications of these materials to a wide variety of fields.[132] Since there is still so much work to be done in these areas, research on nanoparticles and fibrils will continue to increase in future years.

One interesting application of protein nanostructures that has emerged in the past decade is the stabilization of emulsion or foam interfaces. Interface stabilization requires that the nanostructure adsorbs to the interface and confers a resistance to the merging of adjacent interfaces. Nanostructures of greater than 10 nm that adsorb to interfaces may be considered as Pickering stabilizers, which are essentially "stuck" at the interface (*i.e.* detachment energy is pro-hibitive) and, at high surface loads, provide superb resistance to the coalescence (*i.e.* merging) of stabilized emulsion or foam droplets.[141] Protein nanostructures have an added advantage as interface stabil-izers, because many of them possess natural surface activity and are also desirable in food-based formulations as a biodegradable natural product (as opposed to particles based on silica, titanium, or synthetic polymers). This area of research will likely expand in the coming years.

Synthetic peptides have emerged as a powerful method for the creation of fibrous and particulate structures for a wide array of applications, including medicine, sensors, and packaging.[142,143] Synthetic peptides can be readily created in the laboratory to achieve specific amphiphilic properties or a propensity to form secondary structures such as β-sheets.[144] Peptide synthesis allows the scientist to choose the exact material properties necessary to assemble desired structures. Nanostructures from synthesized peptides will likely advance the science of nanostructural development tremendously. However, peptide synthesis is prohibitively expensive for most industrial applications, and the development of nanostructures from protein extracted from natural sources is still a highly valuable endeavor. In particular, nanostructures developed for food use should always be developed from natural materials, and there is then a strong need for continuing development of protein nanostructures from natural sources for biomaterial applications.

4.4 CONCLUSIONS

Nanostructures may be assembled from proteins and polypeptides by careful manipulation of their conformation and intermolecular interactions. Proteins are an essential part of the human diet, a component of all biological systems, and can be utilized as structural materials. Natural sources of proteins for material development include dairy products, vegetables, and muscle tissue. The native shape and structure of these proteins depends strongly on the amino acid composition, which influences intramolecular (*i.e.* conformation) and intermolecular (*i.e.* quarternary structuring) interactions by the physical forces that exist between the component amino acids. By understanding how the peptide components interact within solution under changing conditions of ionic strength, pH, co-solvents, co-solutes, and temperature, it is possible to alter the intra- and intermolecular interactions in such a way as to build desired structures on the nanometer scale.

This chapter has discussed numerous examples of protein nanostructures, and the capacity to form new nanostructures from proteins will likely expand as the theoretical and experimental understanding of colloidal materials grows. Casein micelles are an excellent example of a natural self-assembled protein nanostructure composed of several different caseinate proteins in milk. In the laboratory it is possible to induce assembly of proteins by: introduction of an electrostatically interacting protein or polysaccharide, modification of the protein by a polysaccharide segment to introduce amphiphilicity (*i.e.* self-assembled conjugates), introduction of an ideal-to-poor solvent (*i.e.* simple coacervation), introduction of a very poor solvent or destabilizing co-solutes (*i.e.* desolvated nanoparticles), desolvation in volume-restricted templates (*i.e.* emulsion-templated nanoparticles), or controlled thermal treatment (*i.e.* microgels/ nanogels and fibrils). Considering the huge advances in protein nanostructures in the past several decades, there are likely to be many exciting developments in functional protein nanostructures for food and biomedical applications within the coming years.

REFERENCES

1. C. Tanford and J. Reynolds, *Nature's Robots: A History of Proteins*, Oxford University Press, Oxford, UK, 1st edn, 2004.
2. J. Wisniak, *Chem. Educ.*, 2000, **5**, 343.
3. H. B. Vickery and C. L. A. Schmidt, *Chem. Rev.*, 1931, **9**, 169.
4. US Economic Research Service, Dairy Data, United States Department of Agriculture, 2013.
5. H. E. Swaisgood, in *Handbook of Milk Composition*, ed. R. G. Jensen, Academic Press, San Diego, CA, 1st edn, 1995, pp. 464–468.
6. P. F. Fox and P. L. H. McSweeney, *Advanced Dairy Chemistry*, Springer, New York, 4th edn, 2013.
7. A. Moure, J. Sineiro, H. Domínguez and J. C. Parajó, *Food Res. Int.*, 2006, **39**, 945.
8. D. Fukushima, in *Handbook of Food Proteins*, ed. G. O. Phillips and P. A. Williams, Woodhead Publishing, Philadelphia, PA, 1st edn, 2011, pp. 210–232.
9. L. Day, in *Handbook of Food Proteins*, ed. G. O. Phillips and P. A. Williams, Woodhead Publishing, Philadelphia, PA, 1st edn, 2011, pp. 267–288.
10. F. Seno and A. Trovato, *Physica A: Stat. Mech. Applic.*, 2007, **384**, 122.
11. W. J. Wolf, *J. Agr. Food Chem.*, 1993, **41**, 168.
12. M. A. M. Hoffmann and P. J. J. M. van Mil, *J. Agr. Food Chem.*, 1997, **45**, 2942.
13. M. U. Hammer, T. H. Anderson, A. Chaimovich, M. S. Shell and J. Israelachvili, *Faraday Discuss*, 2010, **146**, 299.

14. S. Nakai, *J. Agr. Food Chem.*, 1983, **31**, 676.
15. S. Damodaran, in *Food Chem*, ed. O. R. Fennema, Marcel Dekker, New York, NY, 3rd edn, 1996, p. 321.
16. A. O. Elzoghby, W. M. Samy and N. A. Elgindy, *J. Controlled Release*, 2012, **161**, 38.
17. G. Wang and H. Uludag, *Expert Opin. Drug Delivery*, 2008, **5**, 499.
18. J. K. Raynes, J. A. Carver, S. L. Gras and J. A. Gerrard, *Trends Food Sci. Tech.*, **37**, 42.
19. C. G. de Kruif, T. Huppertz, V. S. Urban and A. V. Petukhov, *Adv. Colloid Interface Sci.*, 2012, **171–172**, 36.
20. T. Huppertz, in *Advanced Dairy Chemistry*, ed. P. F. Fox and P. L. H. McSweeney, Springer, New York, 4th edn, 2013, vol. 1, pp. 135–160.
21. H. M. Farrell, E. M. Brown and E. L. Malin, in *Advanced Dairy Chemistry*, ed. P. F. Fox and P. L. H. McSweeney, Springer, New York, 4th edn, 2013, vol. 1, pp. 161–184.
22. J. A. O'Mahoney and P. F. Fox, in *Advanced Dairy Chemistry*, ed. P. F. Fox and P. L. H. McSweeney, Springer, New York, 4th edn, 2013, vol. 1, pp. 43–86.
23. J. A. Lucey, M. E. Johnson and D. S. Horne, *J. Dairy Sci.*, 2003, **86**, 2725.
24. J. S. Mounsey, B. T. O'Kennedy and P. M. Kelly, *Lait*, 2005, **85**, 405.
25. T. Huppertz and C. G. de Kruif, *Int. Dairy J.*, 2008, **18**, 556.
26. C. Schorsch, H. Carrie and I. T. Norton, *Int. Dairy J.*, 2000, **10**, 529.
27. S. Lauber, T. Henle and H. Klostermeyer, *Eur. Food Res. Technol.*, 2000, **210**, 305.
28. S. G. Anema, S. Lauber, S. K. Lee, T. Henle and H. Klostermeyer, *Food Hydrocolloids*, 2005, **19**, 879.
29. A.-M. Knoop, E. Knoop and A. Wiechen, *J. Dairy Res.*, 1979, **46**, 347.
30. E. Semo, E. Kesselman, D. Danino and Y. D. Livney, *Food Hydrocoll.*, 2007, **21**, 936.
31. P. Zimet, D. Rosenberg and Y. D. Livney, *Food Hydrocoll.*, 2011, **25**, 1270.
32. C. Schmitt, C. Sanchez, S. Desobry-Banon and J. Hardy, *Crit. Rev. Food Sci. Nutr.*, 1998, **38**, 689.
33. D. Guzey and D. J. McClements, *Food Hydrocoll.*, 2006, **20**, 124.
34. B. L. H. M. Sperber, H. A. Schols, M. A. Cohen Stuart, W. Norde and A. G. J. Voragen, *Food Hydrocoll.*, 2009, **23**, 765.

35. J. L. Xia, P. L. Dubin, Y. Kim, B. B. Muhoberac and V. J. Klimkowski, *J. Phys. Chem.*, 1993, **97**, 4528.
36. O. G. Jones, U. Lesmes, P. Dubin and D. J. McClements, *Food Hydrocoll.*, 2010, **24**, 374.
37. E. Kizilay, A. B. Kayitmazer and P. L. Dubin, *Adv. Colloid Interface Sci.*, 2011, **167**, 24.
38. S. L. Turgeon, C. Schmitt and C. Sanchez, *Curr. Opin. Colloid Interface Sci.*, 2007, **12**, 166.
39. O. G. Jones and D. J. McClements, *Food Biophys.*, 2008, **3**, 191.
40. O. G. Jones and D. J. McClements, *J. Food Sci.*, 2010, **75**, N36.
41. O. G. Jones and D. J. McClements, *Adv. Colloid Interface Sci.*, 2011, **167**, 49.
42. G. Markman and Y. D. Livney, *Food Funct.*, 2012, **3**, 262.
43. J. Li and P. Yao, *Langmuir*, 2009, **25**, 6385.
44. L. Zhang, A. Dudhani, L. Lundin and S. L. Kosaraju, *Eur. Polym. J.*, 2009, **45**, 1960.
45. M. Rubinstein and R. H. Colby, *Polymer Physics*, Oxford University Press, Oxford, UK, 2003.
46. S. A. H. Khalil, J. R. Nixon and J. E. Carless, *J. Pharm. Pharmacol.*, 1968, **20**, 215.
47. B. Mohanty and H. B. Bohidar, *Europhys. Lett.*, 2006, **76**, 965.
48. L. W. J. Holleman, H. G. Bungenberg de Jong and R. S. Tjaden Modderman, *Kolloid-Beih*, 1934, **39**, 334.
49. J. I. Okada, A. Kusai and S. Ueda, *J. Microencaps.*, 1985, **2**, 163.
50. S. Galindo-Rodriguez, E. Allémann, H. Fessi and E. Doelker, *Pharm. Res.*, 2004, **21**, 1428.
51. Q. Zhong, H. Tian and S. Zivanovic, *J. Food Process. Preserv.*, 2009, **33**, 255.
52. Q. Zhong and M. Jin, *Food Hydrocoll.*, 2009, **23**, 2380.
53. A. R. Patel, E. C. M. Bouwens and K. P. Velikov, *J. Agr. Food Chem.*, 2010, **58**, 12497.
54. Y. Luo, Z. Teng and Q. Wang, *J. Agr. Food Chem.*, 2011, **60**, 836.
55. J. W. J. de Folter, M. W. M. van Ruijven and K. P. Velikov, *Soft Matter*, 2012, **8**, 6807.
56. C. Duclairoir, E. Nakache, H. Marchais and A. M. Orecchioni, *Colloid Polym. Sci.*, 1998, **276**, 321.
57. M. C. Mauguet, J. Legrand, L. Brujes, G. Carnelle, C. Larre and Y. Popineau, *J. Microencaps.*, 2002, **19**, 377.
58. İ. Gülseren, Y. Fang and M. Corredig, *Food Hydrocoll.*, 2012, **29**, 258.
59. S. Gunasekaran, S. Ko and L. Xiao, *J. Food Eng.*, 2007, **83**, 31.
60. C. Weber, C. Coester, J. Kreuter and K. Langer, *Int. J. Pharm.*, 2000, **194**, 91.

61. Z. Teng, Y. Luo and Q. Wang, *J. Agr. Food Chem.*, 2012, **60**, 2712.
62. J. Kundu, Y.-I. Chung, Y. H. Kim, G. Tae and S. C. Kundu, *Int. J. Pharm.*, 2010, **388**, 242.
63. H.-B. Yan, Y.-Q. Zhang, Y.-L. Ma and L.-X. Zhou, *J. Nanopart Res.*, 2009, **11**, 1937.
64. A. S. Lammel, X. Hu, S.-H. Park, D. L. Kaplan and T. R. Scheibel, *Biomaterials*, 2010, **31**, 4583.
65. N. Anton, J.-P. Benoit and P. Saulnier, *J. Control. Rel.*, 2008, **128**, 185.
66. B. Shah, S. Ikeda, P. M. Davidson and Q. Zhong, *J. Food Eng.*, 2012, **113**, 79.
67. L. Chen and M. Subirade, *Biomacromolecules*, 2009, **10**, 3327.
68. M. Cascone, L. Lazzeri, C. Carmignani and Z. Zhu, J, *Mater Sci: Mater Med.*, 2002, **13**, 523.
69. L. A. Lyon and A. Fernandez-Nieves, *Annu. Rev. Phys. Chem.*, 2012, **63**, 25.
70. B. R. Saunders, N. Laajam, E. Daly, S. Teow, X. Hu and R. Stepto, *Adv. Colloid Interface Sci.*, 2009, **147–148**, 251.
71. G. Kontopidis, C. Holt and L. Sawyer, *J. Dairy Sci.*, 2004, **87**, 785.
72. T. Phan-Xuan, D. Durand, T. Nicolai, L. Donato, C. Schmitt and L. Bovetto, *Langmuir*, 2011, **27**, 15092.
73. C. Schmitt, C. Moitzi, C. Bovay, M. Rouvet, L. Bovetto, L. Donato, M. E. Leser, P. Schurtenberger and A. Stradner, *Soft Matter*, 2010, **6**, 4876.
74. D. Saglam, P. Venema, R. de Vries and E. van der Linden, *Soft Matter*, 2013, **9**, 4598.
75. O. G. Jones, E. A. Decker and D. J. McClements, *J. Colloid Int. Sci.*, 2010, **344**, 21.
76. D. Sağlam, P. Venema, R. de Vries, L. M. C. Sagis and E. van der Linden, *Food Hydrocoll.*, 2011, **25**, 1139.
77. W. N. Zhang and Q. X. Zhong, *Food Chem.*, 2010, **119**, 1318.
78. J. Greenwald and R. Riek, *Structure*, 2010, **18**, 1244.
79. M. Sunde and C. Blake, *Adv. Protein Chem.*, 1997, **50**, 123.
80. H. Cui, M. J. Webber and S. I. Stupp, *Peptide Sci.*, 2010, **94**, 1.
81. L. M. Ittner and J. Gotz, *Nat. Rev. Neurosci.*, 2011, **12**, 67.
82. L. Wei, P. Jiang, W. X. Xu, H. Li, H. Zhang, L. Y. Yan, M. B. Chan-Park, X. W. Liu, K. Tang, Y. G. Mu and K. Pervushin, *J. Biol. Chem.*, 2011, **286**, 6291.
83. S. B. Padrick and A. D. Miranker, *Biochem.*, 2002, **41**, 4694.
84. N. H. H. Heegaard, *Amyloid*, 2009, **16**, 151.
85. S. A. Priola, B. Chesebro and B. Caughey, *Science*, 2003, **300**, 917.
86. C. Soto and N. Satani, *Trend Mol. Med.*, 2011, **17**, 14.

87. A. Aguzzi and A. M. Calella, *Physiol. Rev.*, 2009, **89**, 1105.
88. E. H. Koo, P. T. Lansbury and J. W. Kelly, *Proc. Natl. Acad. Sci. USA*, 1999, **96**, 9989.
89. J. A. Hardy and G. A. Higgins, *Science*, 1992, **256**, 184.
90. R. Riek, *Nature*, 2006, **444**, 429.
91. R. V. Lewis, *Chem. Rev.*, 2006, **106**, 3762.
92. J. D. van Beek, S. Hess, F. Vollrath and B. H. Meier, *Proc. Natl. Acad. Sci. USA*, 2002, **99**, 10266.
93. K. Liu, H. S. Cho, H. A. Lashuel, J. W. Kelly and D. E. Wemmer, *Nat. Struct. Biol.*, 2000, **7**, 754.
94. G. Marcon, G. Plakoutsi and F. Chiti, *Meth. Enzymol.*, 2006, **413**, 75.
95. C. Akkermans, P. Venema, A. J. van der Goot, H. Gruppen, E. J. Bakx, R. M. Boom and E. van der Linden, *Biomacromolecules*, 2008, **9**, 1474.
96. A. Arora, C. Ha and C. B. Park, *Protein Sci.*, 2004, **13**, 2429.
97. H. A. Lashuel, S. R. LaBrenz, L. Woo, L. C. Serpell and J. W. Kelly, *J. Am. Chem. Soc.*, 2000, **122**, 5262.
98. C. Lara, S. Handschin and R. Mezzenga, *Nanoscale*, 2013, **5**, 7197.
99. R. Ipsen, J. Otte and K. B. Qvist, *J. Dairy Res.*, 2001, **68**, 277.
100. J. F. Graveland-Bikker, G. Fritz, O. Glatter and C. G. de Kruif, *J. Appl. Crystallogr*, 2006, **39**, 180.
101. Z. Liang, A. S. Susha, A. Yu and F. Caruso, *Adv. Mater.*, 2003, **15**, 1849.
102. R. Sadeghi, A. Kalbasi, Z. Emam-Jomeh, S. H. Razavi, J. Kokini and A. A. Moosavi-Movahedi, *J. Nanopart. Res.*, 2013, **15**, 1.
103. L. N. Arnaudov, R. de Vries, H. Ippel and C. P. M. van Mierlo, *Biomacromolecules*, 2003, **4**, 1614.
104. E. P. Schokker, H. Singh, D. N. Pinder and L. K. Creamer, *Int. Dairy J.*, 2000, **10**, 233.
105. P. Aymard, T. Nicolai, D. Durand and A. Clark, *Macromolecules*, 1999, **32**, 2542.
106. C. Akkermans, P. Venema, S. S. Rogers, A. J. van der Goot, R. M. Boom and E. van der Linden, *Food Biophys.*, 2006, **1**, 144.
107. D. E. Dunstan, P. Hamilton-Brown, P. Asimakis, W. Ducker and J. Bertolini, *Protein Eng. Des. Sel.*, 2009, **22**, 741.
108. L. Nielsen, R. Khurana, A. Coats, S. Frokjaer, J. Brange, S. Vyas, V. N. Uversky and A. L. Fink, *Biochem*, 2001, **40**, 6036.
109. W. Dzwolak, *Biochim. Biophys. Acta, Proteins Proteomics*, 2006, **1764**, 470.
110. D. E. Otzen, *Curr. Protein Pept. Sc.*, 2010, **11**, 355.

111. K. Yoshida, T. Yamaguchi, N. Osaka, H. Endo and M. Shibayama, *Phys. Chem. Chem. Phys.*, 2010, **12**, 3260.

112. K. Yamaguchi, H. Naiki and Y. Goto, *J. Mol. Biol.*, 2006, **363**, 279.

113. S. Jordens, J. Adamcik, I. Amar-Yuli and R. Mezzenga, *Biomacromolecules*, 2011, **12**, 187.

114. M. Holley, C. Eginton, D. Schaefer and L. R. Brown, *Biochem. Biophys. Res. Commun.*, 2008, **373**, 164.

115. W. S. Gosal, A. H. Clark and S. B. Ross-Murphy, *Biomacromolecules*, 2004, **5**, 2408.

116. D. Hamada and C. M. Dobson, *Protein Sci.*, 2002, **11**, 2417.

117. W. Swietnicki, M. Morillas, S. G. Chen, P. Gambetti and W. K. Surewicz, *Biochem.*, 2000, **39**, 424.

118. B. A. Vernaglia, J. Huang and E. D. Clark, *Biomacromolecules*, 2004, **5**, 1362.

119. A. Ahmad, I. S. Millett, S. Doniach, V. N. Uversky and A. L. Fink, *Biochem.*, 2003, **42**, 11404.

120. K. Jomova, D. Vondrakova, M. Lawson and M. Valko, *Mol. Cell Biochem.*, 2010, **345**, 91.

121. S. Rivera-Mancia, I. Perez-Neri, C. Rios, L. Tristan-Lopez, L. Rivera-Espinosa and S. Montes, *Chem. Biol. Interact.*, 2010, **186**, 184.

122. B. Raman, T. Ban, K. Yamaguchi, M. Sakai, T. Kawai, H. Naiki and Y. Goto, *J. Biol. Chem.*, 2005, **280**, 16157.

123. M. Innocenti, E. Salvietti, M. Guidotti, A. Casini, S. Bellandi, M. L. Foresti, C. Gabbiani, A. Pozzi, P. Zatta and L. Messori, *J. Alzheimers Dis.*, 2010, **19**, 1323.

124. Y. Porat, A. Abramowitz and E. Gazit, *Chem. Biol. Drug Des.*, 2006, **67**, 27.

125. J. J. Valle-Delgado, M. Alfonso-Prieto, N. S. de Groot, S. Ventura, J. Samitier, C. Rovira and X. Fernandez-Busquets, *FASEB J*, 2010, **24**, 4250.

126. O. G. Jones, J. Adamcik, S. Handschin, S. Bolisetty and R. Mezzenga, *Langmuir*, 2010, **26**, 17449.

127. B. Morel, L. Varela and F. Conejero-Lara, *J. Phys. Chem. B*, 2010, **114**, 4010.

128. R. Nelson, M. R. Sawaya, M. Balbirnie, A. O. Madsen, C. Riekel, R. Grothe and D. Eisenberg, *Nature*, 2005, **435**, 773.

129. T. P. J. Knowles and M. J. Buehler, *Nat. Nanotechnol*, 2011, **6**, 469.

130. J. F. Graveland-Bikker, I. A. T. Schaap, C. F. Schmidt and C. G. de Kruif, *Nano Lett.*, 2006, **6**, 616.

131. R. Giraldo, *Chembiochem*, 2010, **11**, 2347.

132. I. Cherny and E. Gazit, *Angew. Chem. Int. Edit.*, 2008, **47**, 4062.

133. C. Veerman, G. de Schiffart, L. M. C. Sagis and E. van der Linden, *Int. J. Biol. Macromol.*, 2003, **33**, 121.
134. C. Veerman, H. Ruis, L. M. C. Sagis and E. van der Linden, *Biomacromolecules*, 2002, **38**, 869.
135. C. Veerman, L. M. C. Sagis, J. Heck and E. van der Linden, *Int. J. Biol. Macromol.*, 2003, **31**, 139.
136. J. M. Jung and R. Mezzenga, *Langmuir*, 2010, **26**, 504.
137. M. Bokvist, F. Lindström, A. Watts and G. Gröbner, *J. Mol. Biol.*, 2004, **335**, 1039.
138. L. Isa, J. M. Jung and R. Mezzenga, *Soft Matter*, 2011, 7, 8127.
139. K. N. P. Humblet-Hua, G. Scheltens, E. van der Linden and L. M. C. Sagis, *Food Hydrocoll.*, 2011, **25**, 569.
140. L. M. C. Sagis, R. de Ruiter, F. J. R. Miranda, J. de Ruiter, K. Schroen, A. C. van Aelst, H. Kieft, R. Boom and E. van der Linden, *Langmuir*, 2008, **24**, 1608.
141. E. Dickinson, *Curr. Opin. Colloid Int. Sci.*, 2010, **15**, 40.
142. B. D. Briggs and M. R. Knecht, *J. Phys. Chem. Lett.*, 2012, **3**, 405.
143. M. Zelzer and R. V. Ulijn, *Chem. Soc. Rev.*, 2010, **39**, 3351.
144. B. E. I. Ramakers, J. C. M. van Hest and D. W. P. M. Lowik, *Chem. Soc. Rev.*, 2014.
145. T. E. Creighton, *Proteins: Structures and Molecular Properties*, W.H. Freeman and Company, New York, NY, 2nd edn, 1993, p. 507.
146. A. P. Karplus, *Protein Sci.*, 1997, **6**, 1302.
147. D. J. McMahon and B. S. Oommen, *J. Dairy Sci.*, 2008, **91**, 1709.

CHAPTER 5

Lipid Mesophase Nanostructures

CONSTANTINOS V. NIKIFORIDIS

Physics and Physical Chemistry of Foods & Food Process
Engineering Group, Wageningen University, P.O. Box 17, 6700 AA,
Wageningen, The Netherlands
Email: costas.nikiforidis@wur.nl

5.1 INTRODUCTION

Lipid mesophases are organized self-assembled lipid-based materials
with properties intermediate between those of crystalline solids and
isotropic liquids.[1,2] The term lipid mesophase is interchangeably
used with lyotropic liquid crystals (LLC). These materials retain a
degree of organization that resembles an ordered, crystalline solid,
but owing to the dynamic disorder at atomic distances, they behave as
viscous liquids. Therefore, a short definition for accurately describing
these systems is that they are ordered liquids.[3,4] Lyotropic liquid
crystals commonly consist of amphiphilic molecules that interact
through physical forces in solvents. In lipid mesophases the structural units are polar lipids.[5] The geometry of the structures depends
on the chemical composition, along with the morphology of the
amphiphilic molecules. In addition, since only physical forces are
taking place, the mesophases are sensitive to external parameters
such as the composition of the solvent, pressure and temperature.[2,3,6]

Lyotropic lipid mesophases are amongst the most fascinating
structures in nature. Natural amphiphilic phospholipids form a lipid

Edible Nanostructures
Edited by Alejandro G Marangoni and David Pink
© The Royal Society of Chemistry 2015
Published by the Royal Society of Chemistry, www.rsc.org

bilayer for the purpose of generating a selectively permeable supra-molecular complex, which is recognized as a biological membrane.

A part of the biological function, lipid mesophases are present in most edible raw materials. Mesophases can provide important functional properties in food systems, by self-organizing, crystallizing, co-crystallizing and adsorbing at interfaces.[5] Moreover, such systems may be involved in applications that are related with the delivery of biocompounds and aroma, the structuring of edible lipids (oleogels), and interfacial stabilization. Successful application of self-assembly structures requires control of the intermolecular interactions and the spatial distribution. On the other hand, food matrices are complex systems with many different components present. This renders the application of lipid mesophases in edible materials a challenging field, because their functionality may be affected by the forces that the co-existing components imply.[7]

5.2 POLYMORPHISM OF LIPID MESOPHASES

Amphiphilic lipids, such as monoacylglycerols, phospholipids and fatty acids, are molecules with two main domains: a polar head and a hydrophobic organic tail. When they are mixed with water or other polar solvents, physical forces occur and different types of lipid-crystalline phases can be formed. The formation of these highly ordered matrices is encouraged by the shape and amphiphilic character of lipids. The flexible hydrocarbon tails of the lipids interact to form fused hydrophobic regions, while the hydrophilic headgroups (ionic or neutral) form extended hydrophilic regions.[8] Therefore, as we will describe in this section, an understanding of the required molecular architecture is crucial in order to design structures with the desired morphology.[7]

5.2.1 Self-assembled Structures

Depending on the geometry of the self-assembled molecules[9,10] and their interfacial energy,[11] lipids can form a variety of structures from spherical to cylindrical micelles and three-dimensional inter-connected channels. Figure 5.1 illustrates some examples of lyotropic phase structures, which are the spherical micelles, the lamellar (L), the hexagonal (H) and the cubic (I) phases.[7,12]

The lamellar mesophases are considered to have no intrinsic curvature. They are one-dimensional, and the amphiphilic molecules are equally oriented towards both hydrophilic and hydrophobic

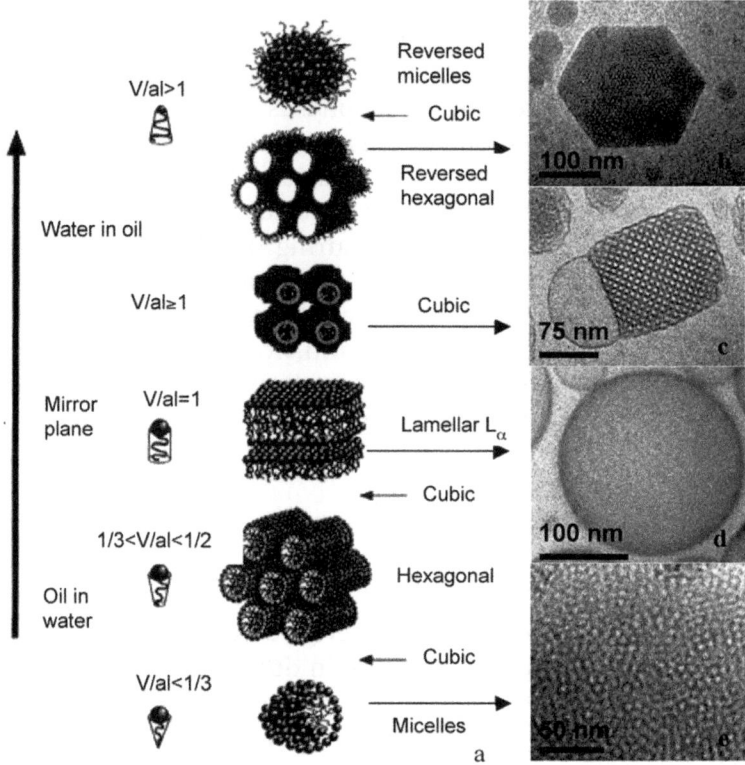

Figure 5.1   (a) Schematic of some of the possible self-assembly structures and their
corresponding packing factors. (b) Cryo-TEM for a dispersed reversed
hexagonal phase. (c) Cryo-TEM for a dispersed reversed bicontinuous
cubic phase of a space group made from Dimodan U. (d) Cryo-TEM of a
vesicle, which can be obtained by dispersion of a lamellar liquid crystalline phase (obtained from mixture of Dimodan U and sodium stearoyl
lactylate). (e) Cryo-TEM of a micelle dispersion (obtained from a polysorbate 80 solution).
(Reprinted with permission from ref. 7.)

regions. This assembly corresponds to an ideal, symmetrical crystal
phase development.[13] Depending on the space between the adjacent
bilayers, the lamellar mesophases are subdivided into L_a, L_{2a}, L_{3a},
etc.[14] The hexagonal phase consists of infinite cylindrical micelles
arranged in a hexagonal lattice,[15] and they are subdivided into two
types. Normal hexagonal (H_I) consists of lipid cylinders arranged in a
continuous polar matrix, and reverse hexagonal (H_{II}) consists of
cylinders with a hydrophilic core, arranged in a continuous lipid
matrix.[16] Lyotropic liquid crystalline mesophases of cubic symmetry
exhibit the most complex organization. They can be bi- or

discontinuous and, like hexagonal structures, they can be normal (I) and have a curvature away from water, or reverse (II) and have a curvature towards water.[17] Besides these more typical liquid crystal architectures, there are also intermediate phases such as ribbon, mesh or nematic (less common).[18,19] Based on their symmetries, these crystalline phases are called rhombohedral, monoclinic or tetrahedral, but their structures still are not well characterized.[20]

The factors that strikingly affect the transitions between the different phases are temperature and the presence of a polar solvent. The exact morphology of a mesophase crystal structure depends on the point they are at in the temperature–water composition phase diagram. A plethora of experimental phase diagrams exists for the most common lipid–water mixtures, such as monoacylglycerols. Monoacylglycerols, being amphiphilic lipids, belong to a class of water insoluble lipids which swell in the water and form various kinds of lyotropic liquid crystal.[4,21,22] Figure 5.2a shows the molecular structure of a widely used monoacylglycerol, glycerol monooleate (GMO), and a typical binary phase diagram of the phases that are formed when GMO is present in lipid–water mixtures (Figure 5.2b).[23] GMO goes through three different one-phase regions. At low water content, a reversed micellar phase co-exists with lamellar liquid crystalline phases (L_{α}) and a cubic phase. At higher water content, the transition to the cubic phase (C) is taking place. This phase consists of two cubic phases belonging to a different cubic space group. Heating the above system can again lead to phase transition and the formation of a reversed hexagonal liquid crystalline phase (H_{II}). All of these phases are in thermodynamic equilibrium, which means they are stable only under certain physicochemical conditions.

5.2.2 Packing Geometry

The molecular shape of a given amphiphilic molecule plays a determining role in predicting its phase transitions. The geometric features of a lipid can be described by the molecular critical packing parameter (CPP).[9] [See equation (5.1)]

$$P = v_0/\alpha l_o, \qquad (5.1)$$

where v is the hydrophobic tail volume, α is the surface area of the hydrophobic core of the assembly expressed per molecule present (area per molecule), and l_o is the length of the aliphatic tail.[9]

The geometry of the assembly is well connected with the properties of the individual molecules, since it determines the favoured

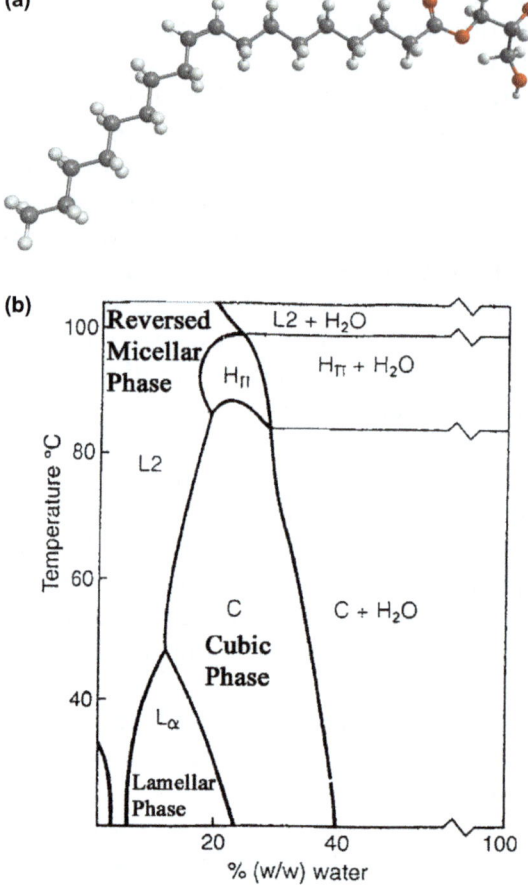

Figure 5.2 The chemical structure of glycerol monooleate (a) and a phase diagram of glyceryl monooleate in water (b).
(The phase diagram is reprinted with permission from ref. 21)

curvature and, therefore, the morphology of the assembly. A spherical micelle with a core radius R, made up of k molecules, has a core volume of $V = kv_o = 4\pi R^3/3$. The surface area of the core is $A = k\alpha = 4\pi R^2$ and hence $R = 3v_o/\alpha$. If the core is fully packed with aliphatic tails without any space, then R cannot exceed the extended length l_o of the tail. This means that there are certain constraints for R; therefore, for cylindrical micelles the packing parameter has to be between 0 and $1/3$ $(0 \leq v_o/\alpha l_o \leq 1/3)$.[24] These simple geometric relations, together with the constraints on R, provide a connection between the molecular packing parameter and possible shapes of the assembly that can be derived: $0 \leq v_o/\alpha l_o \leq 1/3$ for a sphere, $1/3 \leq v_o/\alpha l_o \leq 1/2$ for a cylinder (hexagonal), $1/2 \leq v_o/\alpha l_o \leq 1$ for a bilayer (lamellar) and $v_o/\alpha l_o \geq 1$ for cubic phases (see Figure 5.1).[9] Additionally, this figure illustrates cryo-transition electron microscopy (cryo-TEM) images and schematics of different self-assembly structures. When the packing parameter is 1, planar lamellae phases are formed that, when dispersed, give vesicles (Figure 5.1d). Molecules with a smaller polar head ($P > 1$) form reversed micelles and under specific conditions they form reversed hexagonal phases (H_{II}) (Figure 5.1b).[7]

The chemical structure of the amphiphilic molecules has a strong impact on the packing geometry.[25] The length of the aliphatic chains, along with the chains of the polar head, increases the overall hydrophobicity and subsequently enhances the hydrophobic chain–chain interactions. Stronger interactions between the chains increase the chain-melting temperature and favour the formation of inverse non-lamellar phases.[26,27] On the other hand, the presence of *cis*- or *trans*-double bonds decreases the gel–fluid transition temperature and hinders the formation of inverse non-lamellar phases.[28] The effect of the length of chains of monoacylglycerols on the phase diagrams of aqueous systems is schematically illustrated in Figure 5.3. All monoacylglycerols with chain lengths greater than C8 form lamellar crystalline phases, but cubic phases are formed only when the carbons in the chain number >14. Moreover, the formation of hexagonal phase H_{II} requires an aliphatic chain with >16 carbons.[29]

Apart from hydrophobic chains, polar head groups also have a major impact on the assembly polymorphism. Even minor modifications, such as the existence of hydroxyl groups or replacement of a methyl group, can significantly alter the phase behavior.[30] These small variations in the chemical structure of the polar heads affect the overall polarity of the headgroup, the coulombic interactions, and also the steric effect.[31]

Another important parameter that can influence the structure of the mesophases is temperature. Fluctuations of temperature can lead to a conformational disorder in the hydrocarbon chains, which will probably expand over the covered area per molecule. More specifically, as shown in Figure 5.3, increasing the temperature leads to phase transition from L_α to H_{II}, depending on the chain length, with the cubic phases having an intermediate location in these phase transitions.[5]

Furthermore, a significant factor that may affect the structure of the mesophase, as already mentioned, is the presence of polar compounds such as water. As illustrated in Figure 5.4, an increase in the water content increases the repulsions between the polar

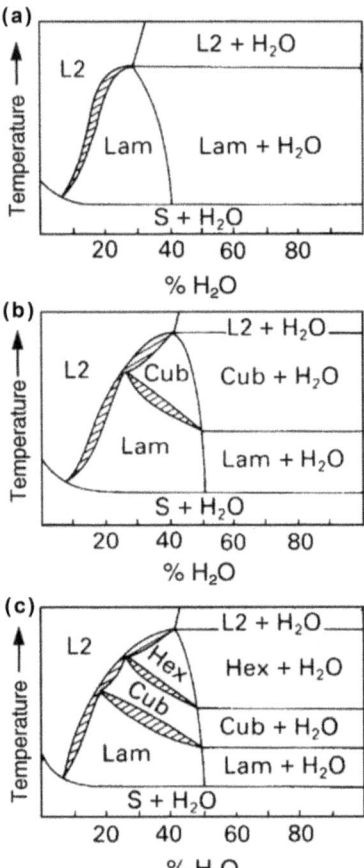

Figure 5.3 Aqueous phase diagrams of monoacylglycerols with different acyl groups. The phase diagram type in (a) corresponds to acyl chains C8–C12, in (b) to chains C14–C18, and in (c) to chains C20 and longer (or unsaturated C18 and longer).
(Reprinted with permission from ref. 5)

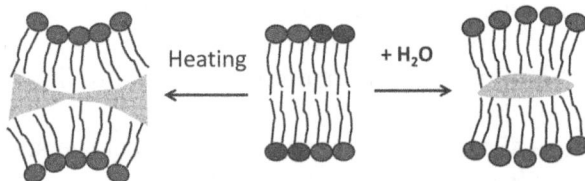

Figure 5.4 Schematic representation of bilayer packing and how it can be influenced by heating or water content.

headgroups[32] and, hence, the interfacial area per molecule. This expansion forces the monolayer to curve and deviate from the ideal, symmetrical lamellar phase (Figure 5.4).[8] Moreover, the presence of a polar solvent, such as water, leads to the formation of additional hydrogen bonds between the polar heads. Stronger interactions between the molecules can have a striking effect on the assembly structure, such as the induction of non-lamellar phase appearances, mentioned in many reports.[33–37]

5.3 IDENTIFICATION OF SELF-ASSEMBLY STRUCTURES

As previously mentioned, mesophases are sensitive to external parameters such as temperature, pressure and the polarity of the solvent, and may appear in several different structures. Therefore, a major issue in the investigation of the lipid mesophases is the appropriate choice of analytical methods to allow structural characterization of the system. Several techniques have been proposed, for example polarized optical microscopy (POM),[38–40] cryo-transition electron microscopy (cryo-TEM),[41–43] cryo field emission scanning electron microscopy (cryo-FESEM),[44] small angle X-ray scattering (SAXS),[42,45,46] small angle neutron scattering (SANS),[46,47] X-ray diffraction (XRD),[38,48,49] nuclear magnetic resonance (NMR),[38,40] atomic force microscopy (AFM),[50] and rheology.[51,52] Details of some of these are given in the current chapter.

5.3.1 Polarized Optical Microscopy

Polarized optical microscopy is a well-established technique to identify lipid crystalline phases. Under cross-polarized filters, mesophases, in most cases, appear with black and white textures. They can also appear with colour, when additional filter plates with strong birefringent properties are used. More specifically, lamellar phases typically display a woven structure and Maltese cross patterns (Figure 5.5a), while hexagonal structures appear with a fan-shaped and angular texture (Figure 5.5b).[38] On the other hand, when cubic mesophases are present, no texture can be observed as they are isotropic and they do not have birefringent properties. The limitation, though, of using this technique in identifying lipid mesophases, is that only microscale and sub-microscale structures can be observed, while characterization of nanocrystals requires the application of more efficient microscopy techniques such as transmission and/or scanning electron microscopy (TEM, SEM).

(a)

(b)

Figure 5.5 Polarized microscopy images recorded for (a) the L_{α}-phase at 20 °C, composition 90 wt% MO and 10 wt% H_2O, magnification 10× and b) the reversed hexagonal phase at 40 °C, composition 65 wt% MO, 25 wt% DO, and 10 wt% H_2O, magnification 20×.
(Reprinted with permission from ref. 38)

5.3.2 Cryo-Transmission Electron Microscopy (cryo-TEM)

Cryogenic transmission electron microscopy has been widely used to characterize nanostructures of lipid mesophases.[41–43] Cryo-TEM provides information on the nano- and microstructure of single particles, such as size and shape, along with precise information on the interplanar distances using fast Fourier transformations (FFT). This information cannot be extracted with the use of polarized microscopy, especially when the particles are in complex systems like foods.[38,53]

As illustrated in Figure 5.6, with the use of cryo-TEM different cubic structures can be identified. The figure shows examples of bicontinuous cubic nanoparticles made of dispersed glycerol monooleate (GMO) and diglycerol monooleate (DGMO) in water, in the presence of surfactant and polymeric stabilizers.[54] Inserted are high magnifications and their Fourier transformations, which yield reflections consistent with the body-centered cubic type in the minimal surface description of bicontinuous cubic phases.[41] Moreover, at a given ratio between the compounds, reversed hexagonal monocrystalline particles are formed.

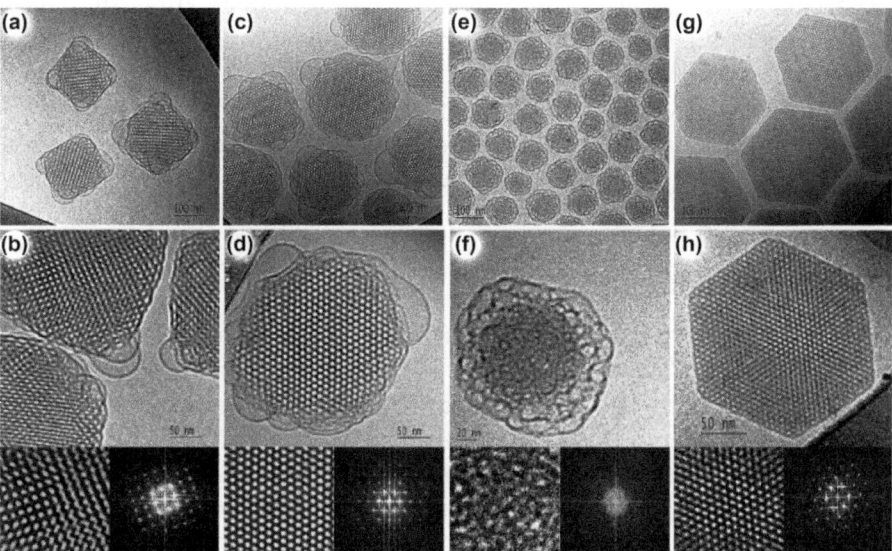

Figure 5.6 Representative cryo-TEM micrographs of different non-lamellar lipid nanoparticles. Panels (a), (b), (c) and (d): Reversed bicontinuous cubic phase particles viewed along different directions. The dispersion was prepared at the weight ratio GMO : F127 : water = 1.88 : 0.12 : 98.0. Panels (e) and (f): Monodisperse "sponge" phase nanoparticles prepared at the weight ratio DGMO : GDO : P80 : water = 2.13 : 2.13 : 0.74 : 95.0. Panels (g) and (h): Reversed hexagonal monocrystalline particles made of lipids at the weight ratio DGMO : GDO : F127 : water = 2.25 : 2.25 : 0.5 : 95.0. Fourier transforms of magnified areas in panels (b), (d), (f), and (h) show the structural periodicity of the different nanoparticles consistent with the mesophase structures indicated above.
(Reprinted with permission from ref. 54)

5.3.3 Cryogenic Field Emission Scanning Electron Microscopy (cryo-FESEM)

Cryo-field emission scanning microscopy is a relatively new method that has been used to obtain structural information on the mesophases. Its main advantage over cryo-TEM is that it allows the reproduction of three dimensions of the structure of the crystalline particles. Furthermore, cryo-FESEM allows analysis of hydrated and native samples at low pressures, in contrast to other microscopy techniques where high-vacuum conditions are compulsory. Cryo-FESEM does not give any information on the internal structure of the particles. It mainly provides information on the geometry of the aggregates and the morphology of the surface.[16] An interesting example of the use of cryo-FESEM is illustrated in Figure 5.7, where a three-dimensional image of a hexagonal structure is obtained and compared with the two-dimensional image of the same particle, but visualized with the use of cryo-TEM.[55] These images show that hexagonal mesophases pose a unique three-dimensional structure. Some of the particles had an edge on the top that can be described as a cylinder, capped with a cone at both ends (Figure 5.7b–d). Additionally, hexagonal morphologies with centrally projected edgy structures were also observed (Figure 5.7e and f).

5.3.4 X-Ray Diffraction (XRD)

X-ray diffraction has wide uses in the investigation of complex food-based materials.[56] At this point, it is important to note that there is confusion between the terms scattering and diffraction, and often they are used interchangeably. As an attempt to clarify the two terms, it is worthwhile mentioning that scattering is a result of the interaction of radiation (electrons or neutrons) with individual atoms, while diffraction is a result of the interference of these initial waves.[57] However, in common phraseology, diffraction describes in general the scattering from crystalline materials, and scattering defines all other cases. Food materials that are analysed with X-ray diffraction should compose crystallites with a uniform distribution of orientations, or at least some degree of preferred orientation.[57] Lipid mesophases are examples of this food-based material.

A typical XRD figure that shows the polymorphs and transitions of lipid crystals is given in Figure 5.8.[58] It is shown that monolinolein in water can form different types of structures depending on temperature and water concentration. By increasing the temperature, the

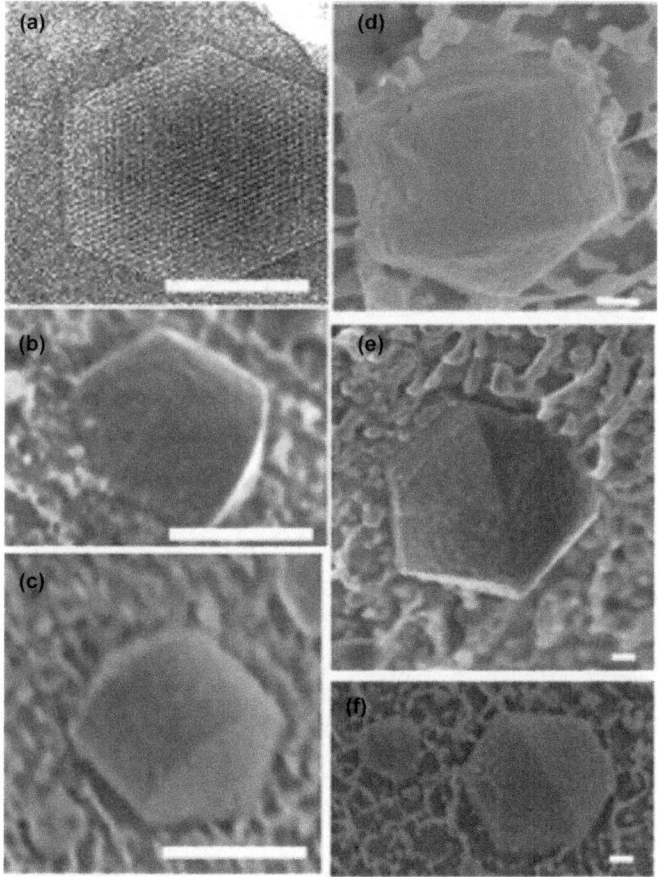

Figure 5.7 Images of hexosomes prepared by the dispersion of phytantriol + 10% vitamin E acetate in 1% Pluronic F-127 solution. (a) Representative image obtained using cryo-TEM. (b–f) Representative images obtained using cryo-FESEM showing the three-dimensional structure of hexosomes, with spinning top (b–d) and spine (e and f) structures on the upper surface. (f) Large and small hexosome structure in the field of view. (Reprinted with permission from ref. 55)

internal structure of the binary monolinolein–water system undergoes a transition from cubic *via* hexagonal to fluid isotropic (L) phase, and *vice versa*.[58] The co-existence of reverse hexagonal and cubic phase between 40 and 45 °C may be attributed to the presence of impurities.

5.3.5 Small Angle X-Ray and Neutron Scattering (SAXS/SANS)

Microscopy techniques are usually applied in characterizing the size and shape of particles and for identifying the self-assembly structure

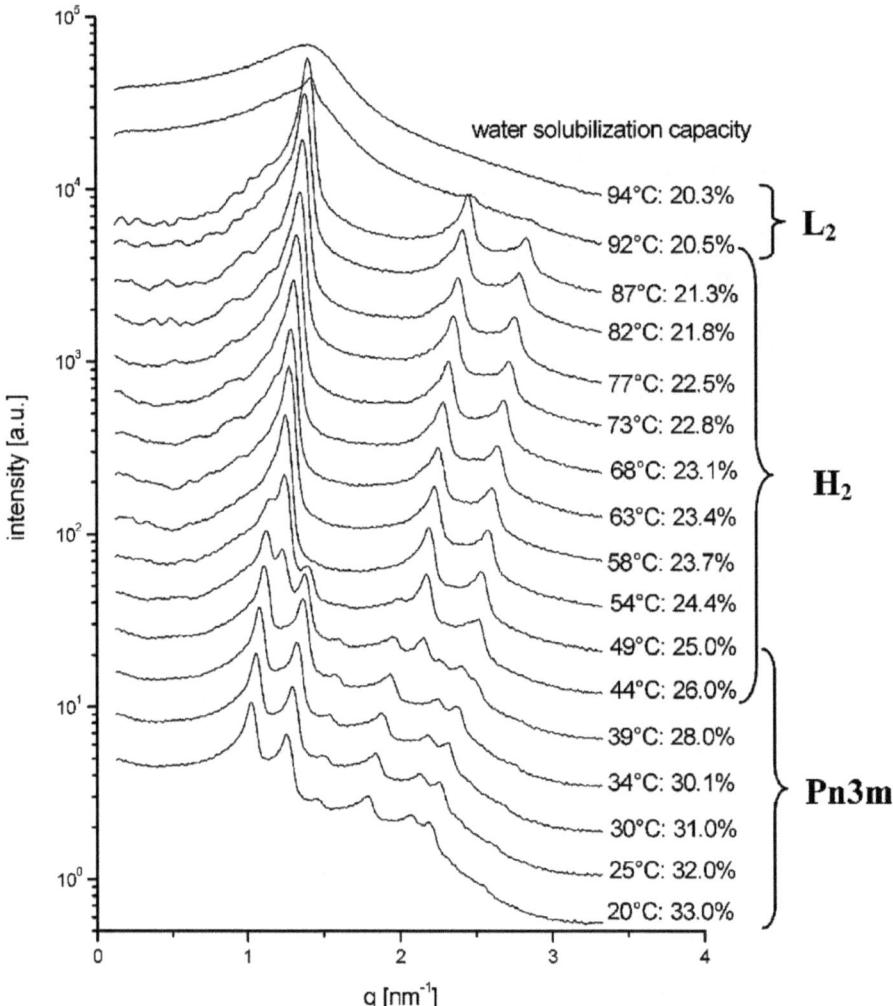

Figure 5.8   Temperature dependence of scattering curves for a non-dispersed sample of MLO with excess water (40 wt% water).
(Reprinted with permission from ref. 58)

type in the bulk phases. However, thorough investigation of the assembly structure requires more sophisticated techniques such as
small angle X-ray scattering. Small angle scattering is a non-invasive
method that enables the study of materials under realistic conditions,
including partial hydration, gels and solutions. Additionally, the
scattering at small angles provides structural information over large
length scales, from approximately one to several hundred

nanometres.[57] With these wide size-scales, small angle scattering is a very sensitive technique. It gives information on the arrangement of assemblies of atoms and enables the determination of the molecular structures and their spatial distribution. Therefore, small-angle scattering, mainly with X-rays (SAXS), and less frequently with neutrons (SANS), has been widely applied to food-based systems and particularly to lipid mesophase characterization.[59] Although foods are complex systems, and sometimes interpreting the signal is difficult, mainly owing to heterogeneity and the obscuration caused by the scattering profile of the surrounding ingredients and solvent, SAXS may provide relatively satisfactory results.[60]

More specifically, SAXS can provide all this valuable information about the spatial structure of the mesophases by giving data on the molecular organization within a lipid aggregate, in terms of the balance of attractive and repulsive forces between the polar headgroups and the aliphatic, non-polar chains. In the headgroups of amphiphilic molecules, including some lipids, an effective attractive force takes place between the polar head atoms and the polar solvent due to hydrogen bonding and the unfavourable contact of the acyl chain (hydrophobic tail) with the solvent.[47] Moreover, some repulsive forces in a shorter extent between the polar head and the solvent may also take place owing to hydration, charge and steric interactions. Another important parameter in defining the structure of mesophases may be the forces that are developed between the acyl hydrophobic chains. These are attractive and repulsive forces, such as van der Waals attractions or lateral pressure.[47]

As previously mentioned, an imbalance in the above forces leads to conformational disorder and spontaneous interfacial curvature (see Figure 5.4). The curvature free energy is associated with the morphology of the formed crystal. Therefore, it is one of the main parameters that are used to describe, theoretically, the structure and behaviour of the liquid lyotropic phase. Despite the theoretical models, to investigate the concept of any energetic description and the impact of external parameters on the mesophase, experimental data are necessary.[47]

A remarkable example of the use of small angle scattering to characterize lipid mesophases is the different lyotropic crystals that distearoylphosphatidylcholine (DSPC) forms in solvents with different polarity (Figure 5.9).[45] Analysis of the electron profile of two different samples with the same DSPC concentration, but different solvents (water and propylglycerol), reveals a decrease in *d*-spacing from 67.0

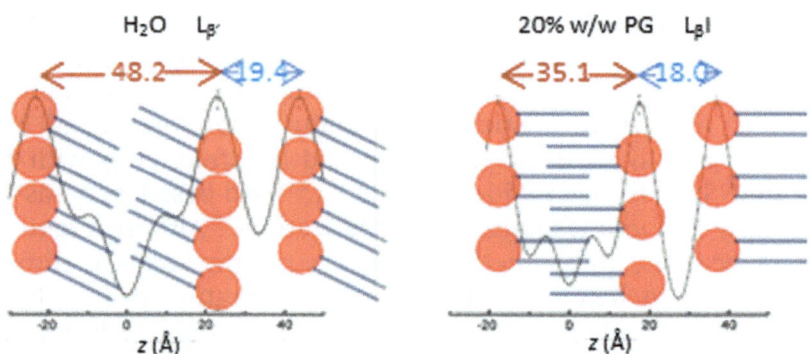

Figure 5.9 Proposed lamellar structures and electron density profiles calculated from small angle X-ray scattering intensity profiles obtained at 25 °C, of DSPC dispersions in (a) water and (b) water with 20% (w/w) propylene glycol. (Reprinted with permission from ref. 45)

to 53.2 Å. The bilayer space for all of the DSPC samples remains roughly constant and unchanged from that observed in water, at ~19.0 Å. Thus, the most apparent effect of the addition of the presence of propyl glycerol instead of water on DSPC bilayer structure is a decrease in layer thickness (from 48.2 Å to 35.1 Å). From this observation, along with complementary data, it was proposed that in different solvents a transition takes place and the acyl chains of the amphiphilic molecules change orientation.[45]

5.3.6 Rheology

Shear rheology appears to be a promising and reliable method for the determination of structural and dynamic aspects of lipid mesophases. Application of rheology in characterizing the structure of lipid mesophases relies on the fact that different crystalline mesophases exhibit different rheological behaviour. For instance, bicontinuous cubic phases are considered to be the most rigid structures amongst all the possible crystalline structures. Reverse hexagonal phases with intermediate viscoelastic properties are next, with lamellar phases being the least rigid structures because they exhibit plastic fluid properties.[53] Apart from the characterization of the inner structure of crystals, interpretation of these differences in viscoelastic behaviour may lead to the detection of structural transition between the mesophases, including cubic to cubic, cubic to hexagonal and hexagonal to isotropic fluid transitions.[16,61,62] This means that dynamic phenomena and phase transition may be detected when observing changes in the slope of storage (G') and loss moduli (G'').[53]

The most reliable rheological method to detect order–disorder transitions in lipid mesophases and, in general, liquid crystalline phases is the interpretation of variations of the longest relaxation time (τ_{max}) *vs.* composition.[53] Longest relaxation time is defined as the inverse of the frequency at which the crossover of G' and G'' takes place ($\tau_{max} = 1/\omega$). The physical meaning that has been attributed to τ_{max} is the characteristic diffusion time of the amphiphilic molecules at the water–lipid interface. Stated in another way, τ_{max} provides the time scale for relaxation to the equilibrium configuration of the hydrophilic–hydrophobic interface, following perturbation by shear deformations.[16]

However, as established in the literature,[52,63] the complexity of the lipid crystalline phases cannot be interpreted by a single relaxation time. Hence, research groups have tried to model the complex rheological frequency response of lipid mesophases by multiple Maxwell models, with several relaxation times. A further expansion of the use of multiple relaxation times was achieved by introducing the time Laplace-transform of G' (ω) and G'' (ω).[51] By doing so, a multitude of relaxation times were obtained and, thus, the relaxation spectra of each mesophase may be considered as its rheological signature, as shown in Figure 5.10.[51] By applying this model, the dominating relaxation time was assigned to τ_{max}, while the residual times were

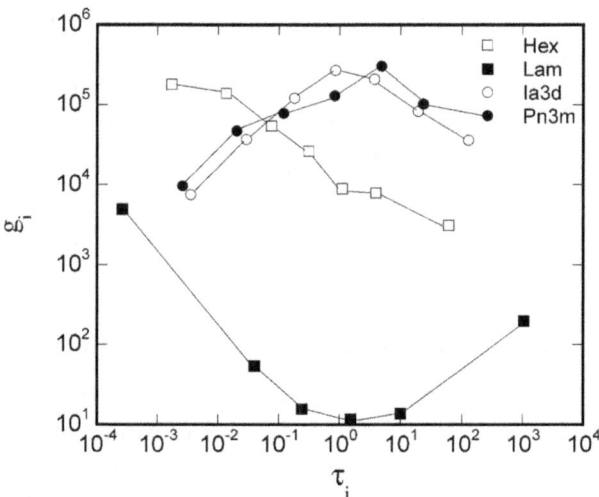

Figure 5.10 Discrete relaxation spectra pairs (g_i, τ_i) for the main liquid crystalline phases investigated. Lamellar phase, 10 wt% water at 40 °C; hexagonal phase, 20 wt% water at 70 °C; Ia3d cubic phase, 20 wt% at 40 °C; Pn3m cubic phase, 20 wt% at 55 °C.
(Reprinted with permission from ref. 51)

attributed to secondary relaxation mechanisms that refer to constrained hydrophobic–hydrophilic phases.[16] More specifically, as demonstrated in Figure 5.10, the lamellar phase mainly showed two relaxation times. The first one was attributed to the relaxation of the lipid–water interface (of the order of 10^3 s), and the second one (of the order of 10^{-3} s) was probably due to the relaxation of water restricted in lamellae. Conversely, the cubic phases have been found to present only one dominant relaxation time and a less important multiple relaxation regime, indicating that this assembly presents a more complex rheological response.[51,64] Lastly, the reverse hexagonal phase exhibited a monotonic distribution of relaxation times.[16,51]

5.4 EDIBLE APPLICATIONS OF LIPID MESOPHASES

Materials that contain lyotropic liquid crystals are a massive playground for researchers. Many different fascinating applications already exist in soft matter nano- and biotechnology.[65] All of the applications, from organic electronics to membrane gas separations, exploit the ability of molecules to self-assemble into structures that exhibit specific nanoscale arrangements. Based on the different behaviour of the crystalline phases formed, new functional materials can be generated. In the current book we are interested in edible materials; therefore, among all the exciting activities only bio-related examples will be discussed. These applications rely on the fact that researchers can take advantage either of the macroscale properties of the material or of the large surface that is formed in the self-assembly crystalline structure.[66] Among all the edible applications that have been proposed or are currently under investigation, this chapter will focus on four general areas: the protection and controlled release of functional compounds, chemical reactivity, the structuring of edible lipid materials, and finally, the emulsifying properties of lipid mesophases.

5.4.1 Protection and Controlled Release of Functional Compounds

One of the main edible applications of lipid crystalline phases is the accommodation and delivery of functional compounds, such as flavours, vitamins and proteins. These types of application have already been widely investigated for medical proposes.[3,23,67,68] More specifically, functional ingredients can be incorporated in edible matrixes by placing them in the lipid crystalline phases.[7,16] wing to the nature of the polar lipids and of the formed mesophases, different types of molecule can be hosted. Figure 5.11 illustrates the possible

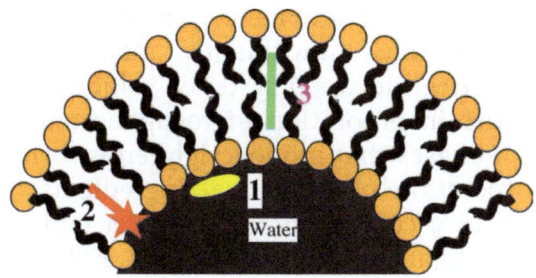

Figure 5.11 Possible localization of (guest) molecules within the inverted bicontinuous cubic phase. For simplicity only part of the lattice is represented. The molecule marked 1 is hydrophilic; the one marked 2 is amphiphilic, and the one marked 3 is lipophilic.
(Reprinted with permission from ref. 7)

localization of hydrophilic, lipophilic and amphiphilic molecules in a lipid mesophase.[7,69] As expected, hydrophilic compounds may be located close to the polar heads of lipids; lipophilic compounds may be located in the matrix of the aliphatic chains; while amphiphilic molecules may act as a building block of the mesophase. It is worth mentioning here that the presence of the "guest" molecule may influence the structure and properties of the mesophase.[7,69] As an example, the addition of β-carotene to phosphatidylcholine–water systems resulted in a decrease of the overall viscosity.[70,71] According to the authors of that report, β-carotene acts as an end-cap-active agent in phosphatidylcholine H_{II} mesophases. Additionally, the presence of sodium oleate in monoolein–water systems induced a transition from bicontinuous cubic to a lamellar liquid crystalline phase.[72]

An important example of the ability of lipid mesophases to host functional ingredients is the accommodation of large proteins, even in their native state. For example, bicontinuous cubic phases formed in a glycerine monooleate–lysozyme–water system can host enzymatically active lysozyme. The lipid mesophase provides protection against any structural changes and, subsequently, enzymatic degradation.[73] Similar protection provided by cubic phases was found in other compounds such as glucose dehydrogenase and vitamin K.[74] Moreover, lipid mesophases have the ability to protect chemically sensitive molecules against oxidation and hydrolysis.[75,76] Additionally, incorporated food-volatile compounds in H_{II} mesophases were approximately two to four times more stable, compared to when they were dissolved in organic solvents.[75]

Protective carriers, lipid self-assembly structures, can also be used to solubilize poorly water soluble ingredients.[16,77–79] This is a very critical application, because food systems usually require relatively high amounts of nutrients. For instance, a recent study[78] showed that tocopherols, lipid-soluble antioxidants, may be solubilized in unsaturated monoglyceride–water systems. Solubilization occurs as a consequence of the incorporation of tocopherol molecules in the reverse hexagonal phases that monoacylglycerols form.

Effective incorporation of functional ingredients in lipid meso-phases also requires a controlled release. As a topic, this has been systematically investigated in pharmaceutics and its purpose is to determine whether the self-assembly and the location of the "guest" molecule influence diffusion. A study of a mixture of eight volatile compounds resulted in the conclusion that the aroma release properties were influenced by the formation of self-assembly structures.[76] The volatiles entrapped in the aqueous phase were released faster and in the highest amounts. On the other hand, aromas incorporated in bicontinuous cubic phases made from unsaturated mono-acylglycerols were released differently from those that were present in the oil phase of an oil-in-water emulsion. This observation leads to the conclusion that the structure of the host plays an important role in the release of the volatile compound. Moreover, this study showed that the release patterns from the cubic phase were not only con-trolled by its composition, but also depended strongly on the lipid interfacial area.[76]

5.4.2 Chemical Reactivity

Self-assembled mesophases can also be used as nanoreactors because they are capable of increasing the yield of Maillard reactions.[59,80] This interesting reaction can be attributed to solubilization phenomena and, thus, can be used to control and enhance the flavour formation in foods. A highly cited example is the reaction of an aminoacid (L-cysteine) with a heterocyclic aldehyde (furfural)[80] to form 2-furfuryl-thiol (FFT). The yield of the model reaction was much higher in microemulsions and in bicontinuous cubic phases, compared to the yield when the reaction took place in an aqueous environment. Similar results were also obtained for the Maillard reaction between a monosaccharide (xylose) and an amino acid (glycine) to produce 4-hydroxy-5-methyl-3 furanone (norfuraneol).[81] The yield of the reaction was much higher in self-assembled structures than in an aqueous environment.

The enhancement of the reaction achieved can be attributed to different causes.[66] First, the concentration of the reactants at the interface may be locally increased. Second, the mobility of the reactants may be lower at the lipid–water interface. The immobility of molecules due to the surroundings is called the "cage effect",[82] and it can increase the efficiency of chemical reactions.

The self-assembly structure also plays an important role in the reaction. For example, the reaction between L-cysteine and furfural has been found to be more efficient in the reverse bicontinuous cubic phases than in the lamellar phases.[80] Additionally, under specific conditions, the production of norfuraneol was found to be approximately two times higher when the reaction took place in the reversed hexagonal phase than in the reversed bicontinuous cubic phase.[81]

At this point, it is important to note that, despite these fascinating results, additional systematic research is necessary to be able to predict chemical reaction yields and pathways as a function of the properties of the self-assembly structures.[66]

5.4.3 Structuring Edible Lipid Material

The texture of many food products, from pastry dough and cookies to sausages, is provided by the texture of the solid oil phase. Traditionally, the physical and chemical characteristics of fat were determined by the composition of saturated and *trans* fatty acids in the triacylglycerol hardstock.[83] Additionally, much evidence exists in the literature to show that their consumption contributes to global epidemics such as the metabolic syndrome and cardiovascular disease.[84,85] Therefore, structuring edible oil has become an active research area in the last decade, mainly owing to legislation pressures to find alternative structurants to saturated and *trans* fatty acids.[86–88] Several strategies have been proposed to create saturated-free and *trans*-free structured oil.[88] All of the strategies include the addition of compounds to oil that can form building blocks leading to its gelation, making this a very challenging field, especially within the constraints imposed by the foreseen use in healthy foods. One of the proposed routes to achieving oil structuring is the addition of polar lipids that can assemble to form lipid mesophases, entangling in a three-dimensional network which is required to solidify the oil.

There are several published works on the use of polar lipids as oil structurants. One large group of these filler compounds is monoacylglycerols,[89–91] which, when present in edible oils, mostly form lamellar phases. Other groups that belong to this category of structurants are fatty acids,[92] fatty acid salts,[92] fatty alcohols,[93] waxes,[94] and sorbitan alkylates.[95] Moreover, the most studied polar lipid that can structure oil is lecithin.[35] Lecithin has been found to form lipid mesophases in an oil phase, either by itself,[96] or after mixing with another compound.[37,97–99]

When lecithin is present solely in the oil or other organic solvents at concentrations below 45 wt%, it mainly forms reversed micellar structures that are not able to entangle to a three-dimensional network.[100] Upon addition of water or another polar solvent, the interactions between the lecithin molecules change and, instead of reversed micelles, long cylindrical micelles are formed and entangled into hexagonal phases (H_{II}).[101] The combination of lecithin with another polar molecule, such as sorbitan tri-stearate[98] or tocopherol (vitamin E)[37] leads to similar results. The shape of the assembly can be estimated by the molecular packing parameter and is a result of the geometry of the molecules and the interactions between them.[37] When only lecithin is present, the small headgroup and the two hydrophobic chains favour the formation of inverse spherical micelles (Figure 5.12a). The addition of a molecule with a smaller

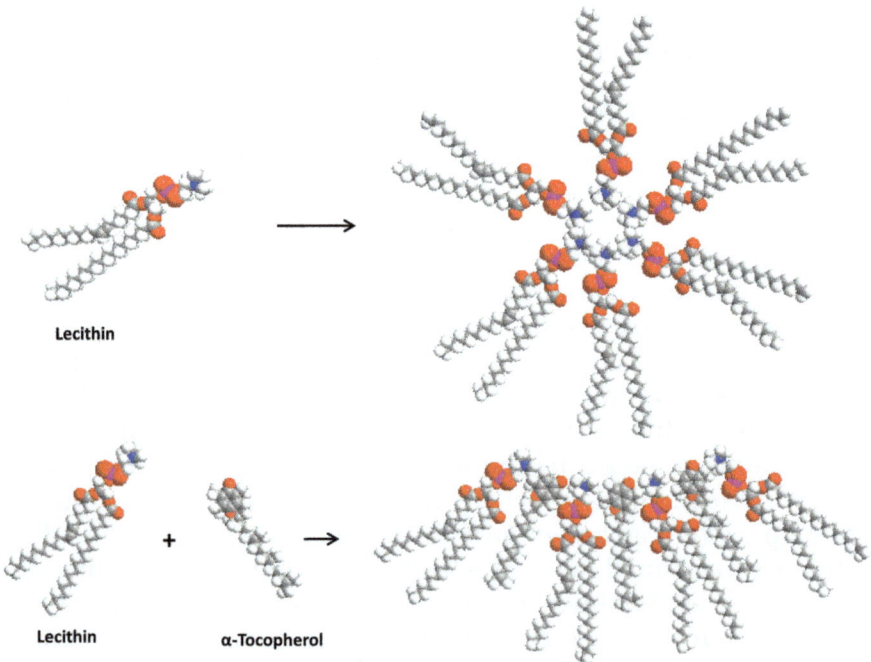

Lecithin

Lecithin   α-Tocopherol

Figure 5.12 Schematic presentation of packing geometry. (a) Inverse micelle formation with large curvature, (b) formation of small curvature favouring inverse cylindrical micelles or lamellar phases.
(Reproduced by permission of The Royal Society of Chemistry from ref. 37)

hydrophobic area, such as tocopherol, leads to an assembly with a smaller curvature that favours the formation of lamellar and/or hexagonal phases (Figure 5.12b). The presence of these self-assembly structures induces the formation of a spanning network and therefore gelation of the system.[37]

5.4.4 Emulsifying Properties

Complex foods are often composed of non-mixable liquid phases that mainly are oil and water. In the presence of an efficient emulsifier, stable oil–water interfaces are formed to create oil-in-water or water-in-oil emulsions.[102] The emulsifiers are mostly amphiphilic molecules that can be absorbed in the interface and form stable interfacial films that prevent the coalescence of the droplets. One important category of edible emulsifiers used in the food industry is the polar lipids.[103] Depending on the compounds that exist in the surrounding environment, lipids can have desirable or undesirable effects on the emulsification process. They may act as effective emulsifying agents that can also create repulsive interactions between droplets, or they can destabilize the emulsion as a result of competitive adsorption at the air–water interfaces.[104,105] Furthermore, if the emulsifier is present in sufficient concentrations, polar lipids themselves may form lipid mesophases at the interfaces that can positively affect the stability of the resultant emulsions, possibly by inhibiting droplet aggregation and coalescence.[106,107] That is, polar lipids not only form a flat film along the interface of the droplet but, owing to cohesive forces, they can also create more organized interfacial structures and form micelles, hexagonal, lamellar and cubic mesophases.[108,109] The additional advantage of the mesostructures formed by polar lipids at oil–water interphases is that, apart from their emulsifying ability, they can provide further stabilization by the steric hindrance mechanism.[110,111]

Liquid crystals, also called mesophases, can be formed either at the interface (after the emulsification process due to interactions between the polar lipids), or they can be initially formed at the bulk phase, and then they can adsorb at the interface during emulsification as ready-formed mesophases.[112,113] If the formed mesophases have interfacial activity, they adsorb at the droplet interface and as common emulsifiers. These mesophases can also be viewed as solid particles owing to their size and nature; therefore, the emulsions they stabilize can fall into the category of the so-called Pickering emulsions. These form a distinct category of emulsions that are stabilized

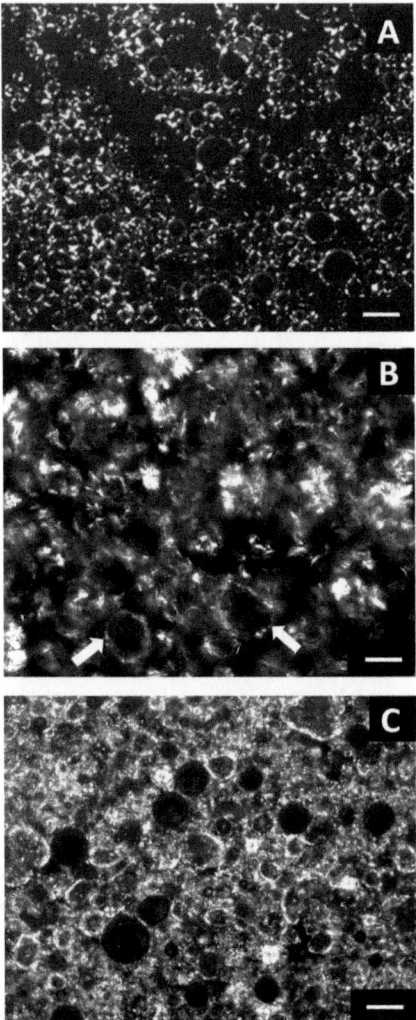

Figure 5.13 Polarized light micrographs of Pickering and network-stabilized water-in-oil emulsions: (A) 4 wt% glycerol monostearate emulsion (Pickering stabilization); (B) 4 wt% glycerol monooleate–10 wt% hydrogenated canola oil emulsion (network stabilization); and (C) 4% glycerol monostearate–10% hydrogenated canola oil emulsion (combined Pickering and network stabilization). The presence of hydrogenated canola oil crystals at the glycerol monostearate–stabilized water droplet surface is shown by arrows in panel B. Scale bar: 40 μm.
(Reprinted with permission from ref. 117)

by solid particles, and they have been found to be remarkably more stable in the presence of destabilization mechanisms than the respective "conventional" small molecular weight surfactant- and protein-stabilized emulsions.[112,114] On the other hand, if the meso-phases formed lack interfacial activity, they can still stabilize the droplets, but only by formation of a plastic network in the continuous phase that entraps the dispersed droplets. In this way, the mobility of the droplets can be minimized, resulting in coalescence and phase separation inhibition.[113] This type of emulsion stabilization is known as network stabilization.[115] The stability of this second type of emulsion can be further enhanced by adding a surfactant.

The most extensively studied emulsifiers that form self-assembled structures at oil–water interfaces are the monoacylglycerols that have HLB (hydrophilic–lipophilic balance) between 2 and 5, and can be used to stabilize water-in-oil and oil-in-water emulsions if water-soluble emulsifiers are also present.[116,117] Furthermore, when they are added to oil-in-water dispersions, they can form highly hydrated lamellar phases (L_α) which cover the oil droplets and form meso-morphic gels (network stabilization).[90] These structures (as are all mesophases) are sensitive to the presence of other compounds, as well as to environment stresses such as temperature and pressure.[118]

Monoacylglycerols stabilize emulsion droplets not only through network stabilization but, as shown in Figure 5.13, through all three possible mechanisms that were previously mentioned.[119] Figure 5.13a shows that glycerol monostearate is present in a water-in-canola oil dispersion. It forms mesophases that are absorbed in the interface (Pickering-type stabilization).[114] When hydrogenated canola oil was used, the presence of another monoglycerol (glycerol monooleate) primarily led to network-stabilized emulsions (Figure 5.13b). Additionally, when glycerol monostearate was added to a water-in-hydrogenated canola oil dispersion, Pickering and network crystals synergistically contributed to the emulsion stability (Figure 5.13c).[119]

5.5 CONCLUSIONS

Edible polar lipids assemble to form lipid mesophases (liquid crystals) that are in an intermediate state between solid and liquid. These structures are present in almost all raw food materials, because the cell membrane is comprised of polar lipids. Apart from forming selectively permeable membranes when the polar lipids are dispersed in solvents, they self-organize to form different types of mesophase with fascinating functional properties. Since it is important to understand the mechanism behind the formation of these structures, this chapter primarily reviews the parameters that determine the type of assembly. These parameters, on one hand, have to do with the characteristics of the molecules (such as the molecular geometry) and the chemical composition of lipids. On the other hand, external parameters such as the solvent composition, pressure and temperature also play a key role. A direct, or indirect, visualization of the mesostructures can be achieved with the use of modern techniques which provide qualitative and quantitative insights.

Owing to their molecular organization, lipid mesophases have many innovative applications, such as the delivery of biocompounds and aroma, the structure of edible lipids (oleogels), and interfacial stabilization. Despite the efforts of researchers, the edible applications are still a challenging field because foods are particularly complex matrixes. Therefore, further investigation is necessary in order to understand the appropriate molecular architectures that are required to design edible liquid crystals with the desired morphology.

ACKNOWLEDGEMENT

I am grateful to Dr Maria Tzoumaki for her support and proofreading of the chapter.

REFERENCES

1. P. J. Collins, *Liquid Crystals: Nature's Delicate State of Matter*, Princeton University Press, Princeton, New Jersey, 1990.
2. D. L. Gin, C. S. Pecinovsky, J. E. Bara and R. L. Kerr, *Liquid Crystalline Functional Assemblies and Their Supramolecular Structures*, ed. T. Kato, Springer, Berlin, 2007, vol. 128.
3. D. Libster, A. Aserin and N. Garti, *Self-Assembled Supramolecular Architectures: Lyotropic Liquid Crystals*, ed. N. Garti, P. Somasundaran, and R.Mezzenga, Wiley, New Jersey, 2012.
4. K. Larsson, *J. Phys. Chem.*, 1989, **93**, 7304.
5. K. Larsson, *Curr. Opin. Coll. Interface Sci.*, 2009, **14**, 16.
6. J. W. Goodby, V. Gortz, S. J. Cowling, G. Mackenzie, P. Martin, D. Plusquellec, T. Benvegnu, P. Boullanger, D. Lafont, Y. Queneau, S. Chambert and J. Fitremann, *Chemical Society Reviews*, 2007, **36**, 1971.
7. L. Sagalowicz, M. E. Leser, H. J. Watzke and M. Michel, *Trends Food Sci. Technol.*, 2006, **17**, 204.
8. C. Fong, T. Le and C. J. Drummond, *Chem. Soc. Rev.*, 2012, **41**, 1297.
9. J. N. Israelachvili, D. J. Mitchell and B. W. Ninham, *J. Chem. Soc. Faraday Trans. Ii*, 1976, **72**, 1525.
10. S. Svenson, *Curr. Opin. Coll. Interface Sci.*, 2004, **9**, 201.

11. S. M. Gruner, V. A. Parsegian and R. P. Rand, *Faraday Discuss.*, 1986, **81**, 29.
12. L. Latypova, W. T. Gozdz and P. Pieranski, *Langmuir*, 2014, **30**, 488.
13. S. T. Hyde, *Handbook of Applied Surface and Colloid Chemistry*, ed. K. Holmberg, Wiley, New York, 2001, 299.
14. S. T. Hyde, *Langmuir*, 1997, **13**, 842.
15. D. Amar-Yuli and N. Garti, *Coll. Surf. B, Biointerf.*, 2005, **43**, 72.
16. I. Amar-Yuli, D. Libster, A. Aserin and N. Garti, *Curr. Opin. Coll. Interface Sci.*, 2009, **14**, 21.
17. R. H. Templer, *Curr. Opin. Coll. Interface Sci.*, 1998, **3**, 255.
18. U. Henriksson, E. S. Blackmore, G. J. T. Tiddy and O. Soderman, *J. Phys. Chem.*, 1992, **96**, 3894.
19. A. M. Squires, R. H. Templer, J. M. Seddon, J. Woenkhaus, R. Winter, T. Narayanan and S. Finet, *Phys. Rev. E*, 2005, **72**, 1550.
20. S. Hyde, B. W. Ninham, S. Andersson, K. A. Larsson, T. Landh, Z. Blum and S. Lidin, in *The Language of Shape*, ed. S. Hyde, B. W. Ninham, S. Andersson, K. A. Larsson, T. Landh, Z. Blum and S. Lidin, Elsevier Science B.V., Amsterdam, 1997, p. 199.
21. K. M. Kumar, M. H. Shah, A. Ketkar, K. R. Mahadik and A. Paradkar, *Int. J. Pharmaceut.*, 2004, **272**, 151.
22. P. Pitzalis, M. Monduzzi, N. Krog, H. Larsson, H. Ljusberg-Wahren and T. Nylander, *Langmuir*, 2000, **16**, 6358.
23. K. Kumar, M. M. H. Shah, A. Ketkar, K. R. Mahadik and A. Paradkar, *Int. J. Pharmaceut*, 2004, **272**, 151.
24. R. Nagarajan, *Langmuir*, 2001, **18**, 31.
25. A. D. Law, M. Auriol, D. Smith, T. S. Horozov and D. M. A. Buzza, *Phys. Rev. Lett.*, 2013, 110.
26. R. Lewis, D. A. Mannock, R. N. McElhaney, D. C. Turner and S. M. Gruner, *Biochemistry*, 1989, **28**, 541.
27. D. A. Mannock and R. N. McElhaney, *Curr. Opin. Coll. Interface Sci.*, 2004, **8**, 426.
28. D. Marsh, *Biochim. Biophys. Act., Biomemb.*, 2010, **1798**, 40.
29. K. Larsson, P. Quinn, K. Sato and F. Tiberg, *Lipids: Structure, Physical Properties and Functionality*, Oily Press, Bridgewater, UK, 2006.
30. M. W. Tate, E. F. Eikenberry, D. C. Turner, E. Shyamsunder and S. M. Gruner, *Chem. Phys. Lipid.*, 1991, **57**, 147.
31. R. Koynova and M. Caffrey, *Chem. Phys. Lipid.*, 1994, **69**, 181.
32. J. F. Sadoc and J. Charvolin, *J. Physiq.*, 1986, **47**, 683.
33. Y. A. Shchipunov and E. V. Shumilina, *Mater. Sci. Engineer. C*, 1995, 3, 43.

34. J. M. Seddon, *Biochim. Biophys. Act.*, 1990, **1031**, 1.
35. Y. A. Shchipunov, *Coll. Surf. A, Physicochem. Engineer. Asp*, 2001, **183**, 541.
36. E. V. Shumilina, Y. L. Khromova and Y. A. Shchipunov, *Colloid J*, 2006, **68**, 241.
37. C. V. Nikiforidis and E. Scholten, *RSC Adv*, 2014, **4**, 2466.
38. J. Borne, T. Nylander and A. Khan, *Langmuir*, 2000, **16**, 10044.
39. T. A. Harroun, M. Koslowsky, M.-P. Nieh, C.-F. de Lannoy, V. A. Raghunathan and J. Katsaras, *Langmuir*, 2005, **21**, 5356.
40. A. Klaus, G. J. T. Tiddy, D. Touraud, A. Schramm, G. Stühler and W. Kunz, *Langmuir*, 2010, **26**, 16871.
41. J. Barauskas, M. Johnsson and F. Tiberg, *Nano Lett.*, 2005, **5**, 1615.
42. A. Yaghmur, L. de Campo, L. Sagalowicz, M. E. Leser and O. Glatter, *Langmuir*, 2004, **21**, 569.
43. L. Sagalowicz, M. Michel, M. Adrian, P. Frossard, M. Rouvet, H. J. Watzke, A. Yaghmur, L. De Campo, O. Glatter and M. E. Leser, *J. Microsc., Oxf.*, 2006, **221**, 110.
44. S. B. Rizwan, Y. D. Dong, B. J. Boyd, T. Rades and S. Hook, *Micron*, 2007, **38**, 478.
45. R. D. Harvey, N. Ara, R. K. Heenan, D. J. Barlow, P. J. Quinn and M. J. Lawrence, *Molec. Pharmaceut.*, 2013, **10**, 4408.
46. V. Lee and T. Hawa, *J. Chem. Phys.*, 2013, **139**, 124905.
47. R. Winter, *Biochim. Biophys. Act. (BBA), Prot. Struct. Molec. Enzymol.*, 2002, **1595**, 160.
48. P. I. Ravikovitch and A. V. Neimark, *Langmuir*, 2000, **16**, 2419.
49. C. A. Faunce and H. H. Paradies, *J. Chem. Phys.*, 2009, **131**, 244708.
50. M. Rittman, M. Frischherz, F. Burgmann, P. G. Hartley and A. Squires, *Soft Matter*, 2010, **6**, 4058.
51. R. Mezzenga, C. Meyer, C. Servais, A. I. Romoscanu, L. Sagalowicz and R. C. Hayward, *Langmuir*, 2005, **21**, 3322.
52. J. Munoz and M. C. Alfaro, *Gras. Aceit.*, 2000, **51**, 6.
53. L. Sagalowicz, R. Mezzenga and M. E. Leser, *Curr. Opin. Coll. Interface Sci.*, 2006, **11**, 224.
54. J. Barauskas, M. Johnsson and F. Tiberg, *Nano Lett.*, 2005, **5**, 1615–1619.
55. B. J. Boyd, S. B. Rizwan, Y. D. Dong, S. Hook and T. Rades, *Langmuir*, 2007, **23**, 12461.
56. M. Michel and L. Sagalowicz, in *Food Materials Science*, ed. J. Aguilera and P. Lillford, Springer, New York, 2008, p. 203–226.

57. E. P. Gilbert, A. Loppez-Rubio, and M. J. Gidley, in *Food Materials Science and Engineering*, ed. B. Bhandaria and Y. H. Roos, Blackwell Publishing Ltd., New Jersey, 1st edn., 2012, p. 52–93.

58. L. de Campo, A. Yaghmur, L. Sagalowicz, M. E. Leser, H. Watzke and O. Glatter, *Langmuir*, 2004, **20**, 5254.

59. N. Garti, A. Spernath, A. Aserin and R. Lutz, *Soft Matter*, 2005, **1**, 206.

60. E. P. Gilbert, A. Lopez-Rubio and M. J. Gidley, in *Food Materials Science and Engineering*, ed. B. Bhandaria and Y. H. Roos, Blackwell Publishing Ltd., New Jersey, 2012, p. 52–93.

61. J. L. Jones and T. C. B. McLeish, *Langmuir*, 1999, **15**, 7495.

62. C. Rodriguez-Abreu, D. P. Acharya, K. Aramaki and H. Kunieda, *Coll. Surf. A, Physicochem. Engineer. Asp.*, 2005, **269**, 59.

63. C. F. Soon, M. Youseffi, T. Gough, N. Blagden and M. C. T. Denyer, *Mat. Sci. Engineer. C, Mater. Biol. Applic.*, 2011, **31**, 1389.

64. K. W. McKay, W. G. Miller, J. E. Puig and E. I. Franses, *J. Dispers. Sci. Technol.*, 1991, **12**, 37.

65. J. P. F. Lagerwall and G. Scalia, *Curr. Appl. Phy.*, 2012, **12**, 1387.

66. M. E. Leser, L. Sagalowicz, M. Michel and H. J. Watzke, *Adv. Coll. Interface Sci.*, 2006, **123**, 125.

67. L. Bitan-Cherbakovsky, D. Libster, D. Appelhans, B. Voit, A. Aserin and N. Garti, *J. Phys. Chem. B*, 2014, **118**, 4016.

68. N. Garti, G. Hoshen and A. Aserin, *Coll. Surf. B, Biointerfac.*, 2012, **94**, 36.

69. M. Leser, M. Mitchel and H. J. Watzke, in *Food Colloids: Biopolymers and Materials*, ed. E. Dickinson and T. van Vliet, RSC, Cambridge, 2003, p. 3–16.

70. Y. A. Shchipunov and H. Hoffmann, *Colloid J*, 1998, **60**, 794.

71. Y. A. Shchipunov and H. Hoffmann, *Langmuir*, 1998, **14**, 6350.

72. J. Borne, T. Nylander and A. Khan, *Langmuir*, 2001, **17**, 7742.

73. B. Ericsson, K. Larsson and K. Fontell, *Biochim. Biophys. Act*, 1983, **729**, 23.

74. E. Nazaruk, E. Górecka and R. Bilewicz, *J. Coll. Interface Sci.*, 2012, **385**, 130.

75. N. Amar-Zrihen, A. Aserin and N. Garti, *J. Agric. Food Chem.*, 2011, **59**, 5554.

76. S. Vauthey, P. Visani, P. Frossard, N. Garti, M. E. Leser and H. J. Watzke, *J. Dispers. Sci. Technol.*, 2000, **21**, 263.

77. L. Bitan-Cherbakovsky, I. Yuli-Amar, A. Aserin and N. Garti, *Langmuir*, 2010, **26**, 3648.

78. L. Sagalowicz, S. Guillot, S. Acquistapace, B. Schmitt, M. Maurer, A. Yaghmur, L. de Campo, M. Rouvet, M. Leser and O. Glatter, *Langmuir*, 2013, **29**, 8222.
79. J. Gurfinkel, A. Aserin and N. Garti, *Coll. Surf. A, Physicochem. Engineer. Asp*, 2011, **392**, 322.
80. S. Vauthey, C. Milo, P. Frossard, N. Garti, M. E. Leser and H. J. Watzke, *J. Agric. Food Chem.*, 2000, **48**, 4808.
81. I. Blank, T. Davidek, S. Devaud, L. Sagalowicz, M. E. Leser, M. Michel, L. P. B. Wender and P. Mikael Agerlin, *Developments in Food Science*, Elsevier, 2006, vol. 43, p. 347–350.
82. I. Martiel, L. Sagalowicz and R. Mezzenga, *Langmuir*, 2013, **29**, 15805.
83. M. Pernetti, K. F. van Malssen, E. Flöter and A. Bot, *Curr. Opin. Coll. Interface Sci.*, 2007, **12**, 221.
84. R. P. Mensink, P. L. Zock, A. D. Kester and M. B. Katan, *Am. J. Clin. Nutr.*, 2003, **77**, 1146.
85. M. C. Cetinkaya, S. Yildiz, H. Ozbek, P. Losada-Perez, J. Leys and J. Thoen, *Phys. Rev. E, Statist., Nonlinear, Soft Matter Phys*, 2013, **88**, 042502.
86. E. Co and A. G. Marangoni, *J. Am. Oil Chem. Soc.*, 2013, **90**, 529.
87. H. S. Hwang, M. Singh, E. L. Bakota, J. K. Winkler-Moser, S. Kim and S. X. Liu, *J. Am. Oil Chem. Soc.*, 2013, **90**, 1705.
88. M. A. Rogers, *Food Res. Int.*, 2009, **42**, 747.
89. N. K. O. Ojijo, E. Kesselman, V. Shuster, S. Eichler, S. Eger, I. Neeman and E. Shimoni, *Food Res. Int.*, 2004, **37**, 385.
90. H. D. Batte, A. J. Wright, J. W. Rush, S. H. J. Idziak and A. G. Marangoni, *Food Biophys.*, 2007, **2**, 29.
91. S. Da Pieve, S. Calligaris, E. Co, M. C. Nicoli and A. G. Marangoni, *Food Biophys.*, 2010, **5**, 211.
92. A. J. Wright and A. G. Marangoni, *J. Am. Oil Chem. Soc.*, 2006, **83**, 497.
93. F. R. Lupi, D. Gabriele, V. Greco, N. Baldino, L. Seta and B. de Cindio, *Food Res. Int.*, 2013, **51**, 510.
94. D. C. Z. Botega, A. G. Marangoni, A. K. Smith and H. D. Goff, *J. Food Sci.*, 2013, **78**, C1334.
95. S. Svenson, *Curr. Opin. Coll. Interface Sci.*, 2004, **9**, 201.
96. Y. A. Shchipunov, S. A. Mezzasalma, G. J. M. Koper and H. Hoffmann, *J. Phys. Chem. B*, 2001, **105**, 10484.
97. K. Hashizaki, N. Watanabe, M. Imai, H. Taguchi and Y. Saito, *Chem. Lett.*, 2012, **41**, 427.
98. M. Pernetti, K. van Malssen, D. Kalnin and E. Floter, *Food Hydrocoll.*, 2007, **21**, 855.

99. T. Tamura and M. Ichikawa, *J. Am. Oil Chem. Soc.*, 1997, **74**, 491.
100. Y. A. Shchipunov, *Uspek. Khim.*, 1997, **66**, 328.
101. R. Angelico, A. Ceglie, U. Olsson and G. Palazzo, *Langmuir*, 2000, **16**, 2124.
102. G. Muschiolik, *Curr. Opin. Coll. Interface Sci.*, 2007, **12**, 213.
103. B. A. Bergenstahl and J. Alander, *Curr. Opin. Coll. Interface Sci.*, 1997, **2**, 590.
104. C. Barbana and M. D. Perez, *Int. Dairy J.*, 2011, **21**, 727.
105. G. Fragneto, S. Alexandre, J. M. Valleton and F. Rondelez, *Coll. Surf. B, Biointerfac.*, 2013, **103**, 416.
106. D. Rousseau, *Food Res. Int.*, 2000, **33**, 3.
107. B. A. Bergenståhl and J. Alander, *Curr. Opin. Coll. Interface Sci.*, 1997, **2**, 590.
108. N. Krog, *Fett. Seif. Anstrichm.*, 1975, **77**, 267.
109. N. Krog, *J. Am. Oil Chem. Soc.*, 1977, **54**, 124.
110. B. P. Binks, *Curr. Opin. Coll. Interface Sci.*, 2002, **7**, 21.
111. A. Macierzanka, H. Szelag, P. Szumala, R. Pawlowicz, A. R. Mackie and M. J. Ridout, *Coll. Surf. A, Physicochem. Engineer. Asp.*, 2009, **334**, 40.
112. Y. Chevalier and M. A. Bolzinger, *Coll. Surf. A, Physicochem. Engineer. Asp.*, 2013, **439**, 23.
113. N. Krog and K. Larsson, *Fett Wissensch. Technol. (Fat Sci. Technol.)*, 1992, **94**, 55.
114. D. Rousseau, *Curr. Opin. Coll. Interface Sci.*, 2013, **18**, 283.
115. E. H. Lucassen-Reynders and M. Van Den Tempel, *J. Phys. Chem.*, 1963, **67**, 731.
116. I. Kralova and J. Sjoblom, *J. Dispers. Sci. Technol.*, 2009, **30**, 1363.
117. N. Krog, *ACA Symp. Ser.*, 1991, **448**, 138.
118. H. D. Batte, A. J. Wright, J. W. Rush, S. H. J. Idziak and A. G. Marangoni, *Food Res. Int.*, 2007, **40**, 982.
119. S. Ghosh, T. Tran and D. Rousseau, *Langmuir*, 2011, **27**, 6589.

CHAPTER 6

Self-assembled Fibrillar Networks of Low Molecular Weight Oleogelators

MICHAEL A. ROGERS[a,b]

[a] Assistant Professor, Department of Food Science, Rutgers University, 65 Dudley, Rd, New Brunswick, NJ 80901, USA; [b] Director of the Center for Gastrointestinal Physiology, New Jersey Institute of Food Nutrition and Health, New Brunswick, NJ, USA
Email: rogers@aesop.rutgers.edu

6.1 INTRODUCING SELF-ASSEMBLED FIBRILLAR NETWORKS

6.1.1 The Gel State

Soft material physics is a burgeoning, and somewhat chaotic, field with little consensus on the central fundamental principles governing the existence and meta-stability of the materials.[1,2] At the core of these debates is the definition of a gel, which has been evolving for more than 150 years! Early definitions relied solely on macro-structural observable elements, leading to conclusions that described "the rigidity" of the crystalline structures shutting out external expression, and the softness of the gelatinous colloid partaking in fluidity, and enabling the colloid to become a medium for liquid diffusion, like water itself.[3] Weiss and Terech highlight the vague nature of this early report where it must "be assumed that Thomas Graham was referring to the ability of a colloidal suspension to undergo

Edible Nanostructures
Edited by Alejandro G Marangoni and David Pink
© The Royal Society of Chemistry 2015
Published by the Royal Society of Chemistry, www.rsc.org

structural changes when it is not in equilibrium".[4] Not until the late 1920s was it recognized that a gel must be composed of two components, a liquid and a solid, and the material, or gel, has the mechanical properties of a solid (defined as a material maintaining its form under the stress of its own weight).[5] Although an important advancement in the definition of gels, a major downfall is that not all gels are colloids, nor are all colloids gels.[6] By the 1950s, a series of seminal observations had led to three major requirements for the gel state to be satisfied: (1) they must be a coherent colloid dispersion of at least two components; (2) they should exhibit the mechanical properties of solids; and (3) both components must be continuous throughout the system.[7] A less rigorous definition was developed to account for outliers of previous attempts, albeit a regression from the gains made from rheological and structural investigations, where a gel "is a substantially diluted system which exhibits no steady state flow".[8] In the most comprehensive review of molecular gels, Weiss and Terech move forward to defining molecular gels by first meeting the overall gel requirement that they must include a continuous microscopic structure with macroscopic dimensions that is permanent on the time scale of an analytical experiment and is solid-like in its rheological properties.[4]

6.1.2   Self-assembled Fibrillar Networks *vs.* Polymeric Gels

The vast majority of gelators discovered to date are polymers and are exclusively responsible for edible gel applications. Polymeric gels are as diverse as they are abundant with the only common underlying characteristic being the fact that they are linear chains comprised of covalently linked monomers. Polymer gels will form if they are suitably soluble in the solvent and if they have a mechanism to promote crosslinking or junction zones that develop *via* covalent interactions (*i.e.*, chemical gels) or physical interactions (*i.e.*, physical gels) and may, or may not, be thermoreversible depending on whether the underlying gelation mechanism is driven by entropy or enthalpy.

Gels with "small" molecular gelators (*i.e.*, oleogels) are thermally reversible, quasi-solid materials comprised primarily of organic liquids that undergo spontaneous formation into self-assembled networks that are usually fibrillar in nature.[4,9,10] These one-dimensional (1D) partly crystalline aggregates form a three-dimensional (3D) network of 1D objects forming a bi-continuous supramolecular self-assembled fibrillar network (SAFiN) in a solution (sol) at low concentrations of gelator molecules (≤2 wt%).[4] As the sol, composed of solvent and monomer, is cooled, the solution becomes supersaturated, causing a change in the chemical potential, which is the driving force for phase separation and nucleation. Gelator molecules begin to self-assemble in stochastic nucleation events, which have highly specific interactions, promoting 1D growth (Figure 6.1). The contrasting, non-covalent, gelator–gelator and gelator–solvent interactions result in fibrillar aggregates that are, in some cases, capable of structuring fluids while preventing macroscopic or observable liquid flow.[11–13] Disassembly of the 1D objects (and their 3D networks) is possible because the monomers are not covalently linked and may be driven to disassemble by application of heat, dilution, shear, or other physical perturbations.[14]

The ability of molecules to self-assemble spontaneously into supramolecular, 1D aggregates requires an intricate balance between contrasting enthalpic and entropic parameters, including solubility and those that control epitaxial growth. Generally, gelation is achieved when the solvent and gelator are unable to form intermolecular non-covalent bonds and the SAFiN is comprised of thin entangled fibers.[15] These fibers, which contain individual monomers, are held together by intermolecular forces that include hydrogen bonding,[16–20] electrostatic interactions, π–π stacking, dipole–dipole,[21,22] and London dispersion forces.[23]

6.2 MECHANISMS GOVERNING SELF-ASSEMBLY IN MOLECULAR GELS

6.2.1 Nucleation (0D–1D Transformations)

Although gels composed of monomers have been phenomenologically observed for decades, it was not until 1974 that Flory, as an afterthought, included in his structural classification of gels particulate disordered structures originating from zero-dimensional (0D) structures.[24] In oleogels, monomers or 0D structures must be driven to phase separate microscopically, facilitating a phase transition (term used loosely) from the sol to the gel state. Originally explained by Gibbs, the solution/sol, in our case a molecular gelator, is cooled and the system becomes supersaturated, representing a metastable state where the 0D elements transition to 1D fibers with a positive change in the chemical potential (σ). Metastability arises because of the necessity to form a phase boundary that has an associated interfacial tension (μ) and a critically sized aggregate or crystal-embryo.[25] Thus, the change in the thermodynamic potential (ΔG) of the newly evolving phase may be expressed as in equation (6.1):[25]

$$\Delta G = -n_\alpha \Delta\mu + \sigma A \qquad (6.1)$$

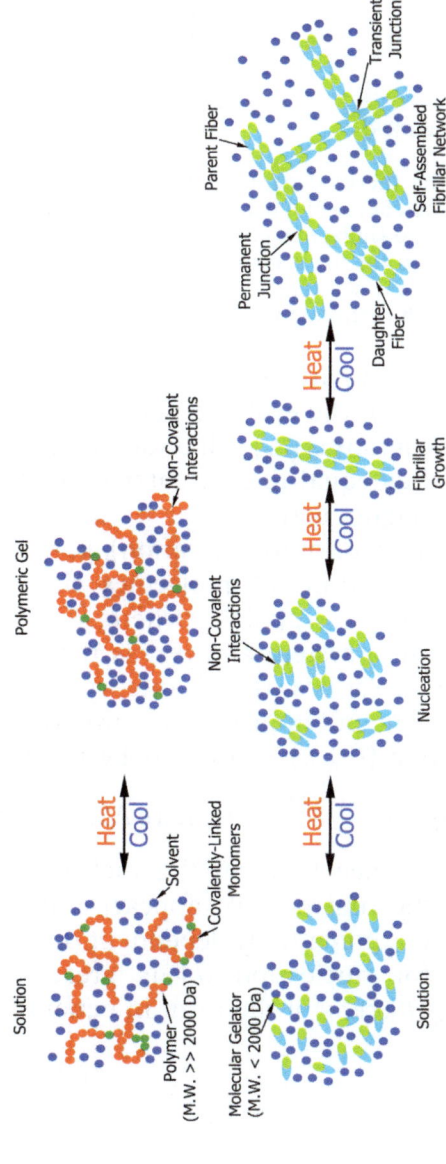

Figure 6.1   Schematic representation of polymeric on monomeric (molecular) gels.

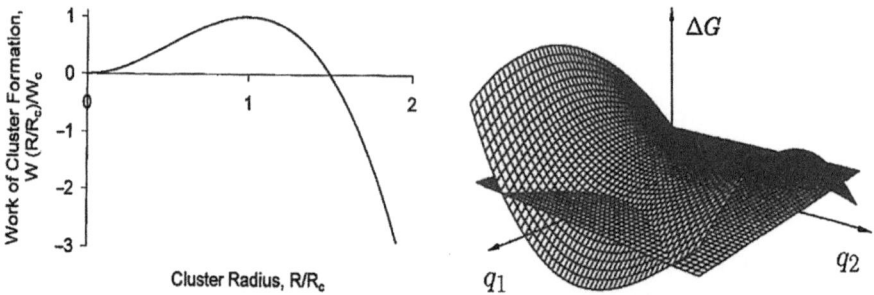

Figure 6.2 Change in the thermodynamic potential ΔG (*i.e.*, work of cluster formation) in the course of cluster formation of a new phase (left). Qualitative shape of a Gibbs free energy surface if several parameters (q_1 and
q_2) are required for the specification of the state of the clusters of the
newly evolving phase (right).
(Reprinted with permission from ref. 25. Copyright 2006 Springer.)

where n_α is the number of particles in the critical cluster and A is the
surface area ($A = 4\pi R_c^2$) of the evolving phase. Clusters with a radius
(R) less than the critical radius (R_c) will undergo dissolution. Once the
cluster reaches a radius equal to or greater than R_c, the nuclei will
grow spontaneously (Figure 6.2).[25,26]

In the metastable state (prior to nucleation), n_α molecules are in the
cluster, while the proportion of molecules that potentially may
nucleate is governed by $(n_\alpha) = p(x)(N_T)$, where N_T is the total concentration of molecules in a system and $p(x)$ is the probability density
function (pdf) that describes the frequency distribution of the
particular event.[27-30] The pdf approach was first developed when the
relative nucleation rate (J/J_{max}), [governed by the time–temperature
cooling trajectory (β) $\left(\beta = \dfrac{1}{2}\dfrac{(\Delta T_C)^2}{\varphi} \right)$ where ΔT_c is the supercooling or
the temperature difference between the melting temperature (the
temperature where the first crystals appear), and φ is the cooling
rate], was found to be exponentially dependent on $\sqrt{\beta}$.[30] Therefore, if
the rate constant for nucleation (k) is greater than zero, the
exponential pdf, $p(\sqrt{(\beta)};k)$, is of the form in equation (6.2):[28]

$$p\left(\sqrt{\beta}\,;\,k \right) = \left\{ \begin{array}{l} ke^{-k\sqrt{\beta}}\,;\; \sqrt{\beta} \geq 0 \\ 0\,;\; \sqrt{\beta} < 0 \end{array} \right\} \tag{6.2}$$

This pdf applies to the randomly distributed variables belonging to
the set $\sqrt{\beta} \in [0;\infty]$. Using this logic, the pdf describes nucleation
kinetics non-isothermally, allowing for an activation energy to be
determined for a SAFiNs, compared to 3D sphereulitic crystal

networks. Molecules, capable of assembling into SAFiNs, have lower activation energies than their counterparts which result in sphereulitic colloidal crystal networks.[29,31] The lower activation energy for SAFiNs compared to sphereulties is due to microscopic phase separation, because of the greater chemical potential differences between the sol and "solid" states in molecular gels, compared to the macroscopic separation observed when triglycerides form sphereulties.[31] Therefore, smaller critical nuclei are required when going from a crystal embryo to a stable nucleus.

6.2.2 Fibrillar Growth of Small Molecules and Crystallographic Mismatches

Nucleation is followed by crystal growth which causes gelator molecules to aggregate into a network of rods, tubes, or sheets forming *via* non-covalent interactions. The aggregation process requires a meticulous balance between the contrasting parameters of solubility and those controlling epitaxial growth.[32] Crystal physics has long been concerned with not only the rate of crystal growth, but also the dimensionality of how materials assemble during crystallization.[33-35] For practical applications, the effect of non-isothermal crystallization conditions on the microstructure and macrostructure is of utmost importance.[29,31] The kinetics of a phase change is frequently modelled using the Avrami equation where it is assumed that the number and size of the crystals is a function of time and temperature [equation (6.3)]:[33]

$$\frac{Y}{Y_{max}} = 1 - e^{-k_{app}(x-x_0)^n} \tag{6.3}$$

where k_{app} is the apparent rate constant, x is the time and x_0 is the induction time, n is the Avrami exponent representing both the dimensionality of growth and mode of nucleation, Y is the new phase volume and Y_{max} is the solid phase volume at $t = \infty$. The Avrami exponent (n) is a function of both the dimensionality of crystal growth and the mode of nucleation (*i.e.*, sporadic or instantaneous).[36] If n is the same for different crystallization rates then, when using an Arrhenius equation, an activation energy may also be determined.

The fibers which radiate from the central nuclei, arising either instantaneously or as a function of time (sporadic) must then interact, forming a 3D network of transient and permanent junction zones.[37] Permanent junction zones are effective at entraining the liquid apolar

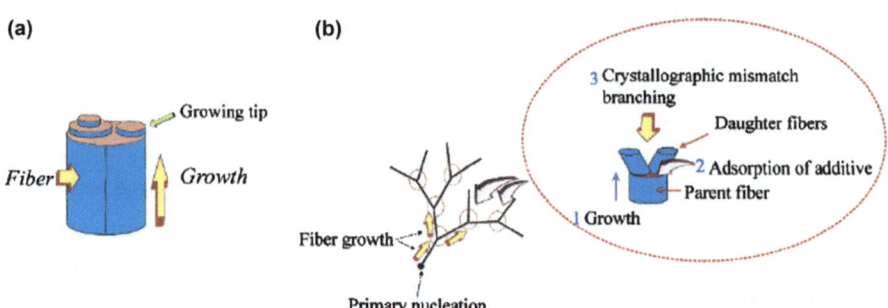

Figure 6.3   The growth of a fiber by deposition of new crystal layers on an existing crystal surface (a). At low supersaturations the layer occurring at the tip of the fiber matches exactly that of the crystal surface, while at high supersaturations crystallographic mismatches develop branch points where two daughters emerge from one parent fiber (b).
(Reprinted with permission from ref. 102. Copyright 2002 American Chemical Society.)

phase in the mesh-like network, arising because of crystallographic mismatches at the interface of the growing 1D crystals and resulting in branch points and truncations along the fiber (Figure 6.3).[37,38] Transient junction zones are thought to arise from non-covalent interactions of adjacent fibers giving rise to thixotropy (*i.e.*, recovery of elastic component upon sequestration of shear), a unique attribute of SAFiNs.[14]

To engineer the characteristic length of the SAFiNs, the isothermal crystallization conditions (or cooling rate of the gelator–solvent system) can be tailored, which results in crystallographic mismatches. These mismatches have been used to explain the branching rate and generation of permanent junction zones. Significant evidence has been provided to support the nucleation–growth–crystallographic mismatch branching (CMB) mechanism (Figure 6.3).[17,31,37,39–42] The CMB method suggests that with low degrees of undercooling, and following nucleation, the fibers grow in one dimension with little branching, interpenetration and entanglement.[37] At low degrees of bulk supersaturation, the crystallographic mismatch nucleation barrier (ΔG^*) is very high, favoring 1D fiber growth with a corresponding large correlation length (ξ).[37] However, when the crystallization temperature is decreased, an increase in the supersaturation causes the crystallographic mismatch barrier to be significantly reduced, increasing the fiber tip branching.[37] The highly branched fibers [*i.e.*, short correlation length (ξ)] coincide with smaller pore sizes. Recently, it has been shown that increased supercooling leads to more elastic gels due to the shorter, highly branched fibers.[39,43]

Upon large degrees of supercooling *via* isothermal or non-isothermal cooling, monomers adhere to the growing crystal surface in a sub-optimal configuration resulting in crystallographic mismatches, further resulting in higher fractal networks ($D_f \sim 1.2$) affecting the hardness and the ability of the network to retain the solvent. As an example, 12-hydroxysteaic acid (12HSA) assembles into short fibers at cooling rates faster than 5 °C/min; while for slower cooling rates longer fibers result with lower fractal values ($D_f \sim 1.0$). Synchrotron based Fourier Transform Infra-Red (FTIR) spectroscopy suggested that crystallographic mismatch (in the case of 12HSA) is a result of the rate of dimerization between carboxylic acid groups, not hydroxyl interactions.[17]

The SAFiNs contain fibers that vary in length and thickness; this is dependent on the crystal history and the types of junction zone present (*i.e.*, transient and/or permanent).[40] Permanent junctions are more effective at entraining the liquid apolar phase into its mesh-like network. The permanent junction zones are formed *via* crystallographic mismatches at the interface of the growing 1D crystals. Transient or non-permanent junction zones occur between individual SAFiN fibers, typically resulting in the formation of weak gels.[44] Non-permanent interactions are subject to the nature and polarity of the solvent, which may either promote or interfere with fiber–fiber interactions. It is thought that the unique thixotropic properties of these gels arise from the non-permanent junctions, which "heal" upon the removal of shear.[45] It has recently been suggested that the gaps in knowledge of what drives thixotropic behaviour in SAFiN are of critical importance; it seems to be influenced by not only the crystalline phase but also the liquid medium.[14]

6.3 FATTY ACID LOW MOLECULAR WEIGHT OLEOGELATORS

Numerous oleogels, composed of different fatty acids, are capable of producing a continuous network of small crystal platelets at significantly lower concentrations than triacylglycerols (TAGs). Examples include long chain fatty acids[46] and combinations of fatty acids and fatty alcohols.[47] Stearic acid or steryl alcohol individually form platelet-like crystals upon cooling. On cooling mixtures of these compounds, needle-shaped crystals are observed, with the smallest crystals observed at a 3 : 7 wt/wt acid : alcohol ratio.[48] These compounds are not effective at forming SAFiNs; their platelet-like crystals have much smaller surface area to volume ratios, fewer crystal–crystal interactions, and construct a much greater space-filling network, diminishing the elastic strength of the material compared to

SAFiNs.[48] However, with only a slight structural modification to the fatty acids (addition of a second hydrogen-bonding group), they may become extremely effective oleogelators owing to a microscopic conformational change from platelet to fiber, driven by the addition of functional groups.

6.3.1 Role of Chirality in Oleogelation

The sense of the supramolecular twist of optically pure 12HSA can be correlated with the chirality of the stereogenic center of the enantiomer (*i.e.*, D-12HSA forms left-handed helices).[49–52] Gelation of enantiopure or entanio-enriched SAFiNs enhances the circular dichroism signal and the surpramolecular spectra correlate with the chirality of the molecule (Figure 6.4).[14] In numerous studies it has been shown that racemic

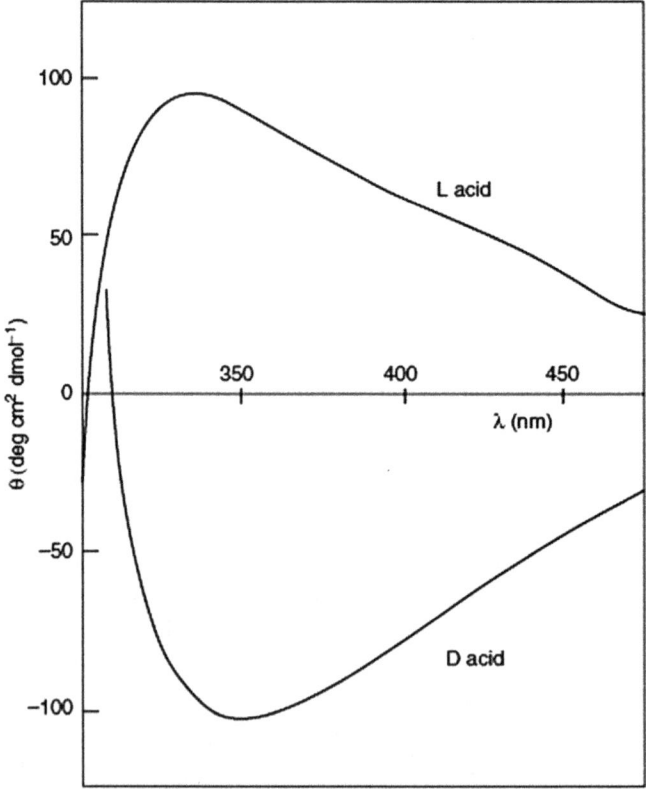

Figure 6.4 Circular dichroism (CD) spectra of enantiomerically pure 12HSA in CCl$_4$ gels: 25.7 mmol l^{-1} for the L-acid; 35.7 mmol l^{-1} for the D-acid. (Reprinted with permission from ref. 103. Copyright 1979 Nature Publishing Group.)

mixtures (critical gelator concentration $\sim$2 wt%, compared to 0.5 wt% for pure enantiopure 12HSA) are far less efficient at entraining oil owing to their inability to form SAFiNs.[53]

Less than 1.0 wt% (optically pure) D-12HSA was needed to form an organogel in mineral oil, and the material was composed of high aspect-ratio fibers, while 2.0 wt% DL-12HSA (racemic) was required to effect gelation which formed platelets. When the D : L ratios of 12HSA are <80 : 20, the Avrami exponent, n, was found to be equal to 3, which corresponds to platelet-like crystals and sporadic nucleation. Conversely, at D : L ratios >80 : 20, the Avrami exponent was 2, indicating that the crystallization process involved fiber-like growth and sporadic nucleation.[53]

The SAFiNs comprised of the enatiomerically pure D-12HSA formed cyclic dimers between carboxyl groups. This was concluded from the observation of a FTIR spectral feature at 1700 cm^{-1} that has been previously ascribed to the cyclic dimer of carboxylic acid head groups.[17,53–58] In these 12HSA SAFiNs, hydrogen bonding of the 12-hydroxyl groups forms along the transverse axis, favoring longitudinal growth (Figure 6.5). The racemic mixture of DL-12HSA organogel was composed of platelet-like interlocking crystals with a molecular arrangement of single in-plane hydrogen bonded acyclic dimers (FTIR peak at $\sim$1720 cm^{-1}), which prevented longitudinal growth and limited the ability of the polar groups to be shielded from the low polarity

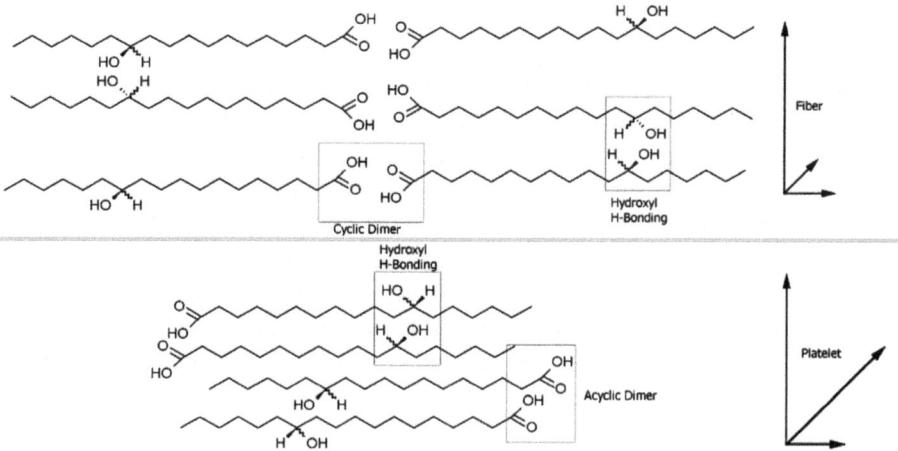

Figure 6.5 Schematic representations of D-12HSA packing and DL-12HSA packing in SAFiNs of mineral oil gels.
(Reprinted with permission from ref. 53. Copyright 2011 Royal Society of Chemistry.)

mineral oil solvent during nucleation. This difference may be the source of the higher activation energy for nucleation and growth of the platelets (~ 40 kJ mol^{-1}) than for fiber formation (~ 10 kJ mol^{-1}).[53]

6.3.2   Role of the Position of Hydroxyl Groups in Organogelation

One challenge in finding SAFiNs to be used in foods is the nuances in chemical structure that affect self-assembly. In an attempt to better understand hydroxy fatty acid assembly, the positional isomers of 12HSA were synthesized. Their supramolecular aggregation forms and nanostructures were affected acutely by the position of the hydroxyl group on the SAFiN. When the hydroxyl group is at position 2, the 2HSA molecules do not dimerize effectively and growth along the secondary axis of the aggregate network is inhibited (Figure 6.6).[54]

These changes in the microscale structural elements arose because of nanoscale interaction between the head groups of 2HSA, resulting in a Bragg distance for the longest dimension of the dimers that is significantly shorter than the sum of the extended lengths of two molecules. In other words, they are partially interdigitated (Figure 6.7).[54] The X-ray diffraction (XRD) data also indicate that 3HSA forms acyclic dimers in which the hydroxyl groups are unable to organize into an effective hydrogen-bonding network along a secondary axis. The SAFiNs of 6HSA, 8HSA, 10HSA, 12HSA, and 14HSA have similar nanostructures and overall packing arrangements, as indicated by FTIR and XRD. These HSAs are able to effectively form both carboxylic cyclic dimers and H-bonding networks *via* their hydroxyl groups.

Along with changes in the microstrucuture and nanostructure, when the carboxylic acid and hydroxy group are separated by fewer than five carbon atoms (*i.e.*, as in 2HSA and 3HSA), there is a much lower activation energy and energy of crystallization.[55] The reduction in the activation energy is associated with the ease of HSA incorporating into the growing crystal lattice or the ease with which the critical nucleus is reached. It is most likely that a combination of these factors is responsible because the critical radius is a balance between the energy associated with creating a new phase and the interfacial free energy of the new surface.[55]

6.3.3   12-Hydroxystearic Acid Oleogels

The most efficient and well studied hydroxy fatty acid is 12HSA, derived naturally from hydrogenated castor seed oil, producing an

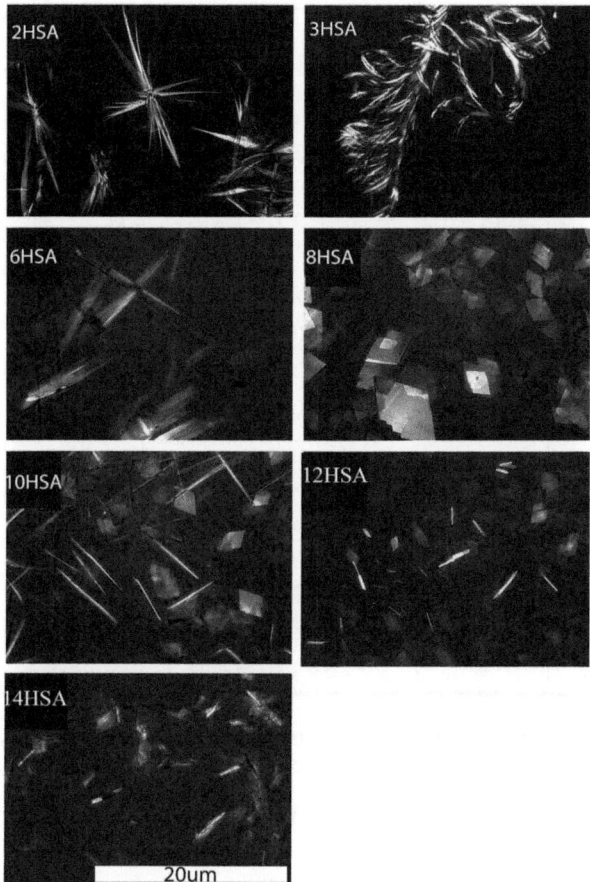

Figure 6.6 Polarized light micrographs of positional isomers of 2 wt% gels and dispersions of HSA in mineral oil at 30 °C. The space bar applies to all panels. (Reprinted with permission from ref. 54. Copyright 2012 American Chemical Society.)

enantiomerically pure D-12HSA. The first study addressing the thickening action of 12HSA occurred in 1972, when 12HSA was added to peanut butter.[59] In this study, low levels (0.5 to 1.0%) of 12HSA were added to peanut butter in an attempt to prevent the syneresis of oil *via* gelation. Using Clark and Ross-Murphy's[60] classical definition of gels (based on the frequency sweeps), it is determined that, above 0.5 wt% 12HSA, the SAFiNs produced were weak gels. However, at 0.5 wt% 12-HSA, a frequency dependence is observed. Hence the sample is not a true gel at this concentration; instead it is classified as a dispersed solution.[61] The final physical properties (such as fiber branching, fiber length and pore size) of 12HSA SAFiNs may be

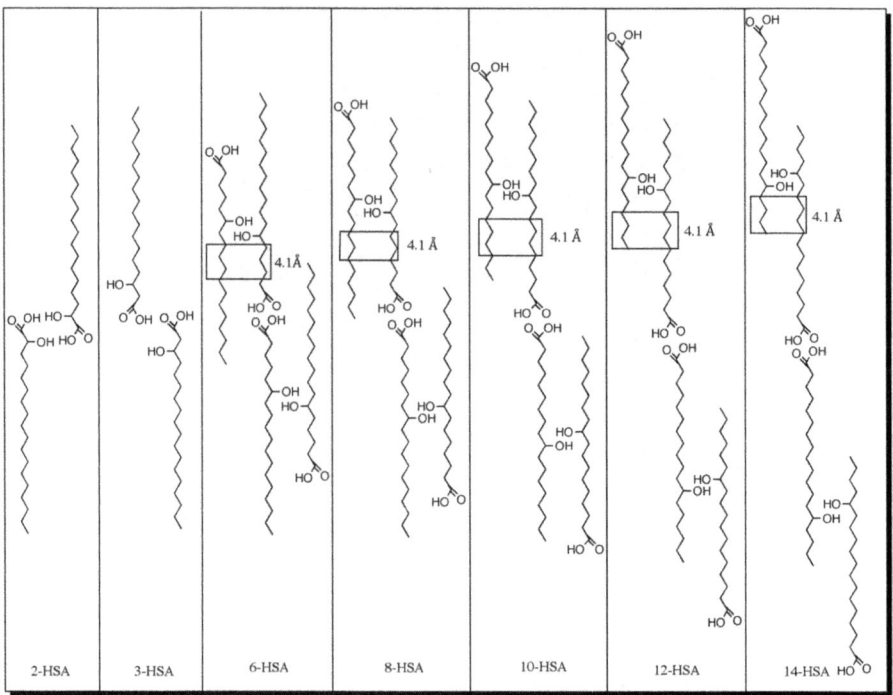

Figure 6.7 Proposed packing arrangements of HSA isomers in the SAFiNs of their gels/dispersions in mineral oil. Distances shown are based on XRD measurements.
(Reprinted with permission from ref. 54. Copyright 2012 American Chemical Society.)

tailored to mitigate problems of oil migration between different phases.[43,61] From the perspective of structuring liquid oils using SAFINs, the yield strain and hardness may be manipulated in a predictable way (*i.e.*, tailoring mass and energy transfer), allowing for the creation of new foods without the need for saturated or *trans* fatty acids.

In a similar fashion to triglycerides, the polymorphic forms of the nanostructure may be tailored by varying the oil conditions.[62] For example, 12HSA in alkanes and thiols forms a hexagonal sub-cell spacing ($\sim$4.1 Å) and a multi-lamellar crystal morphology with a distance between lamella greater than the bi-molecular length of 12HSA has been observed (Figure 6.8). This polymorphic form corresponded to molecular gels composed of a SAFiN and a critical gelator concentration (CGC) at less than 1 wt%.

The 12HSA molecular gels in nitriles, aldehydes and ketones have a triclinic parallel sub cell ($\sim$4.6, 3.9, and 3.8 Å) and interdigitation in

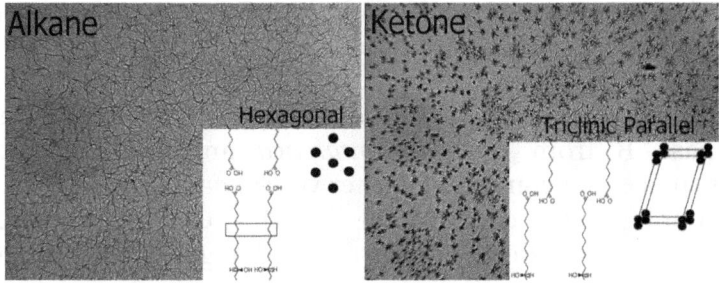

Figure 6.8   Supramolecular network observed using brightfield microscopy and the corresponding nanostructures in each solvent.
(Reprinted with permission from ref. 62. Copyright 2013 American Chemical Society.)

the lamella.[62] This polymorphic form is far less effective at immobilizing solvents with CGCs greater than 1.5 wt% and a sphereultic supramolecular crystalline network. Without continued in-depth study of the physicochemical properties of organogels these systems will continue to be underutilized and their many applications never fully realized.

6.4 LOW MOLECULAR WEIGHT SUGAR-DERIVED OLEOGELATORS

6.4.1 Diversity of Sugar Oleogelators

In practice, simple sugars are often modified by food and pharmaceutical industries to make new functional ingredients such as olestera. Numerous alkyl- and aryl-based monosaccharide gelators have been generated using this strategy.[63] These lactose- and maltose-based hydrogelators are isotropic liquids above 70 °C and, upon cooling, form hydrogels in which one gelator molecule is responsible for the immobilization of more than 6000 water molecules.[63] It is postulated that these alkyl disaccharide amphiphiles entrap water in the large interlamellar spaces as a result of surface tension.[63] These technologies have now advanced to begin structuring oils with sugar–fatty acid esters.

Sugar-based esters have recently been synthesized *via* enzymatic catalysis, which provides tremendous advantages over chemical synthesis that include high selectivity, which can generate a wide range of sugar-based fatty acid esters, and can prepare highly regioselective symmetrical diesters.[64–66] John *et al.* point out that combining the principles of supramolecular chemistry with the selectivity of biocatalysis may represent a new and powerful strategy in developing new molecularly defined and functional materials which may be produced in a food-grade manner.[67,68] Disaccharide-based (sucrose, maltose, lactose and trehalose) diesters have been synthesized using lipase B from *Candida antarctica* and transesterification reactions in acetone containing either vinyl stearate or vinyl butyrate as hydrophobic ester donors (Table 6.1).[67] Of these synthesized compounds, only trehalose diesters resulted in gels; this is thought to be due to the fact that it is a symmetrical disaccharide, resulting in a sole regiospecific isomer.[67] Along with the structure of the sugar molecule, the fatty acid chain length [C2 to C14 (1, 3, 4, and 6 in Table 6.1)] has been observed to affect gelation. In edible oils, trehalose diesters undergo SAFiN (~ 500 μm in length) formation and ultimately gelation when the esterified fatty acid is between 9 and 17 carbons, and gelation efficiency decreases as chain length increases in olive oil. It is also interesting to note that, in most industrial solvents, more effective gelators occur with fatty acids of much shorter chain length.

Of the sugar gelators, D-sorbitol has been shown to be the simplest sugar capable of forming a SAFiN, but it only assembles in ethanol.[69] Assembly of D-sorbitol, like most other organogelators, depends acutely on the mass and energy transfer conditions during assembly. This very weak gelator can be modified into one of the most universal gelators by simply adding two phenyl groups to the sugar, resulting in 1,3:2,4-dibenzylidene sorbitol (DBS). DBS is currently used to nucleate polymers, converting opaque plastics into clear ones by decreasing the crystal size of the polymer blend (Figure 6.9). Such an approach has allowed polyethylene to be substituted with lower concentrations of polypropylene–DBS mixtures.[70-72] Small changes to the molecular structure of DBS often result in more or less efficient nucleators than DBS. They are useful in enhancing polymer blend compositions, including those for food packaging.

6.4.2 Role of Solvent Structure in Oleogel Formation

Given the universal nature of the DBS assembly, it was chosen to highlight the effect of solvent on gelation and SAFiN-forming abilities. Because SAFiN formation requires the establishment of a meticulous balance between contrasting parameters (*i.e.*, solubility and gelator–gelator non-covalent interactions) it is not surprising that solvent is crucially important in dictating the molecular interactions

Table 6.1 Minimum wt% gelation concentration of trehalose-based diesters in different solvents at 25 °C (G indicates gel formation and I indicates insoluble). (Reprinted with permission from ref. 67. Copyright 2006 Wiley InterScience.)

Solvent (log P value)	R=	1 −CH$_3$	2 −(CH$_2$)$_2$CH	3 −(CH$_2$)$_8$CH$_3$	4 −(CH$_2$)$_{12}$CH$_3$	5 −(CH$_2$)$_{16}$CH$_3$	6 −CH=CH$_2$
acetonitrile (−0.34)		G (0.36)	G (0.69)	G (0.18)	G (0.11)	G	G
Acetone (−0.24)		G (0.34)	G (1.0)	G (1.3)	G (1.4)	G	G
isopropanol (0.05)		G (0.54)	G (1.39)	G (2.21)	G (1.31)	G	G
ethyl acetate (0.73)		G (0.04)	G (0.13)	G (0.71)	G (1.1)	G (0.72)	G
methyl methacrylate (1.38)		G (0.05)	G (0.11)	G (0.82)	G (0.85)	G (0.40)	G
p-xylene (3.15)		I	G (0.14)	G (0.18)	G (0.25)	G (0.37)	G (0.72)
olive oil (N/A)		I	I	G (0.09)	G (0.13)	G (0.18)	I

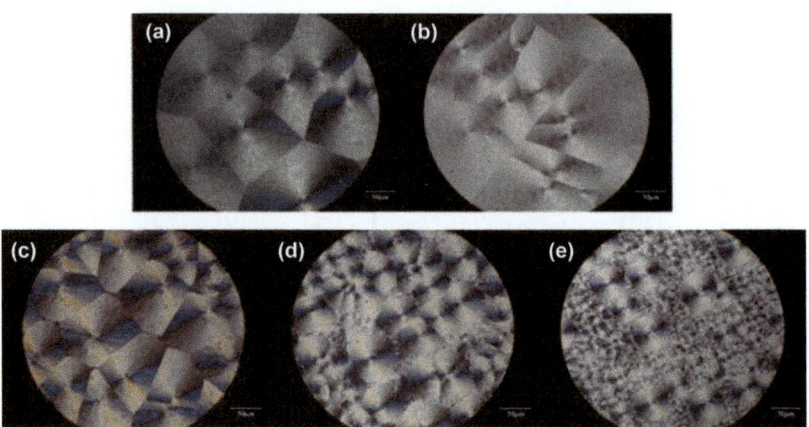

Figure 6.9 Polarized optical microscopy (POM) micrographs of poly (l-lactic acid)
samples with (a) 0, (b) 1, (c) 2, (d) 3, and (e) 4 wt% DBS crystallized at 120 °C.
(Reprinted with permission from ref. 102. Copyright 2013 Elsevier.)

of the gelator, its nanostructures and 3D supramolecular architectures.[4,29,32,73,74] Thicker fibers, resulting in opaque gels, have been
found when the solvent–gelator interactions increase, which corresponds to a decrease in the likelihood of unidirectional gelator–gelator
interactions.[15,73] Optimal gelation is observed when the solvent and
gelator do not form intermolecular non-covalent bonds and the SAFiN
is comprised of thin entangled fibers.[15] Owing to the interplay between the solvent and gelator, numerous attempts have been made to
understand the role solvent plays in dictating SAFiN formation.[29,75–77]
Raynal and Bouteiller recently presented a meta-analysis using
Hansen solubility parameters (HSPs) in evaluating gelation behavior.[75] In an initial study of HSPs and organgels, they revealed that
the solvents which were gelled had similar HSPs (with only a few
exceptions).[75] From an industrial perspective, understanding the
interactions between solvents and gelating molecules is of utmost
importance.[78]

Hildebrand solubility parameters are governed by the gelator–
solvent free energy of mixing, as seen in equation (6.4):

$$\Delta G_\mathrm{m} = \Delta H_\mathrm{m} - T\Delta S_\mathrm{m} \qquad (6.4)$$

where ΔG_m is the Gibbs free energy change during mixing, ΔH_m is the
change in enthalpy during mixing, T is the absolute temperature and
ΔS_m is the entropy change. It is assumed, in polymer physics, that
polymer dissolution is accompanied by a minor entropy increase, and

enthalpy is the deciding factor in the Gibbs free energy change. Therefore, the Hildebrand solubility parameter proposed in two seminal papers, by Hildebrand and Scott[79] and Scatchard,[80] relies solely on the enthalpy term. See equation (6.5):

$$\Delta H_m = V\left(\left[\frac{\Delta E_1^v}{V_1}\right]^{1/2} - \left[\frac{\Delta E_2^v}{V_2}\right]^{1/2}\right)\phi_1\phi_2 \tag{6.5}$$

where V is the mixture volume, ΔE_i^v is the energy of vaporization i, V_i is the molar volume and ϕ_i is the volume fraction of i. Under isothermal vaporization of the saturated liquid, the cohesive energy density (ΔE_i^v) is the energy of vaporization per cm^3, corresponding to the Hildebrand parameter δ_I as seen in equation (6.6).[81]

$$\delta_i = \left(\frac{\Delta E_i^v}{V_i}\right)^{1/2} \tag{6.6}$$

The HSP decomposes the cohesive energy density according to the dispersive (δ_d), polar (δ_p), and hydrogen-bonding interactions (δ_h).[81] Solvents (j) capable of being gelled tend to cluster in a particular region of Hansen space (R_{ij}); see equation (6.7).

$$R_{ij} = \left(4(\delta_{di} - \delta_{ij}^2)^2 + (\delta_{pi} - \delta_{pj}^2)^2 + (\delta_{hi} - \delta_{hj}^2)^2\right)^{1/2} \tag{6.7}$$

HSPs are extremely useful in understanding why certain solvents gel while others form precipitates or sols. They have recently been the basis for new and exciting discoveries pertaining to methods to produce molecular gels from insoluble mixtures,[82] and to understanding the dynamics of gel formation.[78] HSPs not only aid in understanding which solvents are gelled and which are not, but also provide an understanding of why physical aspects of the gels change in certain solvents.[57,58,62,83] In this body of work, it has been shown that certain HSPs, specifically hydrogen-bonding HSP, δ_h, correlate extremely well with the degree of crystallinity, melting temperatures, supramolecular morphology, and polymorphic form (Figure 6.10).

Initial work with 12HSA illustrates that at every level of structure, from the nanoscale to the macroscale, modifications can be made by varying δ_h of the solvent (Figure 6.10).[58] The ability to predict whether 12HSA will assemble in certain solvents is closely associated with the δ_h. Solvents with $\delta_h < 4.7$ MPa$^{1/2}$ produce clear organogels, they form opaque organogels at $4.7 < \delta < 5.1$ MPa$^{1/2}$, and solutions

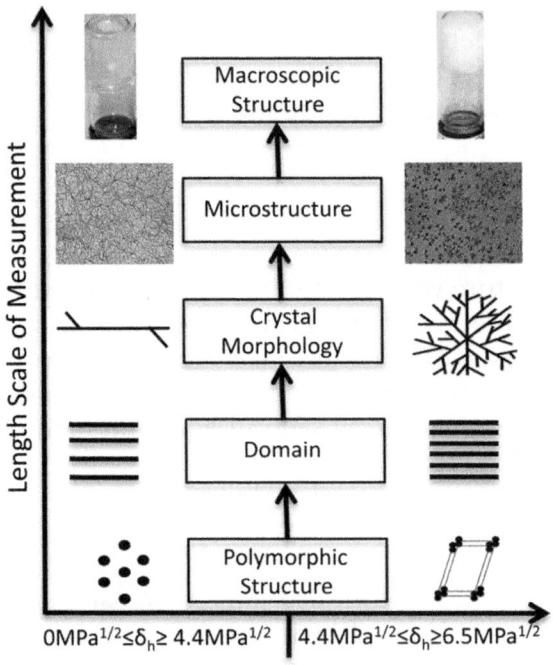

Figure 6.10   Schematic representation of the levels of structure that are altered as δ_h
is varied using solvent selection.
(Reprinted with permission from ref. 58. Copyright 2012 Royal Society
of Chemistry.)

remain when δ_h is >5.1 MPa$^{1/2}$. Amazingly, these trends are being
found for other molecular gelators including the pyrenyl-linker-
glucono low molecular mass organogelator (LMOGs).[82]

Using the HSPs with DBS as the SAFiN, plots were constructed
defining the regions as soluble (S), slow gelation (SG), instant gelation
(IG), and insoluble (I) in the different solvents.[84] It was found that
regions radiate out as concentric shells, *i.e.*, a central solubility (S)
sphere, followed in order by spheres corresponding to SG, IG, and I
regions (Figure 6.11).[84] The distance (R_0) from the origin of the central
sphere quantifies the incompatibility between DBS and a solvent: the
larger this distance the more incompatible the pair. Interestingly,
numerous physical properties of the SAFiN were also found to correlate to R_0, including the elastic modulus.[84] For example, if R_0 is too
small the gels are weak, but if R_0 is too large insolubility occurs – thus
strong gels fall within an optimal window of incompatibility between
the gelator and the solvent.

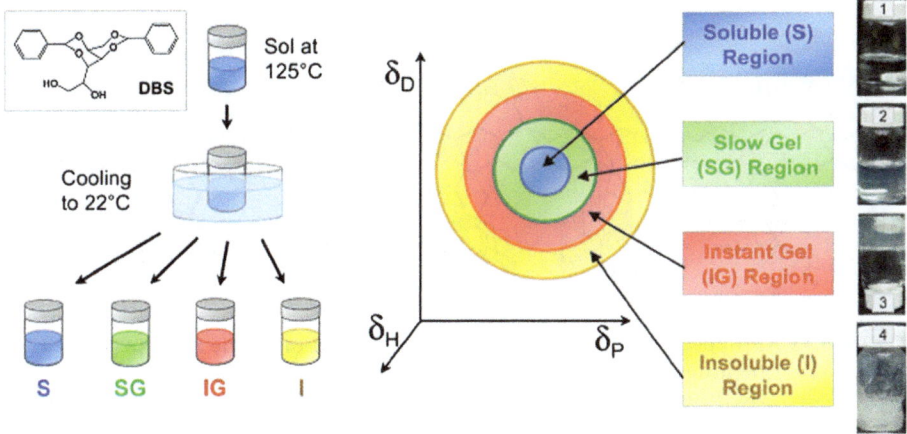

Figure 6.11   Schematic of the experimental setup (left) and the corresponding results in 3D Hansen space (right). Visual observations were used to
classify the samples as soluble or sols (S, blue), slow gels (SG, green),
instant gels (IG, red) and insoluble (I, yellow). Photographs of samples
corresponding to these outcomes are shown on the right.
(Reprinted with permission from ref. 84. Copyright 2014 Royal
Chemistry Society.)

6.5   PEPTIDE-BASED MOLECULAR GELS

6.5.1   Peptide SAFiNs Composed of β-Sheets

It is extremely well known that certain proteins may aggregate into
polymer-based fibers, including gelatin and β-lactoglobulin; however,
it was not until recently that small peptides were found to be capable
of forming SAFiNs. Oligopeptides can be engineered to assemble into
a peptide β-sheet that may be exploited to design SAFiNs in different
solvents.[85] It has been shown that these fibers become entangled at
extremely low concentrations (~ 3 mg ml^{-1}) and the elastic properties
of the gel are determined by both physical and chemical influences.[85]
The first of these peptides that were shown to form SAFiNs were
24-amino acid peptides related to a transmembrane protein. These
were investigated because they readily formed β-sheets in the lipid
bilayer (Figure 6.12).[85] The peptides assemble into β-sheets at the
nanoscale and then into 8 nm-wide fibers that exceed lengths of
0.1 μm, exhibiting rheological properties of highly entangled polymer
gels.[85] A major advantage of these peptide gels is that they are
potentially biocompatible in a similar manner to biopolymer gels
such as gelatin, acitin and agarose.[85]

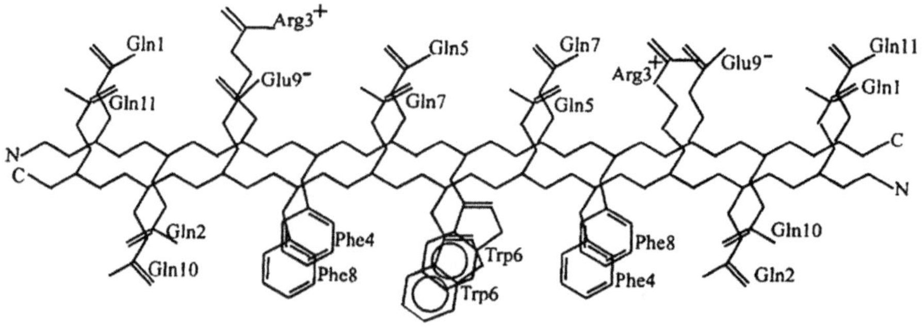

Figure 6.12   Proposed antiparallel β-sheet arrangement of the de novo peptide. The
peptide backbones are drawn as black zig-zag lines.
(Reprinted with permission from ref. 85. Copyright 1979 Nature
Publishing Group.)

Since the first investigations of naturally occurring peptide
fragment gels, numerous peptides have been designed using a
bottom-up approach.[86,87] The smallest of these peptides are termin-
ally protected peptapeptides (Boc–Leu–Val–Phe–Phe–Ala–OMe)
(Figure 6.13).[87] Banerjee et al. eloquently show that engineering these
molecules is as much an art as it is a science, and the replacement of
any of the Phe residues with any other hydrophobic α-amino acid
residue drastically changes the gel-forming properties, indicating that
both Phe residues have an important role in gel formation.[87] The self-
association of the synthetic pentapeptide must be capable of forming
intermolecularly hydrogen-bonded β-sheets that then further
aggregate into SAFiNs in various organic solvents at very low
concentrations.[87] For self-assembly to occur, phenylalanine must be
present because it develops crucial π–π interactions thereby stabilizing the β-sheets.[87]

The smallest known SAFiN-producing peptide is a dipeptide comprised of two l-phenylalanine monomers.[88] It is interesting to note
that the β-amyloid of Alzheimer's disease is composed of a dipheny-
lalanine core, which may be the basis for the driving force for aggregation. The formation of such gels may be driven by the hydrogen
bonds of the peptide main chains and the π–π interactions between
aromatic residues of the peptide (Figure 6.14). These dipeptides are
thermoreversible at concentrations as low as 8 mM and undergo
transition from the sol to gel state between 20 and 70 °C depending
on the concentration.[88] The fibrils form a 3D network with branch
points (permanent junction zones) and entanglements (temporary

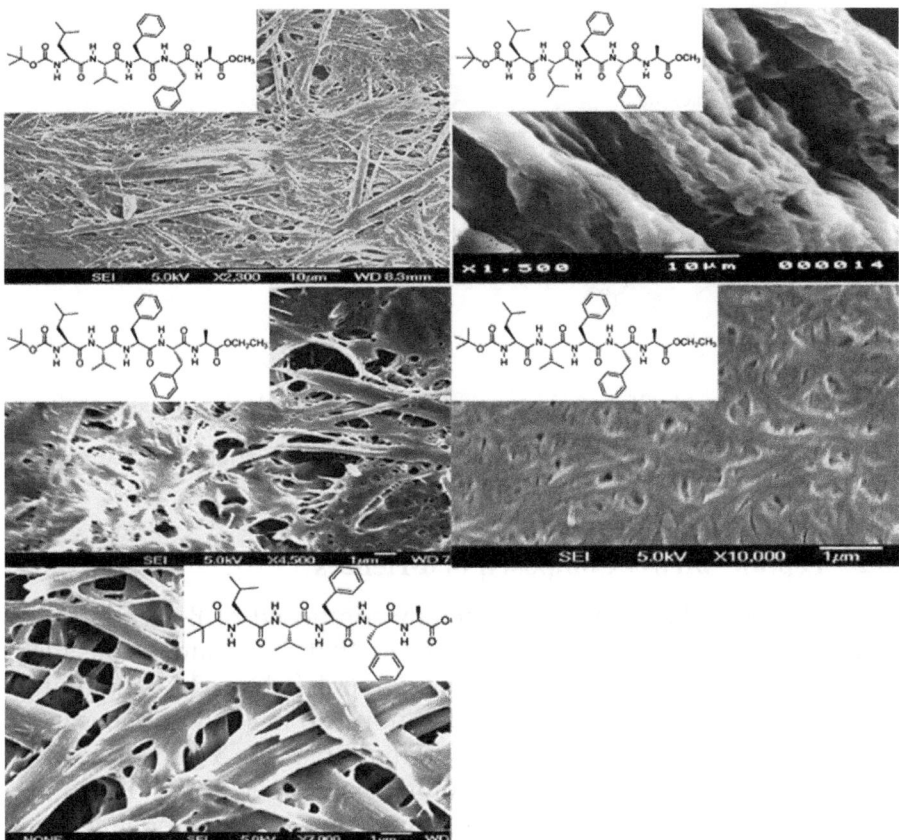

Figure 6.13 Field emission scanning electron micrographs in 1% w/v *m*-xylene of different synthetic pentapeptides.
(Reprinted with permission from ref. 87. Copyright 2008 Royal Chemistry Society.)

junction zones), and grow to have diameters ranging between 10 and 50 nm, reaching micron length scales.[88]

Numerous other organogelators based on different amino acids and various chain lengths have been characterized with respect to their ability to gel safflower oil.[89] In this particular study, tyrosine, tryptophan, and phenylalanine were synthesized and derivatized with various aliphatic chains; tyrosine derivatives formed the strongest gels and tryptophan had the weakest gel properties.[89] Again the peptide derivatives were able to gel safflower oil through van der Waals interactions and hydrogen bonds. The superior gelling ability of tyrosine derivatives has been explained by their well-structured 2D packing.[89]

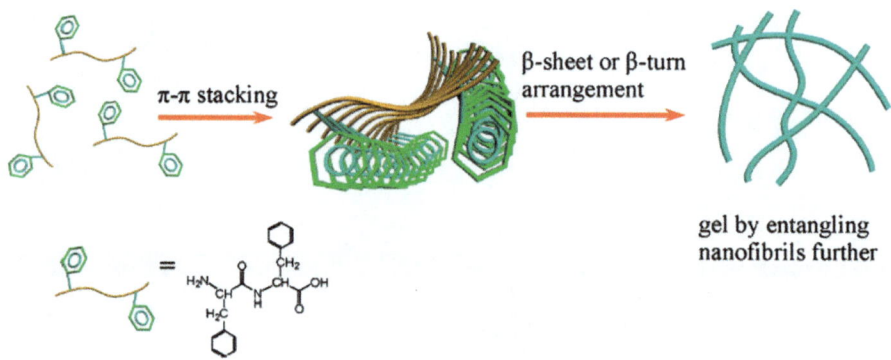

Figure 6.14   Proposed self-assembly mechanism of a dipeptide through π–π stacking and hydrogen bonding.
(Reprinted with permission from ref. 88. Copyright 2013 American Chemical Society.)

6.5.2   Peptide SAFiNs Composed of α-Helices

Peptide-based SAFiN are predominately composed of β-sheets; however, there have been reports of peptides that form supramolecular aggregates *via* α-helices.[90] The polypeptide poly(γ-benzyl-L-glutamate) (PBLG) forms a rod-like polymer in solution and in solid state because of its rigid α-helix (Figure 6.15). The formation of the α-helix has been shown to be a prerequisite for thermoreversible gelation.[90] Based on this mechanism, it is difficult to imagine that block co-polymers of PBLG form a random coil that could generate a gel (at 0.3 wt%) because the block co-polymers cannot be ordered in a liquid crystalline phase in dilute solutions owing to the presence of the bulky random-coil polymer chain at the end of PBLG helix.[90] Therefore, a different mechanism of gelation was required to explain the self-assembly behavior of PBLG–random-coil diblock co-polymers in solution. Along with numerous other observations and the aid of small angle X-ray scattering (SAXS), it is proposed that the nanoribbons arise as a result of the formation of an α-helix.[90]

6.6   SAFINS ARISING FROM MULTI-COMPONENT SYSTEMS

6.6.1   Phytosterols and γ-Oryzanol

Pioneering work by Bot and Agterof[91] investigated mixtures of γ-oryzanol and phytosterols designed to act as a structuring agent in edible liquid oil such as sunflower oil. Numerous phytosterols and/or cholesterols, mixed in appropriate ratios with γ-oryzanol, form

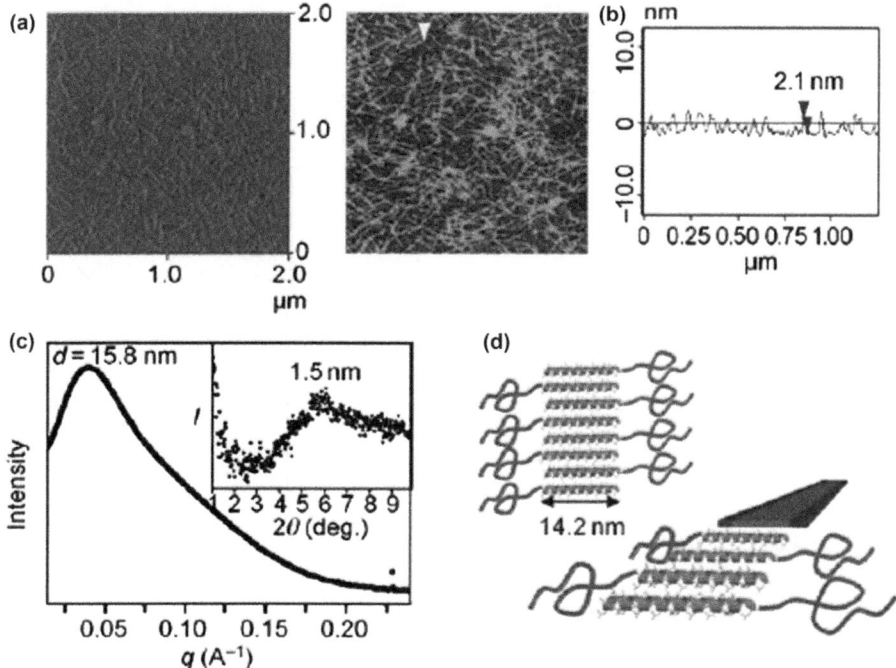

Figure 6.15 a) AFM phase (left) and height (right) images of the toluene gel of 2 (0.1 wt%). b) AFM height profile of the nanoribbons. c) SAXS profile of the dried gel of 2a showing the width of the ribbon (15.8 nm). The inset denotes the diffraction at 1.5 nm, indicating the distance between two PBLG helices obtained by PXRD. d) A schematic presentation of the nanoribbon formed in the network structure of the toluene gel of 2; 14.2 nm indicates the calculated length of the PBLG helix.
(Reprinted with permission from ref. 90. Copyright 205 Wiley-VCH Verlag GmbH & Co.)

SAFiN.[19,91–96] These mixed systems are capable of trapping oil *via* capillary action between the self-assembled γ-oryzanol and β-sitosterol co-crystals.[92] In their work, Bot and Agterof showed that dihydrocholesterol, cholesterol, β-sitosterol, and stigamsterol formed firm gels when mixed with γ-oryzanol.[91] The transparency of the gels implies that the building blocks making up the gel are smaller than the wavelength of visible light.[97] Microstructural studies showed that γ-oryzanol and β-sitosterol self-assemble upon mixing and form nanoscale tubules (Figure 6.16) in the liquid oil phase. The diameter of these tubules ranges between 6.7 and 8.0 nm, while the wall thickness ranges between 0.8 and 1.2 nm.[93] Upon aggregation of these nanoscale tubules a 3D SAFiN is formed, immobilizing the liquid oil phase into a gel. Using a free energy minimization

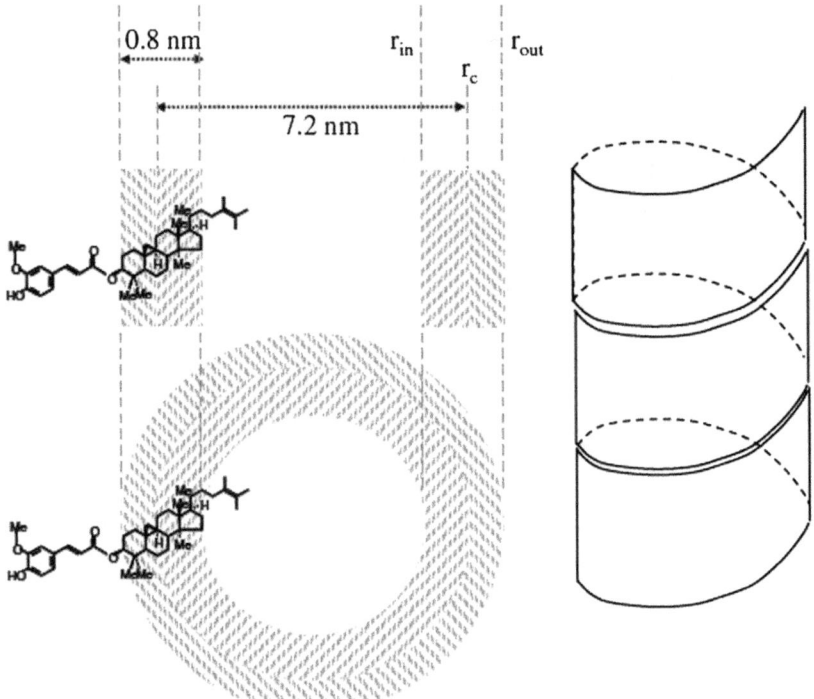

Figure 6.16 Schematic representation of the sitosterol + oryzanol tubule. The up-
ward hatched and downward hatched areas in the ring have the same
size and are meant to explain the meaning of r_{in}, r_c, and r_{out} (left). The
molecules are thought to be tilted relative to the axis and radial plane of
the tubule in such a way that their wedged stacking leads to the formation of a helical ribbon (right).
(Reprinted with permission from ref. 103. Copyright 2009 Elsevier.)

calculation, the mechanism by which the tubules form is believed to
be a result of the ring moieties of the phytosterol and γ-oryzanol
stacking one on top of the other while the sterane cores of both
structurants occupy the tubule wall. The ferulic acid unit of the γ-
oryzanol is situated on the exterior of the formed tubule.[95,98]

The presence of a hydroxyl group on the C-3 position of the sterol
ring stabilizes the stacked ring structure and is an essential element
for SAFiN formation to occur. This hydroxyl group forms a hydrogen
bond with the carbonyl oxygen of the γ-oryzanol.[95] In addition, highly
specific hydrogen bonds formed between the hydroxyl groups of γ-
oryzanol and phytosterol have a minimum free energy if their intermolecular distance follows an optimal specific bond length.[93] At this
optimal bond length, the stacking of the phytosterol units follows a

slightly wedged pattern instead of them being perfectly parallel to one another. As a result, helical tubules or ribbons are formed.[93]

Bot and Agterof have elucidated numerous molecular properties within the phytosterol units that govern gelation within the system.[91] A hydroxyl (–OH) group must be present in the phytosterol for gelation to occur. The properties of the ring within the phytosterol influence the ability to form a gel: the absence of double bonds in the phytosterol ring structure contributes to accelerated gel formation and the presence of a single double bond decelerates the rate of gelation. However, the presence of more than one double bond completely inhibits gelation. The alkyl group composition has been shown to have no major influence on SAFiN formation.

The presence of both γ-oryzanol and the phytosterols are essential to the formation of the nano-fibers. In the absence of either structurant, crystals are formed in place of the nano-fibers, and a hazy slurry is formed in place of the gel network (Figure 6.17).[99] At 8 wt% cholesterol in mineral oil (without γ-oryzanol), long tube-like fibers were formed and at 8 wt% γ-oryzanol in mineral oil (without cholesterol) small plate-like crystals were formed.[98,100] Upon mixing the two components together at 8 wt% : 8 wt% in mineral oil, a new crystal structure was observed, in which a central nucleus forms and fine tubules radiate from it; this could be attributed to the hylotropic behavior (a solid consisting of equal γ-oryzanol and phytosterol ratios) of this ratio combination. Interestingly, at other structurant ratios, both the new crystal structure (a central nucleus from which fine tubules were radiating) and the highly birefringent crystals were seen.

FTIR indicated that the mixtures of the cholesterol : γ-oryzanol in mineral oil showed a broad peak between 3200 and 3600 cm^{-1}, indicative of non-specific hydrogen bonding, followed by a sharp peak at 3450 cm^{-1}, indicative of highly specific interaction between the two structurants. The presence of only a broad peak indicates the presence of either pure phytosterol or γ-oryzanol crystals (Figure 6.18). The presence of the sharp peak (3450 cm^{-1}) after the broad peak (between 3200 and 3600 cm^{-1}) is illustrative of the formation of the 3D tubular structure, which is only apparent when cholesterol and γ-oryzanol are mixed, resulting in tubule formation (sharp peak at 3450 cm^{-1}) with the remaining excess crystallized out (Figure 6.19).[98,100] Upon overlaying the hydrogen-bonding area on the light micrographs (Figures 6.18 and 6.19), it was shown that the hydrogen bonds are evenly distributed, indicating very limited contribution of the solvent to the tubule network formed. Further, upon

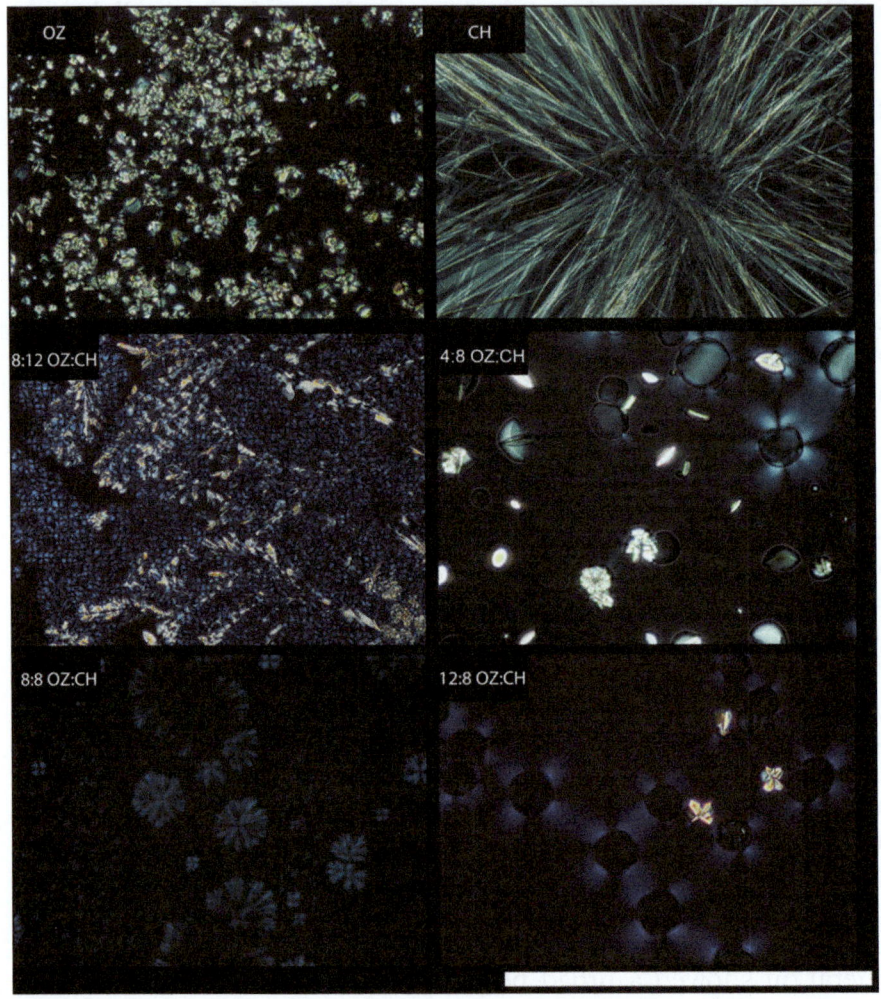

Figure 6.17 Polarized light micrographs of different ratios of cholesterol (CH) and
γ-oryzanol (OZ) imaged at 30 °C. Magnification bar = 100 μm.
(Reprinted with permission from ref. 19. Copyright 2010 Royal
Chemistry Society.)

integration of the 3450 cm^{-1} peak it was deduced that the highly
birefringent crystals consist of either 100% γ-oryzanol or 100%
cholesterol. Concurrently, the fine fibers are in fact the co-crystals of
the two mixed structurants.[98]

Upon fitting the hydrogen-bonding area for the cholesterol :
γ-oryzanol system to the Avrami equation [equation (6.3)],[100] the
dimensionality of growth (n) was found to be equivalent to 2. This,
combined with the fact that short angle X-ray scattering showed

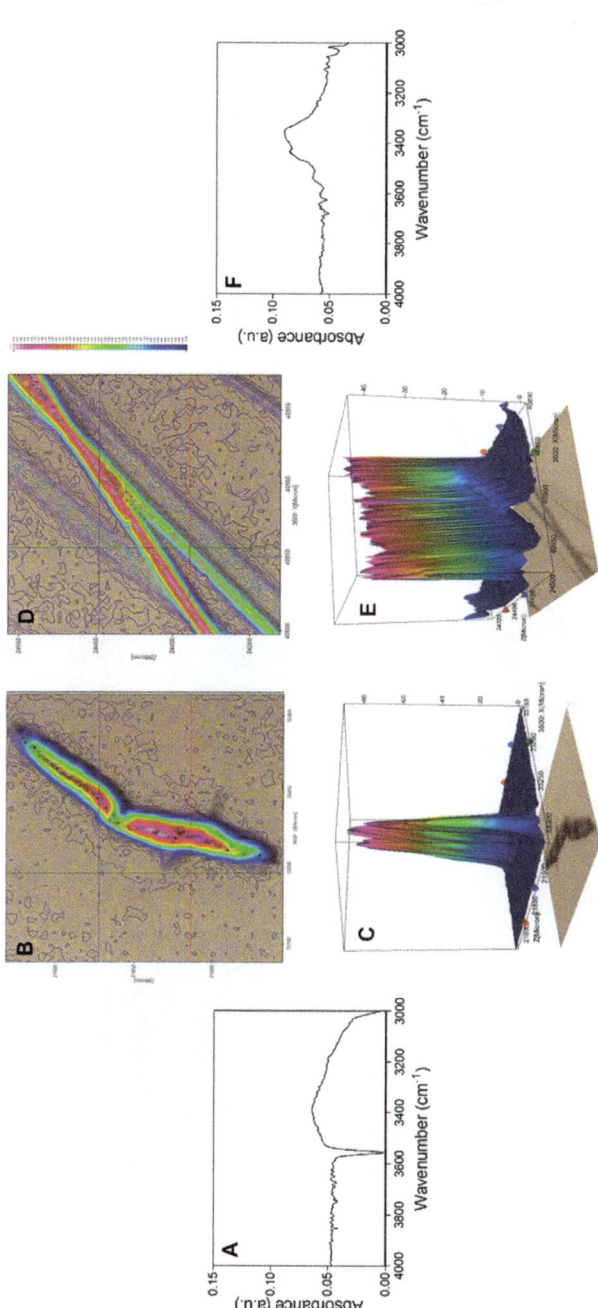

Figure 6.18 FTIR spectrum of 8 wt% γ-oryzanol in mineral oil (A) and 2D (B) and 3D (C) spectromicrographs for 8 wt% γ-oryzanol in mineral oil. FTIR spectrum of 8 wt% cholesterol in mineral oil (F) and 2D (D) and 3D (E) spectromicrographs for 8 wt% cholesterol in mineral oil. Spectromicrographs represent the integrated area under the peak associated with hydrogen bonding (3000 to 3550 cm^{-1}). (Reprinted with permission from ref. 19. Copyright 2010 Royal Chemistry Society.)

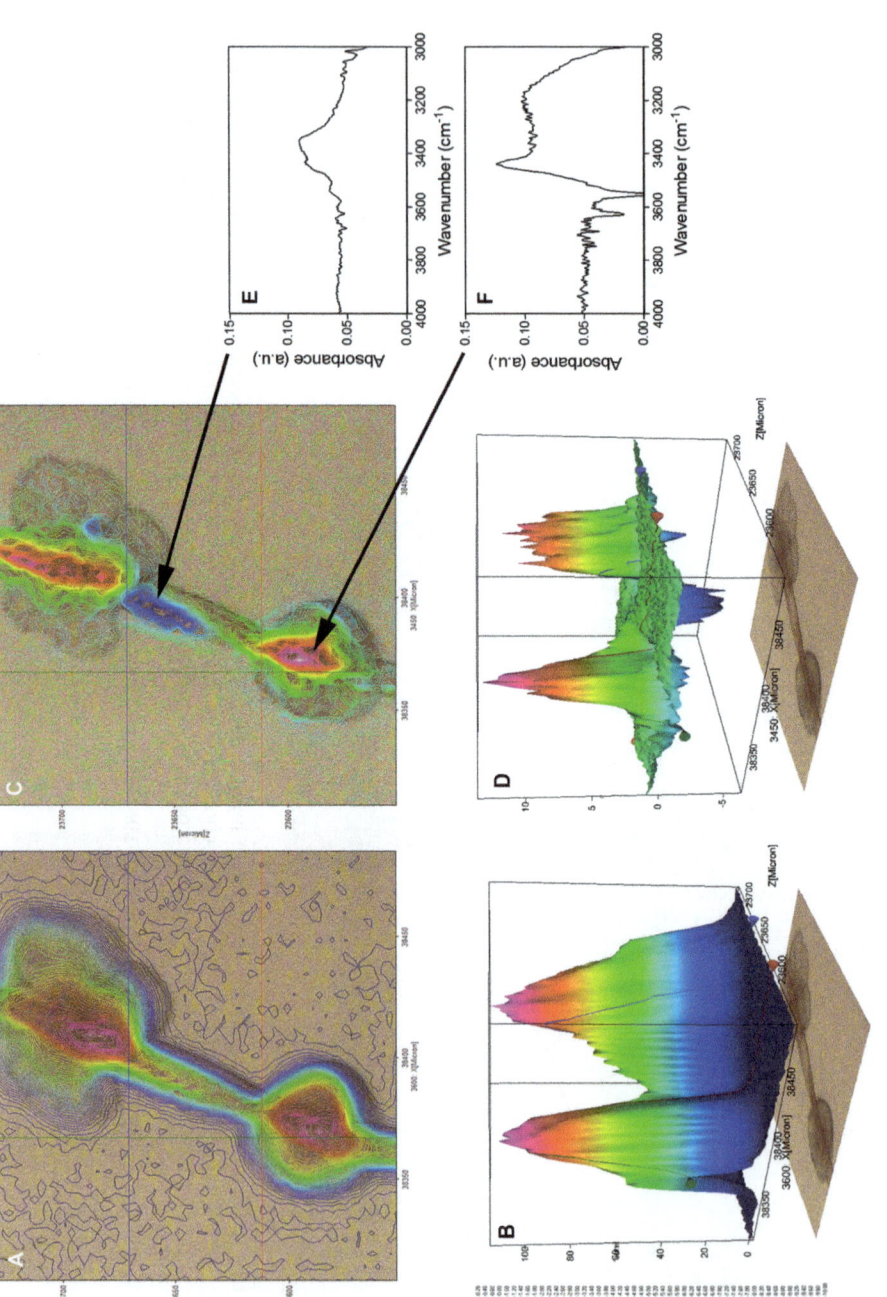

Figure 6.19 FTIR spectrum of 4% γ-oryzanol–8% cholesterol in mineral oil (E, F) and 2D (A, C) and 3D (B, D) spectromicrographs representing the area integrated between 3000 and 3550 cm⁻¹ (A, B) and between 3400 and 3475 cm⁻¹ (C, D). Arrows represent the approximate area from which the spectra were extracted. (Reprinted with permission from ref. 19. Copyright 2010 Royal Chemistry Society.)

that these mixtures form 1D hollow fibers,[93,101] confirms that the nano-fibers formed *via* sporadic nucleation.[100] Also, a positive correlation between turbidity and activation energy was seen as the γ-oryzanol or cholesterol ratios increased beyond 1 : 1.[98] This would imply that the tubules are thicker or longer, creating more junction zones (either permanent or temporary) capable of diffracting light.[100] The location of the co-crystals on the ends or surfaces of the cholesterol fibers (Figure 6.19) suggests that the fibers serve as a surface of nucleation for the co-crystals. As a result, less activation energy is needed to initiate the co-crystal formation.[98]

6.7 CONCLUSIONS

For SAFiNs to reach their potential in edible applications, a fundamental understanding of interactions between the chemical structure, solvent structure, mass and energy transfer are required. These systems are hindered, in part, because of our inability to: (1) identify generally regarded as safe (GRAS). gelators, (2) control their structure/function, and 3) prevent coalescence arising from their meta-stability. With more foundational studies, new and exciting compounds will be identified, allowing liquid oils to be structured with the same precision that researchers are able to use to control polymeric gelators in hydrogels. In the past decade, strides have been made towards these goals; however, a coordinated effort to consult on methods will better advance the field and allow not only edible applications of SAFiNs, but also medical, pharmaceutical and cosmetic applications.

REFERENCES

1. S. Ifuku, R. Nomura, M. Morimoto and H. Saimoto, *Materials*, 2011, **4**, 1417.
2. O. G. Jones, J. Admamcik, S. Handschin, S. Bolisetty and R. Mezzenga, *Langmuir*, 2010, **26**, 17449.
3. T. Graham, *Phil. Trans. Roy. Soc.*, 1861, **151**, 183.
4. R. G. Weiss and P. Terech, in *Molecular Gels: Materials with Self-Assebled Fibrillar Networks*, ed. R. G. Weiss and P. Terech, Springer, Dordrecht, The Netherlands, 2006, pp. 1–13.
5. D. Lloyd-Jones, in *Colloid Chemistry*, ed. J. Alexander, The Chemical Catalog Co., New York, 1926, vol. 1, pp. 767–786.
6. A. G. Marangoni and N. Garti, in *Edible Oleogels*, ed. A. G. Marangoni and N. Garti, AOCS Press, Urbana, IL, 2011, pp. 1–19.

7. P. H. Hermans, in *Colloid Science*, ed. H. R. Kruyt, Elsevier, Amsterdam, 1949, vol. II, p. 484.
8. J. D. Ferry, *Viscoelastic Properties of Polymers*, Wiley, New York, 1961.
9. V. A. Mallia, P. D. Butler, B. Sarkar, K. T. Holman and R. G. Weiss, *J. Am. Chem. Soc.*, 2011, **133**, 15045.
10. M. George and R. G. Weiss, *Acc. Chem. Res.*, 2006, **39**, 489.
11. M. Fahrländer, K. Fuchs, R. Mülhaupt and C. Friedrich, *Macromolecules*, 2003, **36**, 3749.
12. B. Isare, L. Petit, E. Bugnet, R. Vincent, L. Lapalu, P. Sautet and L. Bouteiller, *Langmuir*, 2009, **25**, 8400.
13. E. A. Wilder, C. K. Hall, S. A. Khan and R. J. Spontak, *Langmuir*, 2003, **19**, 6004.
14. K. L. Caran, D.-C. Lee and R. G. Weiss, in *Soft Firillar Materials: Facrication and Applications*, ed. X. Y. Liu and J.-L. Li, Wiley-VCH Verlag GmbH & Co. KGaA, Weinheim, Germany, 2013, pp. 1–67.
15. G. Zhu and J. S. Dordick, *Chem. Mater.*, 2006, **18**, 5988–5995.
16. T. Kuwahara, H. Nagase, T. Endo, H. Ueda and M. Nakagaki, *Chem. Lett.*, 1996, **25**, 435.
17. R. Lam, L. Quaroni, T. Pederson and M. A. Rogers, *Soft Matter*, 2010, **6**, 404.
18. J. L. Li, X. Y. Liu, R. Y. Wang and J. Y. Xiong, *J. Phys. Chem., B*, 2005, **109**, 24231.
19. M. A. Rogers, A. Bot, R. Lam, S. H. T. Pedersen and T. May, *J. Phys. Chem.*, 2010, **114**, 8278.
20. M. A. Rogers, T. Pedersen and L. Quaroni, *Cryst. Growth Des.*, 2009, **9**, 3621.
21. T. Brotin, J. P. Devergne and F. Fages, *Photochem. Photobiol.*, 1992, **55**, 349.
22. P. Terech, I. Furman and R. G. Weiss, *J. Phys. Chem.*, 1995, **99**, 9558.
23. J. F. Toro-Vazquez, J. Morales-Rueda, V. A. Mallia and R. G. Weiss, *Food Biophys.*, 2010, 5, 193.
24. P. J. Flory, *Faraday Disc. Chem. Soc.*, 1974, **57**, 7–18.
25. J. W. P. Schmelzer, in *Molecular Gels: Materials with Self-Assembled Fibrillar Networks*, ed. R. G. Weiss and P. Terech, Springer, Dordrecht, The Netherlands, 2006, 131–160.
26. A. G. Marangoni, in *Fat Crystal Networks*, ed. A. G. Marangoni, Marcel Dekker, New York, 2005, pp. 21–82.
27. R. S. H. Lam and M. A. Rogers, *Cryst. Engineer. Commun.*, 2010, **13**, 866.

28. R. S. H. Lam and M. A. Rogers, *Cryst. Growth Des.*, 2011, **11**, 3593.
29. M. A. Rogers and A. G. Marangoni, *Langmuir*, 2009, **25**, 8556.
30. A. G. Marangoni, T. C. Aurand, S. Martini and M. Ollivon, *Cryst. Growth Des.*, 2006, **6**, 1199.
31. M. A. Rogers and A. G. Marangoni, *Cryst. Growth Des.*, 2008, **8**, 4596.
32. M. Suzuki, Y. Nakajima, M. Yumoto, M. Kimura, H. Shirai and K. Hanabusa, *Langmuir*, 2003, **19**, 8622.
33. M. Avrami, *J. Chem. Phys.*, 1939, **7**, 1103.
34. M. Avrami, *J. Chem. Phys.*, 1940, **8**, 212.
35. M. Avrami, *J. Chem. Phys*, 1941, **9**, 177.
36. A. Sharples, *Introduction to Polymer Crystallization*, Edward Arnold, Ltd., London, 1966.
37. R. Y. Wang, X. Y. Liu, J. Narayanan, J. X. Xiong and J. L. Li, *J. Phys. Chem., B*, 2006, **10**, 25797.
38. X. Y. Liu and P. D. Sawant, *ChemPhysChem*, 2002, **4**, 374.
39. J. L. Li, R. Y. Wang, X. Y. Liu and H. H. Pan, *J. Phys. Chem., B*, 2009, **113**, 5011.
40. R. Wang, X.-Y. Lui, J. Xiong and J. Li, *J. Phys. Chem., B*, 2006, **110**, 7275.
41. R.-Y. Wang, P. Wang, J.-L. Li, B. Yuan, Y. Liu, L. Li and X.-Y. Liu, *Phys. Chem. Chem. Phys.*, 2013, **15**, 3313.
42. B. Yuan, J.-L. Li, X. Y. Liu, Y.-Q. Ma and Y. J. Wang, *Soft Matter*, 2012, **8**, 5187.
43. M. A. Rogers, A. J. Wright and A. G. Marangoni, *Soft Matter*, 2008, **4**, 1483.
44. P. Terech and R. G. Weiss, *Chem. Rev.*, 1997, 3133.
45. M. Lescanne, P. Grondin, A. d'Aleo, F. Fages, J.-L. Pozzo, O. M. Monval, P. Reinheimer and A. Colin, *Langmuir*, 2003, **20**, 3032.
46. J. Daniel and R. Rajasekharan, *J. Am. Oil Chem. Soc.*, 2003, **80**, 417.
47. F. G. Gandolfo, A. Bot, E. Flöter, *J. Am. Oil Chem. Soc.*, 2004, **81**, 1.
48. H. M. Schaink, K. F. van Malssen, S. Morgado-Alves, D. Kalnin and E. van der Linden, *Food Res. Int.*, 2007, **40**, 1185.
49. T. Tachibana and H. Kambara, *J. Colloid Interface Sci.*, 1968, **28**, 173.
50. R. Oda, I. Huc, M. Schmutz, S. J. Candau and F. C. MacKintosh, *Nature*, 1999, **399**, 566.
51. A. Brizard, D. Berthier, C. Aime, T. Buffeteau, D. Cavagnat, L. Ducasse, I. Huc and R. Oda, *Chirality*, 2009, **21**, E153.
52. A. Brizard, R. Oda and I. Huc, *Topics Curr. Chem.*, 2005, **256**, 167.

53. D. A. S. Grahame, C. Olauson, R. S. H. Lam, T. Pedersen, F. Borondics, S. Abraham, R. G. Weiss and M. A. Rogers, *Soft Matter*, 2011, 7, 7359.
54. S. Abraham, Y. Lan, R. S. H. Lam, D. A. S. Grahame, J. J. H. Kim, R. G. Weiss and M. A. Rogers, *Langmuir*, 2012.
55. M. A. Rogers, S. Abraham, F. Bodondics and R. G. Weiss, *Cryst. Growth Des.*, 2012, 12, 5497.
56. M. A. Rogers, S. Abraham, F. Bodondics and R. G. Weiss, *Cryst. Growth Des.*, 2012, 12, 5497.
57. S. Wu, J. Gao, T. Emge and M. A. Rogers, *Cryst. Growth Des.*, 2014, Accepted.
58. S. Wu, J. Gao, T. Emge and M. A. Rogers, *Soft Matter*, 2013, 9, 5942.
59. C. A. Elliger, D. G. Guadagni and C. E. Dunlap, *J. Am. Oil Chem. Soc.*, 1972, 49, 536.
60. A. H. Clark and S. B. Ross-Murphy, *Adv. Polymer Sci.*, 1987, 83, 55.
61. M. A. Rogers, A. J. Wright and A. G. Marangoni, *Curr. Opin. Colloid Interface Sci.*, 2009, 14, 223.
62. J. Gao, S. Wu, T. Emge and M. A. Rogers, *Cryst. Engineer. Commun.*, 2013, 15, 4507.
63. S. Bhattacharya and S. N. G. Acharya, *Chem. Mater.*, 1999, 11, 3504.
64. P. V. Vassil, R. W. Malcolm and W. Chi-Huey, *Chem. Commun.*, 1998, 1865.
65. Y. Yan, U. T. Bornscheuer, G. Stadler, S. Lutz-Wahl, R. T. Otto, M. Reuss and R. D. Schmid, *Eur. J. Lipid Sci. Technol.*, 2001, 103, 583.
66. O. Park, D.-Y. Kim and J. S. Dordick, *Biotechnol. Bioengineer.*, 2001, 70, 208.
67. G. John, G. Zhu, J. Li and J. S. Dordick, *Angew. Chem. Int. Ed.*, 2006, 45, 4772.
68. J. Kim, D. H. Altreuter, J. S. Clark and J. S. Dordick, *J. Am. Oil Chem. Soc.*, 1998, 75, 1109.
69. S. Grassi, E. Carretti, L. Dei, C. W. Branham, B. Kahrd and R. G. Weiss, *New J. Chem.*, 2011, 35, 445.
70. Y. Feng, X. Jin and J. N. Hay, *J. Appl. Polymer Sci.*, 1998, 69, 2089.
71. T. L. Smith, D. Masilamani, L. K. Bui, Y. P. Khanna, R. G. Bray, W. B. Hammond, S. Curran, J. J. Belles, Jr. and S. Binder-Castelli, *Macromolecules*, 1994, 27, 3147.
72. M. Tenma, N. Mieda, S. Takamatsu and M. Yamaguchi, *J. Polymer Sci., B: Polymer Phys.*, 2008, 46, 41.
73. Y. Wu, S. Wu, G. Zou and Q. Zhang, *Soft Matter*, 2011, 7, 9177.

74. A. R. Hirst, I. A. Coates, T. R. Boucheteau, J. F. Miravet, B. Escuder, V. Castelletto, I. W. Hamley and D. K. Smith, *J. Am. Chem. Soc.*, 2008, **130**, 9113.
75. M. Raynal and L. Bouteiller, *Chem. Commun.*, 2011, **47**, 8271.
76. K. Fan, L. Niu, J. Li, R. Feng, R. Qu, T. Liu and J. Song, *Soft Matter*, 2013, **9**, 3057.
77. L. Niu, J. Song, J. Li, N. Tao, M. Lu and K. Fan, *Soft Matter*, 2013, **9**, 7780.
78. K. K. Diehn, H. Oh, R. Hashemipour, R. G. Weiss and S. R. Raghavan, *Soft Matter*, 2014, Accepted.
79. J. H. Hildebrand and R. L. Scott, *The Solubility of Nonelectrolytes*, Dover Publications, Reinhold, NY, 3rd edn, 1959.
80. G. Scatchard, *Chem. Rev.*, 1949, **44**.
81. E. A. Grulke, ed., *Solubility Parameter Values*, John Wiley & Sons, New York, 2005, 4th edn.
82. N. Yan, Z. Xu, K. K. Diehn, S. R. Raghavan, Y. Fang and R. G. Weiss, *Langmuir*, 2013, **29**, 793.
83. J. Gao, S. Wu and M. A. Rogers, *J. Mater. Chem.*, 2012, **22**, 12651.
84. K. K. Diehn, H. Oh, R. Hashemipour, R. G. Weiss and S. R. Raghavan, *Soft Matter*, 2014, **10**, 2632.
85. A. Aggeli, M. Bell, N. Boden, J. N. Keen, P. F. Knowles, T. C. B. McLeish, M. Pitkeathly and S. E. Radford, *Nature*, 1997, **386**, 259.
86. R. Afrasiabi and H.-B. Kraatz, *Chem. Eur. J.*, 2013, **19**, 15862.
87. A. Banerjee, G. Palui and A. Banerjee, *Soft Matter*, 2008, **4**, 1430.
88. X. Yan, Y. Cui, Q. He, K. Wang and J. Li, *Chem. Mater.*, 2008, **20**, 1522.
89. G. Bastiat and J.-C. Leroux, *J. Mater. Chem.*, 2009, **19**, 3867.
90. K. T. Kim, C. Park, G. W. M. Vandermeulen, D. A. Rider, C. Kim, M. A. Winnik and I. Manners, *Ange. Chem.*, 2005, **117**, 8178.
91. A. Bot and W. G. M. Agterof, *J. Am. Oil Chem. Soc.*, 2006, **83**, 513.
92. E. Daniel Co and A. G. Marangoni, *J. Am. Oil Chem. Soc.*, 2012, **89**, 749.
93. A. Bot, R. den Adel and E. C. Roijers, *J. Am. Oil Chem. Soc.*, 2008, **85**, 1127.
94. A. Bot, Y. S. J. Veldhuizen, R. den Adel and E. C. Roijers, *Food Hydrocoll.*, 2009, **23**, 1184.
95. M. Pernetti, K. F. van Malssen, E. Floter and A. Bot, *Curr. Opin. Colloid Interface Sci.*, 2007, **12**, 221.
96. F. AlHassawi and M. A. Rogers, *J. Am. Oil Chem. Soc.*, 2013.
97. M. Zinic, F. Vogtle and F. Fages, *Top Curr. Chem.*, 2005, **256**, 39.
98. M. A. Rogers, *CrystEngComm*, 2011, DOI: 10.1039/C1031CE05818E.

99. R. den Adel and P. Heussen, *J. Phys.*, 2010, **247**, 12.
100. M. A. Rogers, A. Bot, R. S. H. Lam, T. Pedersen and T. May, *J. Phys. Chem. A.*, 2010, **114**, 8278.
101. A. Bot, R. den Adel, E. Roijers and C. Regkos, *Food Biophys.*, 2009, **4**, 266.
102. W.-C. Lai and J.-P. Liao, *Mater. Chem. Phys.*, 2013, **139**, 161.
103. A. Bot, R. den Adel, E. Roijers and C. Regkos, *Food Biophys.*, 2009, **4**, 266.

CHAPTER 7

Nanoemulsions

DONGMING TANG*[a] AND KENNETH J. CHOMISTEK[b,c]

[a] Litehouse Inc., 1109 N Ella Ave., Sandpoint, ID, USA, 83864;
[b] BioConvergence LLC, 4320 W Zenith Dr., Bloomington, IN, USA, 47404;
[c] Microfluidics Corp., 90 Glacier Dr., Westwood, MA, USA, 02090
*Email: tangd81@gmail.com

7.1 INTRODUCTION

A nanoemulsion is an emulsion with droplet sizes in the nano-scale. Similar to macroemulsions, all nanoemulsions contain at least three parts: two immiscible liquids (dispersed phase and continuous phase) and emulsifier(s).[1] The dispersed phase of a nanoemulsion is the immiscible liquid that forms the droplets. The continuous phase is the immiscible liquid that surrounds the outside of the droplets. An example of an emulsion used in the food industry is Italian salad dressing (oil and vinegar), with olive oil as the discrete phase. The salad dressing is typically hand agitated to break up the coalescence of the olive oil that has phase separated and risen to the top of the emulsion. Hand agitation breaks the olive oil into smaller droplets; these droplets become dispersed within the vinegar. However, this emulsion is not stable, nor is it a nanoemulsion. Hand agitation does not constitute high energy processing. Most oil and vinegar salad dressings also lack sufficient stabilizers (*e.g.* surfactants, surface modifiers, emulsifiers, co-surfactants) to assist in preventing or delaying coalescence for an extended period of time.

Edible Nanostructures
Edited by Alejandro G Marangoni and David Pink
© The Royal Society of Chemistry 2015
Published by the Royal Society of Chemistry, www.rsc.org

Different droplet size ranges have been proposed by different researchers, from 1 nm to 100 nm,[2] to 500 nm or less.[3] In this chapter, a nanoemulsion is defined as an emulsion with a droplet size of 100 nm or less, with some exceptions of droplet sizes of 200 nm or less. This is because nanoemulsions exhibit unique characteristics, such as optical transparency and long-term stability over long-term storage, only in this droplet size range. In order to produce nanoemulsions, specific preparation methods such as ultra-high pressure (69 MPa or above) homogenization and spontaneous emulsification techniques are necessary.

Nanoemulsions have drawn a lot of attention from the scientific community owing to their unique advantages over traditional macro-emulsions in various industrial applications. For example, because of their small droplet size, nanoemulsions have a large surface area which makes them ideal chemical reaction vessels, or nano-reactors, especially for polymerization reactions.[4] Nanoemulsions show improved oral delivery of poorly soluble drugs, significantly enhance transdermal delivery, and bioaccessibility of the dispersed phase to the digestive system, and thus have found many uses in the pharmaceutical, healthcare, cosmetic and food industries.[5] Certain nano- and microemulsions of edible oils demonstrate antimicrobial and bactericidal effects on enteric and other pathogens and biofilms. This makes them a great candidate for decontamination or sterilizing agents in the food industry.[6] The optical transparency or translucent characteristics of nanoemulsions have been used in the food and nutraceutical industries to deliver omega-3, co-enzyme Q_{10}, and other lipid-soluble nutrients into clear or semi-clear food beverage products.[5,7]

Nanoemulsions can be prepared using two different approaches: top-down and bottom-up. The top-down or high energy method can utilize high pressure, shear, collision, turbulence, and cavitation to create nano-droplets finely dispersed within a continuous liquid phase. The low energy methods rely on the chemistry of the emulsion rather than mechanical forces to assist in creating nano-droplets. Phase transitions and physicochemical properties of surface modifiers, emulsifiers, co-surfactants, and the discrete phase itself are all factors which influence nanoemulsion manufacture using the low energy processing method.[8]

Some researchers argue that nanoemulsions made by the bottom-up approach, such as the phase inversion temperature (PIT) method, are actually microemulsions. Only nanoemulsions made by applying external shear in order to rupture large droplets into smaller ones are considered true nanoemulsions.[9] However, in this chapter, nano-emulsions are emulsions with droplet sizes in the nano-scale and both top-down and bottom-up approaches will be discussed. The bottom-up approach is closely related to microemulsion formation. Many of the principles and concepts used in microemulsion manufacture, including co-emulsifiers and co-solvents, and microemulsion phase diagrams are used to assist in the manufacture of nano-emulsions using the bottom-up approach. These relevant micro-emulsion concepts and mechanisms will be introduced in this chapter.

When compared to other industries, formation of edible nano-emulsions in the food industry is more challenging because of the limited selection of emulsifiers, co-emulsifiers, and co-solvents, and the low price margin of food products. The non-polar nature and the bigger geometrical volume of edible oils, especially saturated long chain triacylglycerides (TAGs), makes them more difficult to encapsulate by either the bottom-up approach or top-down approach.

There has been much debate about the safety of nanomaterials and nanoemulsions and whether, or how, they should be regulated in food applications. European organizations such as the European Union (EU) Directorate and The Institute of Food Science and Technology, a United Kingdom-based professional association, tend to consider nanoparticles as potentially harmful until proven to be safe. They have also recommended that special regulations on usage of nanomaterials in foods need to be established and enforced.[3] The position of the Food and Drug Administration in the United States is that the agency regulates products, not technologies, so that nano-emulsions will not be evaluated as an individual category for food safety. The safety of food containing nanoemulsions will be assessed case by case, which includes the effects of nano-scale particles or droplets.[10]

In this chapter, several aspects of edible nanoemulsions will be introduced. The first section, "The Molecules", introduces chemical and physical properties of emulsifiers and nanoemulsions, including the hydrophilic–lipophilic balance (HLB) concept, spontaneous curvature and critical packaging parameters (CPP), microemulsion phase behavior (for bottom-up or low energy emulsification methods), and an instability mechanism (mainly Ostwald ripening) of nanoemulsions. The second section, "The Sources of Molecules", will talk about the detailed chemical structure and properties of examples of emulsifiers in each group, and their application in edible nano-emulsions. The third section, "Methods of Preparation", will address the two groups of nanoemulsion preparation methods: the bottom-up or low energy emulsification method, and the top-down or high energy emulsification method.

7.2 THE MOLECULES

Similar to macroemulsions, nanoemulsions are thermodynamically unstable. The total free energy to form a nanoemulsion, ΔG, is positive, and $\Delta G = \Delta A \gamma - T\Delta S$, where ΔA is the increase in interfacial area during emulsification, γ is the interfacial tension, and $T\Delta S$ the increase in entropy before and after emulsification. The main role of emulsifiers is to lower the interfacial tension between the dispersed phase and the continuous phase, to assist in the formation of emulsion droplets and the reduction of emulsion droplet size. The most important considerations when selecting emulsifiers are the HLB, spontaneous curvature, the CPP, phase behavior and Ostwald ripening.

7.2.1 The HLB Concept of Emulsifiers

Hydrophilic–lipophilic balance (HLB) is a numerical value which represents the degree of hydrophilic- and lipophilic-like properties of the emulsifiers (Table 7.1). The HLB value of a non-ionic emulsifier is calculated[11] as the molar percentage of hydrophilic groups in

Table 7.1 HLB values of selected food emulsifiers.

Type	Example of emulsifier	HLB value
Non-ionic emulsifiers		
Glycol fatty acid esters	Glycol distearate	1
	Glycol stearate	2.9
Glyceryl fatty acid esters	Glyceryl stearate	3.8
	Glyceryl laurate	5.2
Sorbitants	Sorbitan trioleate	1.8
	Sorbitan oleate	4.3
	Sorbitan laurate	8.6
Polysorbates	Polysorbate 85	11
	Polysorbate 60	14.9
	Polysorbate 20	16.7
Propylene glycol esters	Propylene glycol isostearate	2.5
Anionic emulsifiers	Diacetyl tartaric acid esters (DATEM)	8.3
	Sodium oleate	18
	Sodium lauryl sulfate	40
Zwitterionic emulsifiers	Lecithin	4–9

the surfactant molecule divided by 5. A completely water-soluble hydrophilic molecule has a HLB value of 20, whereas a completely oil-soluble lipophilic molecule has a HLB value of 0. For anionic emulsifiers, the above method is not applicable and their HLB value must be determined by experimental methods; values can be as high as 40 for sodium lauryl sulfate.[12] According to Bancroft's rule, the emulsion phase in which the emulsifier is most soluble becomes the continuous phase. Emulsifiers with an HLB of 4 to 6 form water-in-oil (W/O) emulsions, emulsifiers with HLB values of 8 to 18 form oil-in-water (O/W) emulsions, and emulsifiers with HLB values of 7 to 9 are used as wetting agents. Owing to the fact that most of edible nano-emulsions are of the O/W type, the emulsifiers used are mainly high-HLB value emulsifiers, with the exception of lecithin, which has a HLB value around 7, yet can still promote both W/O and O/W emulsion formation.

The HLB values of a mixture of emulsifiers can be calculated from the HLB values of each individual emulsifier and their weight percentage,[11] as in equation (7.1):

$$\mathrm{HLB_{mix}} = \frac{(W_A \mathrm{HLB_A} + W_B \mathrm{HLB_B})}{(W_A + W_B)} \tag{7.1}$$

where $\mathrm{HLB_{mix}}$ is the HLB value of a mixture of emulsifiers A and B, $\mathrm{HLB_A}$ and $\mathrm{HLB_B}$ are the HLB values of emulsifiers A and B separately, and W_A and W_B are the weights of emulsifier A and B separately.

7.2.2 Critical Packing Parameter (CPP) and Spontaneous Emulsification

The CPP is a geometric value expressed as the ratio between the hydrocarbon chain volume, V, and the product of length, L, and the interfacial area occupied by the head group, A,[13]

$$\mathrm{CPP} = \frac{V}{LA} \tag{7.2}$$

When $1/2 < \mathrm{CPP} < 1$, emulsifiers form spherical bilayers; when CPP is approximately 1, emulsifiers form planar bilayers; when $\mathrm{CPP} > 1$, inverted micelles are produced. Table 7.2 shows the CPPs, critical packing shapes, and the structures formed by some emulsifiers used in the detergent industry.[14]

Table 7.2 CPP of selected emulsifiers and their preferred self-assembled structures. (Reprinted with permission from ref. 14)

Lipids	Critical packing paramenter	Critical packing shape	Strurctures formed
Single-chained lipids (surfactants) with large headgroup areas: • SDS[a] in low salt	<1/3	Cone	Spherical micelles
Single-chained lipids with small headgroup areas: • SDS and CTAB[b] in high salt • Non-ionic lipids	1/3–1/2	Truncated cone	Cylindrical micelles
Double-chained lipids with large headgroup areas, fluid chains: • Phosphatidyl choline (lecithin) • Phosphatidyl glycerol • Phophatidyl inositol • Phophatidic acid • Sphingomyelin, DGDG[c] • Dihexadecyl phosphate • Dialkyl dimethyl ammonium • Salts	1/2–1	Truncated cone	Flexible bilayers, vesicles
Double-chained lipids with small headgroup areas, anionic lipids in high salt, saturated frozen chains: • Phosphatidyl ethanaiamine • Phosphatidyl serine + Ca^{2+}	~1	Cylinder	Planar bilayers

Table 7.2 (*Continued*)

Lipids	Critical packing paramenter	Critical packing shape	Strurctures formed
Double-chained lipids with small headgroup areas, non-ionic lipids, poly(*cis*)unsaturated chains, high *T*: • Unsaturated phosphatidyl ethanolamine • Cardiolipin + Ca^{2+} • Phosphatidic acid + Ca^{2+} • Cholesterol, MGDG^d	>1	Inverted truncated cone or wedge	Inverted micelles

^{a}Sodium dodecyl sulfate.
^{b}Cetyltrimethylammonium bromide.
^{c}Digalactosyl diglyceride, diglucosyldiglyceride.
^{d}monogalactosyl diglyceride, monoglucosyl diglyceride.

7.2.3 Microemulsion Phase Behavior

A typical so-called Kahlweit fish diagram (Figure 7.1) is commonly used to describe the phase behavior of microemulsions. It demonstrates how the structure of a non-ionic emulsifier in a microemulsion changes when compared to varying emulsifier concentrations and temperatures at constant water to oil ratios.[15]

At low emulsifier concentrations, with increased temperature, the interaction between the hydrophilic group of the emulsifiers and water is decreased, and thus the preferred curvature of the emulsifiers is decreased. As a consequence, the HLB value of the emulsifiers is decreased and the microemulsion formed changes from an O/W microemulsion with lower excess of water (Winsor phase II), to a middle bi-continuous microemulsion with an upper excess of oil and a lower excess of water (Winsor phase III), then to a lower W/O microemulsion with excess of oil (Winsor phase I). When the concentration of emulsifier is increased to a value higher than the starting point of the Kahlweit fish tail, a homogeneous single phase O/W or W/O microemulsion is formed. A further increase in emulsifier concentration will form a liquid crystal phase of emulsifiers.

The phase with water as the continuous phase is desired for edible nanoemulsions, but because nanoemulsions normally use

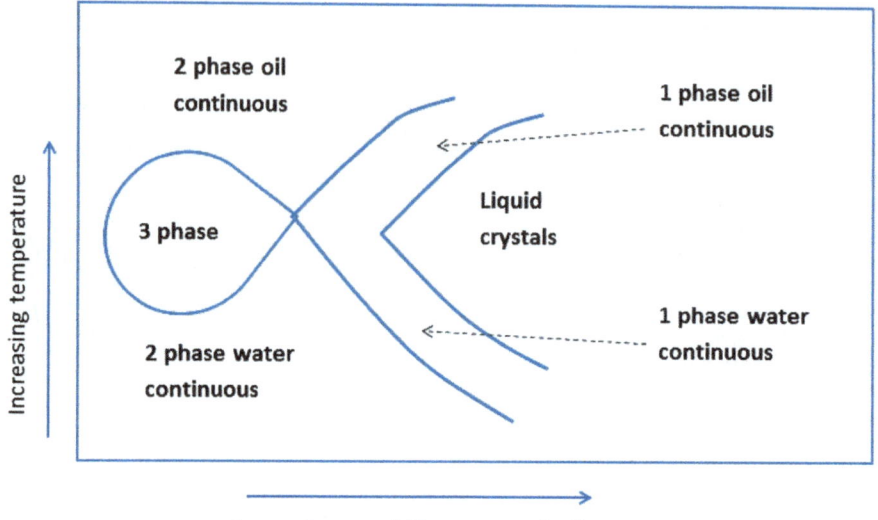

Figure 7.1   Kahlweit fish microemulsion phase diagram.

much less emulsifier than microemulsion, many edible nanoemulsions, especially the edible nanoemulsions formed by the bottom-up approach such as the PIT method, start at Winsor I phase. By adjusting environmental conditions, such as increasing temperature, using co-solvent(s), or simply applying high energy shear, a homogeneous one phase edible nanoemulsion can be formed. Certain bottom-up approaches such as the emulsion inversion point method[16] will require starting from the bi-continuous phase of microemulsion.[17]

7.2.4   Stability of Nanoemulsions and Ostwald Ripening

Conventional emulsions can destabilize *via* several different mechanisms, which include creaming, aggregation, coalescence, flocculation, partial coalescence, and Ostwald ripening.[18] The very small droplet size in nanoemulsions prevents creaming, sedimentation, or flocculation from occurring. The main mechanism of nanoemulsion instability is thus Ostwald Ripening.[19]

Ostwald ripening is the phenomenon where larger oil droplets grow over time at the expense of smaller droplets, which is caused by the difference in Laplace pressure, γ/r, between small and large droplets. Ostwald ripening can be quantitatively assessed following

Lifshitz–Slesov–Wagner (LSW) theory[20] by plotting the cube of the radius *vs.* time[17] [equation (7.3)]:

$$\omega = \frac{\mathrm{d}r^3}{\mathrm{d}t} = \frac{8}{9}\left[\frac{c_\infty \gamma V_m D}{\rho R T}\right]$$ (7.3)

Where ω is the rate of Ostwald ripening, r is the droplet size, t is time, c_∞ is the bulk phase solubility, γ is the interfacial tension, V_m is the molar volume of the oil, D is the diffusion coefficient of the oil in the continuous phase, ρ is the density of the oil, R is the gas constant, and T is the absolute temperature. The rate of Ostwald ripening increases by increasing interfacial tension between the dispersed phase and continuous phase and decreases with increases in droplet size. For example, in a fatty acid methyl ester (methyl decanoate)-in-water nanoemulsion,[17] the slope of the r^3 of the nanoemulsion as a function of time t was found to decrease with decreasing oil to emulsifier ratio from 0.8 to 1.2.

A common method to reduce Ostwald ripening is to add a secondary dispersed phase component, which has a lower solubility or is insoluble in the continuous phase, such as adding an oil of lower polarity (*e.g.* a medium or long chain TAG) into essential oils. Modification of the interfacial film (including reducing interfacial tension) by increasing the concentration of emulsifiers below their critical micelle concentration (CMC) and deposition of a layer of emulsifiers or polymers at the interface (which do not desorb during ripening) can significantly reduce Ostwald ripening as well.[19]

7.3 SOURCES OF EMULSIFIERS FOR EDIBLE NANOEMULSIONS

Emulsifiers are amphipathic molecules that consist of a non-polar hydrophobic portion and a polar or ionic hydrophilic portion. Based on the nature of the hydrophilic group, emulsifiers can be divided into four classes: anionic, cationic, amphoteric, and non-ionic emulsifiers. Polymeric emulsifiers are usually considered separately from the above groups. In edible nanoemulsions, the most important group of emulsifiers are non-ionic, amphoteric (Zwitterionic), and food or generally recognized as safe (GRAS) grade polymeric emulsifiers.

The non-ionic emulsifiers most commonly used for edible nanoemulsions are high HLB value emulsifiers, which may include polyoxyethylene (20) sorbitan esters or Tweens, sucrose esters, and polyoxyethylene glycol alkyl ethers (Brij). Sorbitan fatty acid esters, or

Spans, and monoglycerides have a low HLB value. Sometimes they can be used together with high HLB value emulsifiers in order to obtain the desired HLB value of the emulsifier mixtures, especially nanoemulsions formed by the bottom-up approach through spontaneous emulsification and self-assembly of non-ionic emulsifiers.

7.3.1 Polyoxyethylene (20) Sorbitan Esters (Polysorbates or Tweens)

Polysorbates are oily liquids derived from PEG-ylated sorbitan esterified with fatty acids. The number following the polysorbate part is related to the type of fatty acid associated with the polyoxyethylene sorbitan part of the molecule. The chemical structure of one of the polysorbates, Polysorbate 80, is shown in Figure 7.2.[21]

Polysorbates are water soluble and have high HLB values, from 10.5 for Polysorbate 65 to 16.7 for Polysorbate 20. At low concentrations, interfacial tension between the dispersed phase and continuous phase is decreased by increasing polysorbate concentration. Once the solution is saturated with emulsifiers, additional polysorbate will form micelles with very little effect on interfacial tension. The polysorbate concentration above which micelles are formed is the critical micelle concentration (CMC). The size of polysorbate micelles ranges from 8.5 nm for Polysorbate 20 micelles to 10.7 nm for Polysorbate 80 micelles.[22] Polysorbates and other non-ionic emulsifier micelles are often used to solubilize lipophilic components and form O/W nanoemulsions. The hydrophilic–lipophilic nature of polysorbates and other non-ionic emulsifiers strongly depends on temperature.

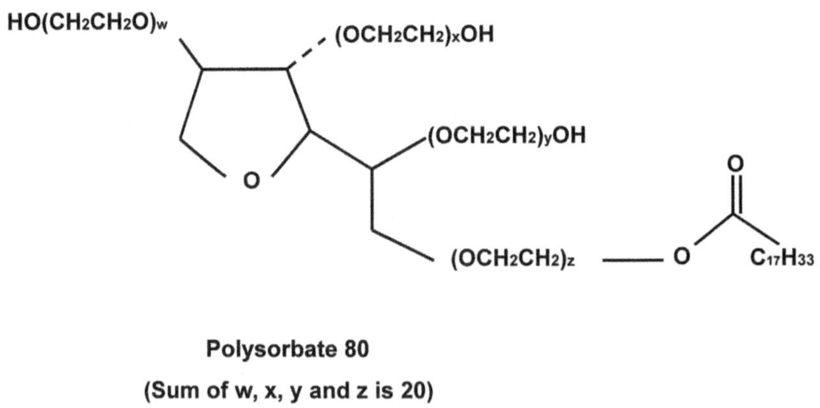

Polysorbate 80

(Sum of w, x, y and z is 20)

Figure 7.2 Chemical structure of Polysorbate 80.
(Reprinted with permission from ref. 21)

As the temperature increases, the polyoxyethylene chain is dehydrated and polysorbates become more lipophilic, so their HLB value is decreased.

Polysorbates have been used to form edible nanoemulsions by both top-down and bottom-up approaches. The US patent application 20130064954 disclosed a food flavor oil-in-water nanoemulsion consisting of polyoxyethylene sorbitan monolaurate (Polysorbate 20), sugar alcohol, and water. Nanoemulsions with mean droplet size of 60 nm to 80 nm were obtained by spontaneous emulsification of polysorbates, which were diluted to obtain nanoemulsions with slightly larger droplet sizes (80 nm to 100 nm) through phase inversion.[23] However, a large amount of emulsifier has to be used (5 : 1 ratio or greater of polysorbate to flavor oil), which may limit its usage in food applications. In another study, a rice bran oil nanoemulsion using sorbitan monooleate (Span 80) and PEG 30 [polyethylene glycol (30) hydrogenated castor oil] was produced using the PIT method.[24] The lowest emulsifier to oil ratio of 1 : 1 was obtained with a mean droplet size of 80 nm. Considering the nature of rice bran oil TAGs compared to the much smaller geometric volume of essential oils, the 1 : 1 ratio of emulsifier to oil is impressive. In addition to using the low energy bottom-up approach, polysorbates are often used to make nanoemulsions using a high energy shearing device through a top-down approach. One example is a silicone O/W nanoemulsion prepared using Microfluidizers™ (Westwood, MA) at 50–150 MPa.[25] An oil droplet size of about 120 nm was obtained after 1–5 passes, depending on the homogenization pressure.[25]

7.3.2 Sucrose Esters

The second group of commonly used non-ionic emulsifiers is sucrose esters. Sucrose esters are non-ionic compounds synthesized by esterification of fatty acids or natural glycerides with sucrose. Sucrose esters can be mono-, di-, and tri-esters of fatty acids linked with sucrose molecules on one, two, or all of the three primary hydroxyls (C6, C1', C6'). Figure 7.3 shows the chemical structure of one of the sucrose esters, sucrose monopalmitate.[26]

Depending on the degree of esterification and ester substitution, sucrose esters can have a wide range of HLB values from 18.5 for sucrose monopalmitate to 1 for sucrose polystearate. Depending on their HLB value and purity, sucrose esters are partially soluble in water and polar solvents such as glycol and propylene glycol, but are still soluble in ethanol.

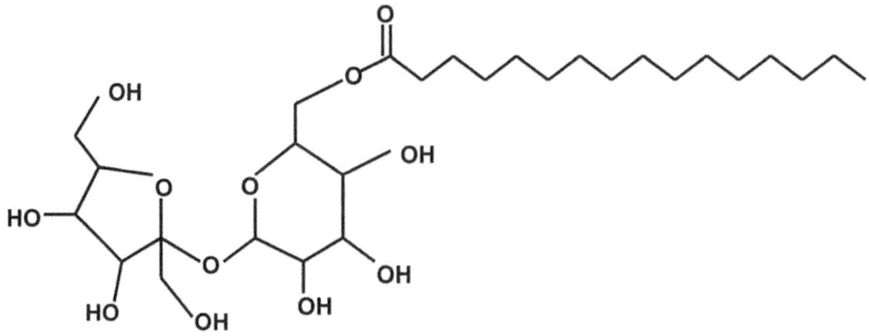

Figure 7.3 Chemical structure of sucrose monopalmitate.
(Reprinted with permission from ref. 26)

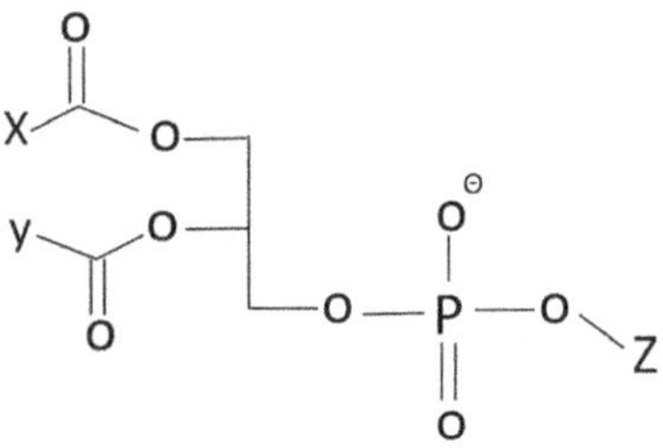

Figure 7.4 Chemical structure of phospholipids.
(Reprinted with permission from ref. 29)

As food-grade non-ionic emulsifiers, sucrose esters have been widely used in nano- and microemulsion preparation. Nanoparticles of astaxanthin with a particle size from 73 nm to 144 nm were produced by high-pressure homogenization combined with the solvent evaporation method, using sucrose laurate, sucrose palmitate, and sucrose stearate.[27] In another study, sucrose laurate, sucrose palmitate, and sucrose oleate were used to form olive oil in water nanoemulsions with the help of glycerol. Sucrose esters were dissolved in glycerol using heat. Oil was then added slowly to the sucrose/glycerol mixture, while stirring with a glass rod. A transparent nanoemulsion with droplet size of 129 nm was formed at a ratio of 1:2 sucrose laureate:olive oil.[28]

7.3.3 Lecithin

Lecithin is normally a mixture of phospholipids, which has a basic structure as shown in Figure 7.4.[29]

Either one, or both, X and Y are fatty acids. Individual species of glycerophospholipids can be distinguished by the structure of the alcohol "Z", which is esterified to the phosphate. The three most common phospholipids species in food lecithin are PC (phosphatidylcholine), PE (phosphatidylethanolamine), and PI (phosphatidylinositol), with headgroups of choline, ethanolamine, and inositol, respectively. Lecithin is known to be able to form single and multilayer liposomes, reverse micelles, and other ordered structures, which can be used to encapsulate both lipophilic and hydrophilic components. In edible nanoemulsions, lecithin is normally used as a co-emulsifier. In the formation of flaxseed oil and medium-chain triglyceride (MCT) nanoemulsions using an ultrasonication method,[30,31] lecithin was used together with other emulsifiers such as polysorbates and sorbitants and/or with solvents. In patent WO 2007/026271A1, a microemulsion of flavor oil was formed using a mixture of high HLB value sucrose esters and soy lecithin.[32] The soy lecithin is presented at 0.5% to 5% by weight in the microemulsion and is used to convert the milky solution into a transparent microemulsion. In another study, a grape marc extract in a sunflower oil in water nanoemulsion was formed using soy lecithin and high pressure homogenization.[33] Lecithin can be used as an oil-soluble emulsifier to form O/W emulsions together with water-soluble emulsifiers. In a palm oil nanoemulsion, lecithin was used together with the water-soluble emulsifiers such as polyethylene glycol 400 and Cremophor EL. A droplet size of 100 nm was obtained using a high shear mixer and high pressure homogenization at 80 MPa.[34] In a study on hydrocarbon microemulsions using lecithins (1, 2-dialkanoyl-*sn*-glycero-3-phosphocholine) with different carbon chain lengths, the effects of lecithin hydrocarbon chain length and the effects of alcohols as co-solvents were investigated.[35] It was found that, for preparing microemulsions with short chain lecithins, one has to add butanol as a co-solvent. For long chain lecithins, propanol is added as the co-solvent. The effects of salts on the phase behavior of different lecithin–alcohol systems are also different. Although the formulation is not edible, the effects of lecithin carbon chain length, alcohol chain length, and salts are of interest for edible lecithin nanoemulsions, as well as non-edible nanoemulsions.

7.3.4 Food- or GRAS-Grade Polymeric Emulsifiers

Food-grade polymeric emulsifiers include gum arabic, modified starches, especially starch sodium octenyl succinate (OSA-starch), food proteins and peptides, and others.

Normally, food polymeric emulsifiers can only produce nano-emulsions (<100 nm) for flavor oils and MCT,[36] owing to their small geometric volume. High energy emulsification methods such as microfluidization or high pressure valve homogenization are often applied in order to produce long chain TAG nanoemulsions.

Many studies have been carried out using microfluidization and ultrasonication for various lipophilic materials such as essential oils,[37] β-carotene,[38] and vitamin E.[39] Whey protein concentrate (WPI) and modified starch (HI-Cap®) were employed to make *d*-limonene nanoemulsions using microfluidization and ultrasonication. Both microfludization and ultrasonication were capable of producing nanoemulsions with a median droplet diameter range of 150–700 nm.[37] MCT-in-water nanoemulsions (5% w/w) were made with OSA modified starch. The smallest mean droplet diameter obtained was 169 nm at 22 000 psi after three passes through a Microfluidizer. The OSA modified starch showed superior emulsification capacity and gave smaller droplet sizes when compared to gum arabic.[36]

7.3.5 Recent Advances in Nanoemulsion Emulsifiers

In addition to conventional emulsifiers such as non-ionic emulsifiers, lecithin, and modified starches, some new emulsifiers have been added recently for their unique characteristics and properties, such as *Quillaja* saponin, and some GRAS pharmaceutical grade emulsifiers like d-α-tocopheryl polyethylene glycol succinate (vitamin E TPGS).[7]

Quillaja saponins are natural and are extracted from the *Quillaja saponaria* Molina tree (native to Chile). It is a small molecular weight emulsifier capable of forming micelles and stabilizing O/W emulsions. The molecular structure of *Quillaja* saponin is shown in Figure 7.5.[40]

During the preparation of an 8% to 50% flavor O/W emulsion, Q-Naturale® (a quillaja extract in a liquid form containing ∼14% saponin on average), shows a superior emulsification capability to gum arabic and starch sodium octenyl succinate (OSA-starch), at a typical oil to Q-Naturale® ratio of 4 : 1 to achieve about 250–300 nm droplet size, or 1.5 : 1 to achieve droplet size of 100–150 nm.[41] In another study, on 10% MCT O/W nanoemulsions, the emulsification capability of *Quillaja* saponin was compared with that of Tween 80.[38] Both emulsifiers gave a similar interfacial tension and were able to produce nanoemulsions with droplet sizes <200 nm, but Tween 80 gave smaller droplet size (<150 nm) than *Quillaja* saponin.

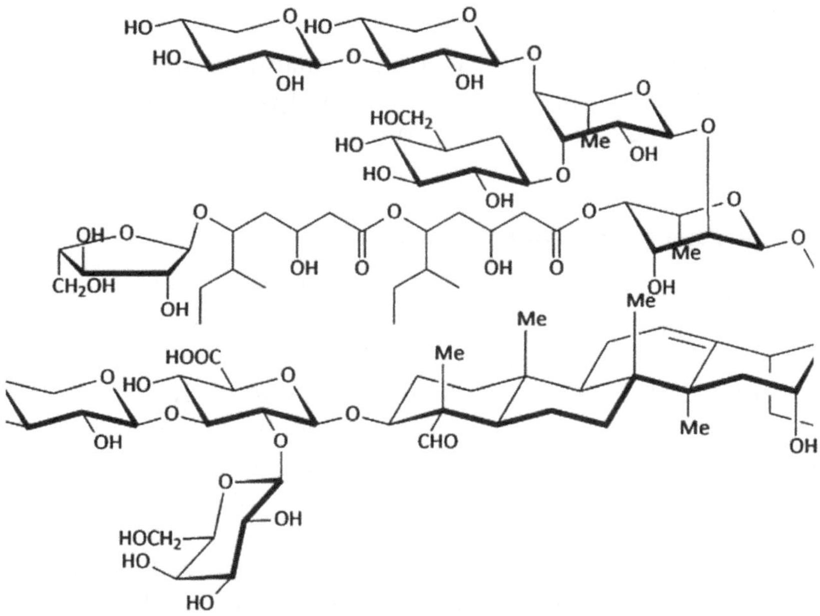

Figure 7.5 Chemical structure of *Quillaja* saponin. (Adapted with permission from ref. 40)

Vitamin E TPGS is prepared by esterification of the acid group of crystalline d-α-tocopheryl acid succinate with polyethylene glycol 1000 and can be used as an emulsifier and drug solubilizer. An omega-3 fish O/W nanoemulsion was made using vitamin E TPGS with a droplet size of 72 nm at surfactant to oil ratio (SOR) as low as 0.6 with high pressure homogenization at 34 MPa for up to 10 passes.[40] In addition to *Quillaja* saponin and vitamin E TPGS, other food or pharmaceutical-grade emulsifiers such as polyethylene lauryl ether (Brij 30), polyoxyethylene (10) oleyl ether (Brij 97), and polyoxyl 40 hydrogenated castor oil (Cremophor® RH 40)[16,42] have been used to make edible nanoemulsions.

Another interesting area is the formation of self-assembled nano-complexes between lipophilic materials and food polymeric emulsifiers. Studies have shown that there is interaction among amylose, protein and free fatty acids (FFA), resulting in the formation of a soluble high molecular weight nano-complex with a diameter of 20–70 nm. The important characteristics of this nano-complex are its solubility, its ability to carry valuable lipid-based compounds such as conjugated linoleic acid (CLA), and its slow starch digestion rate in beverage-based food products.[43]

7.4 METHODS OF PREPARATION

There are two main methods utilized in manufacturing nanoemulsions: the top-down approach or high energy processing, and the bottom-up approach or low energy processing.

7.4.1 Low Energy Emulsification Methods

7.4.1.1 Phase Inversion Temperature (PIT) Method. The concept of PIT was developed by Shinoda, who found that many O/W emulsions made with non-ionic emulsifiers undergo a process of phase inversion from O/W to W/O at a critical temperature (PIT).[44,45] At the PIT, extremely low interfacial tensions can be achieved and thus emulsion droplets in the nano-scale range can be formed. However, at temperatures close to the PIT, the coalescence rate is high. By rapid cooling or heating the emulsion to a temperature at least 25–30 °C from its PIT,[46] relatively stable nanoemulsions can be formed. Emulsion droplet size and stability are dependent on the PIT of the whole emulsion system rather than the PIT of the emulsifier.[14] In the investigation of the mechanism of nanoemulsion formation by the PIT method,[47–50] the droplet size and polydispesity index were decreased by increasing surfactant concentration. The smallest droplet sizes were obtained when the starting emulsion was at a bi-continuous phase or at two-phase region, having both bi-continuous and liquid crystalline (Lα) phases. It shows that, in addition to changing the temperature of the system, starting from a microemulsion in the bi-continuous phase is the mechanism used to obtain a small droplet size by the PIT method. When the nanoemulsion system is cooled during the PIT method, the hydration of the polar group of the non-ionic emulsifiers is dramatically increased, and thus the bi-continuous microemulsion structure is disrupted and nano-scale droplets are formed through change in the curvature of the emulsifier monolayer.[46]

The PIT method has been used to prepare soybean O/W nano-emulsions. Ratios in the range 1 : 2 to 3 : 1 Brij 97 (polyoxyethylene-10-oleyl ether) to soybean oil were used.[51] The PIT of the soya oil–Brij 97–water system with different oil and emulsifier content was found to be between 74 °C and 88 °C. Most transparent nanoemulsions were formed under rapid cooling at 5–16 °C min^{-1}. At emulsifier concentrations of 7.5%, rapid cooling produced nanoemulsions with a droplet size of about 30 nm, while slow cooling at 2 °C min^{-1} produced much larger droplet sizes of approximately 120 nm. It must be emphasized that Brij 97 is not a food-grade emulsifier, but has been used in some pharmaceutical microemulsions.[52]

Polysorbates (Tweens) are non-ionic emulsifiers commonly used in edible nanoemulsions. A flavor O/W (lemon oil) nanoemulsion with a droplet size <100 nm was created, similar to the PIT method.[53] The mixture of lemon oil, Tween 80, and water was heated to 90 °C and held for 30 min. Depending on the surfactant to oil ratio (SOR), a macroemulsion (SOR < 1), nanoemulsion (1 < SOR < 2) or microemulsion (SOR > 2) can be created.

7.4.1.2 Spontaneous Emulsification or Emulsion Inversion Point Method. In contrast to the PIT method, the spontaneous emulsification method is carried out at a constant temperature. A mixture of water, oil, lipophilic and/or hydrophilic emulsifiers and a water-miscible co-surfactant is mixed to form an O/W microemulsion. When water is continuously added to this O/W microemulsion, it leads to changes in the curvature of the emulsifier film. At the emulsion inversion point, where an O/W emulsion is about to change into a W/O emulsion, the microemulsion is in its bi-continuous phase and a minimum interfacial tension is achieved, which results in the formation of nano-scale emulsion droplets. This spontaneous emulsification is also called the emulsion inversion method,[16] or the catastrophic phase inversion (CPI) method.[54] In addition to surfactant structure and concentration, the interfacial tension between the dispersed phase and the continuous phase, the interfacial and bulk viscosity, and the starting microemulsion phase all have effects on the spontaneity of the emulsification process.[55]

In non-edible nanoemulsions, addition of co-emulsifiers such as lower alcohols (butanol, pentanol and hexanol) to milky emulsions can lower the interfacial tension and produce a transparent microemulsion,[56] from which a nanoemulsion can be made.

In the preparation of Acetem (acetic acid esters of mono- and diglycerides) in water nanoemulsions, droplets with diameters of 100–200 nm were formed following the standard CPI procedure. The emulsion obtained was further treated by direct homogenization. The process time to reach 100 nm droplet size was reduced 12-fold compared to starting from direct emulsification.[54] Compared to the high-energy emulsification method, spontaneous emulsification often requires higher SOR to produce droplets at the nano-scale. During fabrication of ultrafine edible emulsions of MCT–Tween 80 or Tween 85–water systems, the spontaneous emulsification method requires SOR > 0.5, while the high-energy method (microfluidization method) requires much less emulsifier (SOR < 0.1).[57]

In spontaneous emulsification, the composition of the co-solvents plays an important role in determining emulsification efficiency and droplet size. In a vitamin E-enriched nanoemulsion made by spontaneous emulsification, the smallest droplet size was obtained when either 30% propylene glycol or 20% ethanol was present in the aqueous phase. However, the resulting nanoemulsion is unstable towards droplet growth during storage, especially at elevated temperatures.[58] These co-solvents can increase the critical micelle concentration (CMC) of the emulsifiers and penetrate into the emulsifier interface to minimize the liquid crystal phase in the microemulsion phase diagram, and thus enlarge the bi-continuous phase, which is the ideal starting phase to make nanoemulsions by many methods, including both low energy and high energy techniques.[59]

7.4.2 High Energy Processing

Within the high energy processing method, there are two main subgroups, high pressure homogenization and ultrasonication, for the production of nanoemulsions.[8] Both methods use high energy densities to create sub-micron mixing pockets, called eddies. These eddies can promote intense mixing at the nanometer scale and expose the discrete phase of the emulsion to the stabilizer(s) (*e.g.*, surfactant, co-solvent) within the formulation.[60,61] Using the example of high pressure homogenization, a decrease in the cross-sectional area of the flow path and/or an increase in the processing velocity can cause the eddy size to decrease. As the intensity of the mixing increases, the theoretical size of the eddies decreases through molecular diffusivity and thus smaller nano-droplets can be formed.[61,62] Owing to the intense level of mixing, the molecular diffusivity of the added stabilizers in the formulation is enabled. The stabilizers coat the nanodroplets and create a kinetically stable nanoemulsion.[61] A solvent evaporation method is a modification of the high energy homogenization method. In a solvent evaporation process, the dispersed phase material can be dissolved in food or pharmaceutical grade solvents to form a hydrophobic phase which is dispersed in the continuous phase and emulsified using either high pressure homogenization or ultrasonication to make nanoemulsions. A detailed process of solvent evaporation will be given in the discussion below. The ultrasonication method is mainly used on the laboratory scale, such that the focus will be on high pressure homogenization in the following discussions.

The most common types of high pressure homogenizers to create edible nanoemulsions are Microfluidizers and valve homogenizers. Both processing technologies begin by using high pressure (as high as 45 000 psi) to accelerate a pre-emulsion (coarse emulsion) through an orifice. The pre-emulsion can experience the forces of shear, collision, turbulence, and cavitation during processing. If the desired characteristic goal(s) (*e.g.* droplet size or stability) are not achieved after one processing cycle, the formulation can be processed for additional cycles until the goals are achieved.

7.4.2.1 Microfluidizer Processors. The distinguishing feature between Microfluidizer processors and valve homogenizers is the interaction chamber (IXC). Inside an IXC is a fixed geometry microchannel design that does not move or oscillate during processing. There are two different geometries of IXC available: Z-type IXC (ZIXC) are generally used to reduce the size of solid particles dispersed in a liquid media and to produce cell disruption and reverse emulsions, and Y-type IXC (YIXC), which are generally used to process nanoemulsions and liposomes, and to produce polymer encapsulation.[61,63]

When the pre-emulsion enters the YIXC, it is forced to split into two separate streams and these streams subsequently collide with each other under high energy conditions inside a microliter volume, utilizing impinging jet technology.[60] The micro-channels' depth and width range from 75 to 200 µm, and the fluid velocities may reach 500 m s^{-1}. Owing to the small cross-sectional area and high velocity, the fluid flow inside the IXC is intensely turbulent; typical values for Reynolds numbers are >20 000.[64,65] The unique geometry and high Reynolds number promotes uniform mixing at the nanometer scale inside the IXC with uniform processing conditions. These conditions allow for greater droplet size reduction and more uniform and monodispersed droplet size distributions. The technology of impinging jet mixing has been successfully used in the past in the food industry and in other applications.[64,65]

In many instances, the impinging jet YIXC can be used in series with an auxiliary processing module (APM) to create nanoemulsions (Figure 7.6). An APM is placed downstream of the YIXC to act as an intermediate pressure relief between the high processing pressures and the atmosphere. An APM can also be added downstream of a ZIXC to create reverse emulsions or upstream of a ZIXC to pre-process suspensions.

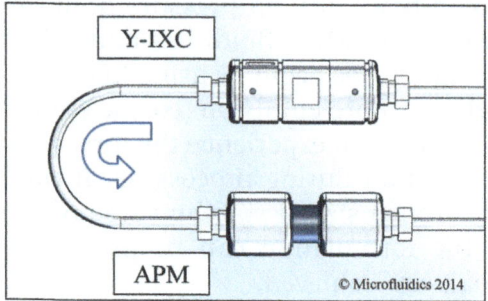

Figure 7.6 Microfluidizer chamber configurations used to create nanoemulsions. The arrow shows the flow profile of the pre-emulsion as it travels through the YIXC and then the APM. The role of the APM is to stabilize the fluid flow by acting as an intermediate pressure relief between the high pressure YIXC and the atmosphere.

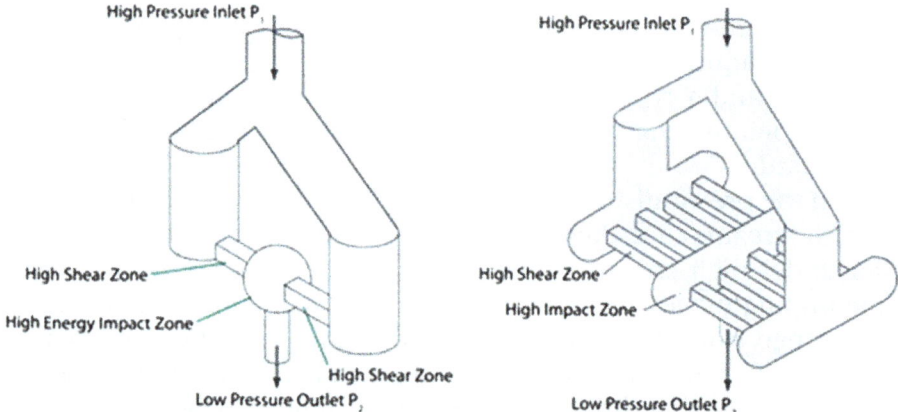

Figure 7.7 Schematic of the impinging jet design (YIXC) of the interaction chamber. On the left is a single slot micro-channel used for laboratory and pilot scale, and on the right is an example of a multi-slotted design for production scale Microfluidizer processors.

Single slotted IXCs (Figure 7.7) have a single micro-channel and are used for laboratory-scale testing/processing. Multi-slotted IXCs are composed of parallel multiple micro-channels in order to increase the volumetric flow rate through the IXC while maintaining the levels of shear, impact, and energy dissipation. The increase in flow rate allows for larger volumes of product to be continuously processed in an efficient amount of time.[63]

The Microfluidizer processor (Figure 7.8) is composed of an inlet reservoir, intensifier piston, YIXC, APM, and heat exchanger. The high

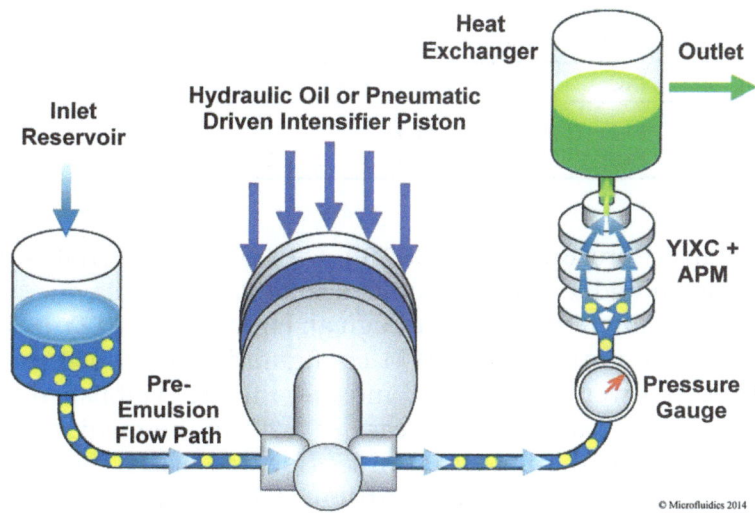

Figure 7.8 A drawing of a Microfluidizer processor and its critical components. The basic drawing is applicable to all laboratory and industrial processors.

velocities through the YIXC are created by applying high pressure to the pre-emulsion. Pressures of 20 000 psi (138 MPa) or more are required to generate velocities necessary to create nanoemulsions.[8] The processing pressure is created using hydraulic oil or pneumatic driven pressure multiplier/accumulator, referred to as an intensifier piston. After the pre-emulsion has travelled through the YIXC and APM (if appropriate), it may pass through a heat exchanger for excess heat removal. If necessary, the emulsion can be passed through the Microfluidizer processor for additional passes.

An example of a nanoemulsion processed on a Microfluidizer processor is a flavor O/W nanoemulsion to be used as an additive in food and beverages. The proprietary formulation was an O/W emulsion that was stabilized by natural surfactants. The processing goal was to produce a transparent and stable nanoemulsion that would not degrade, separate, or alter the color of the product to which it was added. Droplet size characterization was by mean diameter using static laser light scattering (SLS). A Horiba LA-910 (Horiba Instruments Inc., Irvine, CA) was used with measuring capabilities of 20–1000 nm. A laboratory-scale Microfluidizer processor (M-110EH-30) was used to create a median (d_{50}) droplet size of 84 nm and a d_{90} droplet size of 110 nm after two cycles through the F12Y (75 μm)–H30Z (200 μm) IXC and APM configuration at 30 000 psi (Table 7.3). The same formulation was scaled up in volume with an industrial

Table 7.3 Processing conditions for Microfluidizers used to create flavor oil nanoemulsions and the resulting droplet size measurements.

Processor	IXC	Pressure (psi)	Passes	d_{50} (nm)	d_{90} (nm)
M-110EH	F12Y–H30Z	30 000	2	84	110
M-7250-30 CP	F12Y-8–T50Z	30 000	2	82	108

Microfluidizer processor (M-7250-30 CP). The industrial volume nanoemulsion had a median droplet size of 82 nm and a d_{90} droplet size of 108 nm after two cycles through an eight-slotted F12Y-8 (75 μm × 8)–T50Z (550 μm) IXC and APM configuration at 30 000 psi. The nanoemulsions were transparent and met the pre-established stability criteria. There was also a decrease of >2% in the median and d_{90} droplet size when the industrial processor was used. The improvement in the droplet size can be attributed to two factors: adding a second intensifier piston and processing the pre-emulsion with "constant pressure" using the industrial processor.[7]

Nanoemulsions created using a single intensifier piston laboratory processor can be scaled up in volume using an industrial Microfluidizer processor. As previously described, the YIXC can be scaled up by adding parallel multiple slots to provide the same high energy processing conditions at higher volumetric flow rates. The motor on the industrial Microfluidizer can also be scaled up from a laboratory 1.5–5 hp motor to a 25–100 hp motor. Also, a second intensifier piston can be added to nearly double the volumetric flow rate when compared to a single intensifier piston processor. When operating under "constant pressure" with a dual intensifier piston industrial processor, the two pistons are synchronized using proprietary electronic software that senses and times the position of the intensifiers. Sensors force one piston to draw the pre-emulsion into the industrial Microfluidizer processor, while the second piston applies high pressure to process the sample that was previously drawn in. Then the two pistons alternate roles until the entire batch is processed. The timing of the two intensifier pistons creates a pressure profile that is "constant" at the desired processing pressure (Figure 7.9).

The constant pressure profile can assist in reducing droplet size more than a laboratory processor because it almost eliminates any of the pre-emulsion from being processed at suboptimal pressures. As previously stated, the laboratory Microfluidizer processors have only one intensifier piston and create a synchronous pressure profile (Figure 7.9).

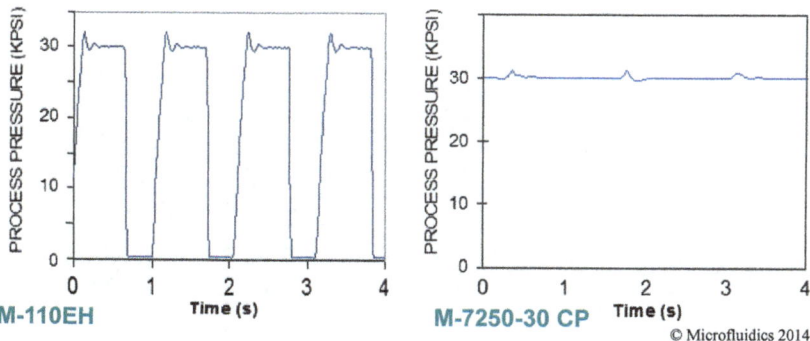

Figure 7.9 The graph on the left is a synchronous pressure profile from a single intensifier piston laboratory Microfluidizer processor. The graph on the right is a constant pressure profile from a dual intensifier piston industrial Microfluidizer processor.

7.4.2.2 Valve Homogenizers. High pressure valve homogenizers consist of a high pressure generator and a homogenizing valve assembly. A mixture of immiscible liquids is accelerated rapidly when it enters the gap between the valve and the seat, and a significant pressure drop is created due to this sudden increase in velocity. It causes turbulent flow and localized pressure differences. Depending on the geometry of the valves and the mode of operation, turbulence, laminar shear or cavitation can be an effective disruption mechanism to break down the dispersed phase into small droplets. Extreme shear and high energy input have to be applied to reduce droplets to nanoscale range. The homogenization pressure is typically in the range of 5–50 MPa for a conventional emulsion. Ultra high pressure up to 400 MPa[66] has been used to make nanoemulsions. According to their droplet disruption mechanism, high pressure valve homogenizers can be divided into two groups: disruption due to inertial forces in turbulent flow and disruption due to shear forces in laminar elongational flow. Each group can be further categorized based on their nozzle geometry,[67] as shown in Figure 7.10.

The effect of emulsification and the resulting droplet size of high pressure valve homogenizers largely depend on the geometry of the homogenization valves, product properties such as the viscosity of continuous and dispersed phases, and operating parameters including the flow rate, pressure, number of cycles and temperature.

In a coenzyme Q_{10} nanoemulsion produced by high pressure homogenization at 150 MPa using valves with different geometry, different droplet sizes of coenzyme Q_{10} nanoemulsion were obtained,

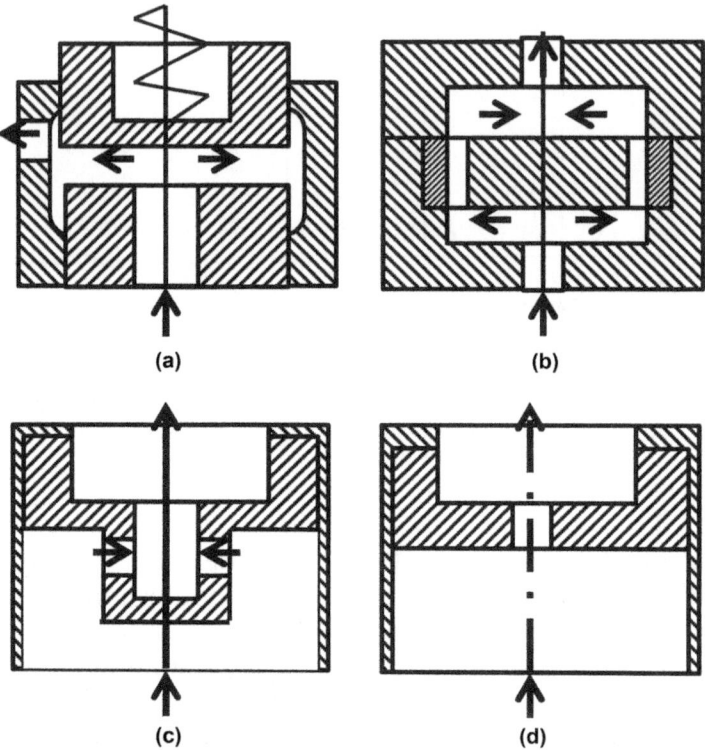

Figure 7.10   Cross section of high pressure homogenizer valves showing their inside design. Disruption predominantly due to inertial forces in turbulent flow: (a) standard nozzle, (b) Microfluidizer. Disruption predominantly due to shear forces in laminar elongational flow: (c) jet disperser, (d) orifice valve.
(Reprinted with permission from ref. 67)

ranging from 40 nm to >200 nm. The homogenization valve with more shear force and turbulent flow gave a smaller droplet size.[68] In another study on O/W and W/O nanoemulsions produced by Microfluidizer and a high pressure valve homogenizer, the Microfluidizer had higher emulsification efficiency in reducing droplet size for O/W nanoemulsions, with fewer cycles and smaller droplet size. However, for W/O nanoemulsions, the droplet size of the nanoemulsions made by Microfludizer and high pressure homogenizer was the same.[69] This effect was explained by the hypothesis that droplet break-up and stabilization are a function of the energy dissipation and are less dependent on the geometries of the high pressure device used.

High pressure homogenization can be combined with certain low-energy emulsification methods (*e.g.* the PIT method) and a spontaneous emulsification method to produce a microemulsion in Winsor II or Winsor III phase, which can be further treated by high pressure homogenization to create a homogeneous one phase O/W nanoemulsion. In the SoluClear™ process,[26] first a citrus flavor O/W microemulsion is created, in a similar way to the PIT method. Then it is homogenized at 30 MPa to produce a nanoemulsion with a droplet size of approximately 55 nm. In this citrus flavor O/W nanoemulsion, propylene glycol and glycerol were used as co-solvents and sunflower oil lecithin was used as the co-emulsifier in the oil phase.[26]

Normally, small molecule emulsifiers (including non-ionic and anionic emulsifiers) have to be used to make nanoemulsions by high pressure valve homogenizers. Food polymer emulsifiers, (such as proteins and modified starches) are not very efficient in reducing oil droplet size, owing to their slower absorption rate to the interface than small molecule emulsifiers. Emulsions made with food polymer emulsifiers can be over-emulsified at ultra high pressures when using a high pressure valve homogenizer.[70]

7.4.2.3 Solvent Extraction/Displacement Method.

The solvent extraction method is also called the solvent displacement method. It is one of the top-down nanoemulsion approaches, and uses high energy shearing devices such as a high-shear mixer, homogenizer, and Microfluidizer to break down the dispersed phase into small droplets. Organic solvents are used to help mix the dispersed phase, followed by the high energy emulsification process.

The solvent extraction process normally consists of three steps: (1) dissolution or dispersion of the dispersed phase into an organic solvent, which may or may not contain another polymer shell material; (2) use of a high-energy shearing device to emulsify this organic phase to form a coarse emulsion; and (3) extraction of the solvent from the dispersed phase by the continuous phase or evaporate solvent by reduced pressure or other methods (*e.g.*, rotor-evaporator) to form a finer emulsion such as a nanoemulsion.[71] Figure 7.11 shows detailed steps of the solvent extraction process.[72]

In addition to making nanoemulsions, polymer shell materials can be added to the organic solvent and thus used to make nanoparticles coated with the polymeric shell materials.[71] The HLB value of the emulsifiers or the mixture of emulsifiers used in the solvent extraction method plays a critical role in determining the spontaneity of the emulsification process and the final droplet size. The droplet size of caprylic/capric triglyceride–water–acetone nanoemulsions[55] is

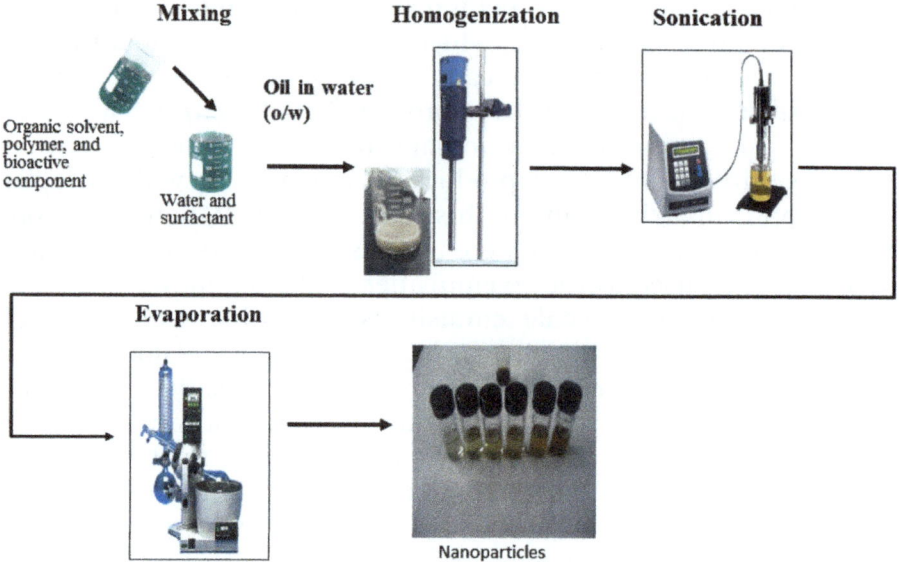

Figure 7.11   Solvent extraction process.
            (Adapted with permission from ref. 72)

decreased from 725 nm to 124 nm when the HLB of the mixture of emulsifiers is increased from 10.98 to 21.23.

7.5   AN EXAMPLE OF NANOEMULSION PREPARATION: A NANO-CLEAR OMEGA-3 OIL-IN-WATER EMULSION

An omega-3 O/W nanoemulsion was prepared by emulsifying a mixture of the emulsifier, the dispersed phase, an organic solvent, and a mono- and/or di-saccharide in a water solution.[7]

Vitamin E TPGS was melted using a water bath set at 60 °C. The melted TPGS, propylene glycol, glycerol and fish oil were then mixed together. The mixture was added slowly to an aqueous sugar solution. The mixture was stirred for 5 min and then homogenized at 5000 psi using a microfluidizer for 10 passes. Storage was at 4 °C. The droplet size of the emulsions was about 72 nm, as determined by a Particle Analyzer Beckman Coulter LS230. The emulsions remained clear after dilution by water in any ratio.[7]

The clarity of most of the emulsion samples[7] was tested by light absorbance of a 1.8% solution in deionized water (1.8 g emulsion : 100 g water to give 50 mg of eicosapentaenoic acid (EPA)+docosahexaenoic acid (DHA) per 250 g serving) at 400 nm, and varied from 0.01 (deionized water used as a reference) to 0.1.

In addition to vitamin E TPGS, other emulsifiers that can be used in this omega-3 O/W emulsion include polysorbate 80, polysorbate 85, monoglyceride, and lecithin. The emulsifier was first melted, if necessary, and then mixed with the components of the dispersed phase and co-solvent(s). The ratio of emulsifier to dispersed phase is about 1.5 or less.[7] The concentration of the emulsifier in this first mixture is above the emulsifier's CMC, so nanoscaled emulsifier micelles containing the dispersed phase are formed. The co-solvent also helps the emulsifier micelles to disperse and prevents gelation of some of the emulsifiers when they are in contact with water. The co-solvents used are propylene glycol and glycerol.[7]

The mono- and/or disaccharide used in the preparation of this omega-3 nanoemulsion has two functions. First, the sugar solution increases viscosity and facilitates particle size reduction by homogenization in the next step. Second, the sugar acts as a stabilizer for the emulsion and prevents the formation of liquid crystals, which cause aggregation of droplets and cloudiness in the emulsion.

Antioxidants can be added in either the continuous or the dispersed phase to protect the oils. The antioxidants can be added at any step during the emulsion formation process. When the antioxidant is added, and at what amount, will depend on the particular antioxidant.

The nano-clear omega-3 oil-in-water emulsions were used to deliver omega-3 oils or other lipophilic functional materials such as co-enzyme Q_{10} in several beverages. All beverages were pasteurized. The emulsion was added before pasteurization at a dosage of 32 mg EPA+DHA per 500 g serving. For example, the vitamin E TPGS-based omega-3 O/W clear emulsion was used to fortify lemon–lime and orange GATORADE™ at 32 mg EPA+DHA per 500 g serving. Moreover, the sensory properties of the fortified GATORADE™ were acceptable after storage at 4 °C for up to 56 days.[7]

7.6 CONCLUSIONS

In this chapter, emulsions with droplet sizes of 200 nm or less are considered nanoemulsions. The emulsifiers used in nanoemulsions include non-ionic emulsifiers such as polysorbates and sucrose esters, Zwitterionic emulsifiers such as lecithin, and food polymeric emulsifiers such as OSA-modified starch. Recently some new emulsifiers have been used to make nanoemulsions, such as *Quillaja* saponin and some pharmaceutical emulsifiers, *e.g.*, vitamin E TPGS. To optimize the formulation and process of nanoemulsion formation, it is important to understand the properties of emulsifiers, such as their HLB values, CPP, and their phase behavior. There are two approaches to the preparation of nanoemulsions: top-down or high energy shearing methods and bottom-up or low energy emulsification methods.

REFERENCES

1. T. Panagiotou and R. J. Fisher, *Funct. Foods Hlth Dis.*, 2013, **3**, 274.
2. National Nanotechnology Initiative, http://www.nano.gov/nanotech-101/what/definition (accessed on 31 March 2014).
3. T. Tarver, *Food Technol.*, 2011, **11**(6), 23.
4. B. H. Robinson, Application of microemulsions, *Nature*, 1986, **320**(27), 309.
5. P. Shah, D. Bhalodia and P. Shelat, *Syst. Rev. Pharm.*, 2010, **1**(1), 24.
6. P. C. Teixeira, G. M. Leite, R. J. Domingues, J. Silva, P. A. Gibbs and J. P. Ferreira, *Int. J. Food Microbiol.*, 2007, **118**(1), 15.
7. D. Tang and S. Cloutier, *International Application No. PCT/IB2011/002743*, publication No. WO/2012/032416, 2011.
8. A. A. Date, N. Desai, R. Dixit and M. Nagarsenker, *Nanomedicine*, 2010, **5**, 1595.
9. T. G. Mason, J. N. Wilking, K. Meleson, C. B. Chang and S. M. Graves, *J. Phys.: Condens. Matter*, 2006, **18**, R635.
10. US Food and Drug Administration, FDA's Approach to Regulation of Nanotechnology Products, http://www.fda.gov/ScienceResearch/SpecialTopics/Nanotechnology/ucm301114.htm (accessed on 31 March 2014).
11. W. C. Griffin, *J. Soc. Cosmet. Chem.*, 1949, **1**(5), 311.
12. W. C. Griffin, *J. Soc. Cosmet. Chem.*, 1954, **5**(4), 249.
13. J. N. Israelachvili, J. N. Mitcheli and B. W. Ninham, *J. Chem. Soc., Faraday Trans. II*, 1976, **72**, 1525.
14. T. Tadros, *Applied Surfactants: Principles and Applications*, Wiley-VCH Verlag Gmbh & Co., KgaA, Weinheim, 2005.
15. M. Kahlweit, R. Strey, D. Haase, H. Kunieda, T. Schmeling, B. Faulhaber, M. Borkovec, H. F. Eicke, G. Busse, F. Eggers, T. Funck, H. Richmann, L. Magid, O. Soderman, P. Stilbs, J. Winkler, A. Dittrich and W. Jahn, *J. Colloid Interface Sci.*, 1987, **118**, 436.
16. D. J. McClements and J. Rao, *Crit. Rev. Food Sci. Nutr.*, 2011, **51**, 285.
17. L. Wang, X. Li, G. Zhang, J. Dong and J. Eastoe, *J. Colloid Interface Sci.*, 2007, **314**, 230.

18. P. Walstra, Dispersed systems: basic consideration, in *Food Chemistry*, ed. O. R. Fennema, Marcel Dekker, Inc., 1996, New York, NY.
19. T. Tadros, P. Izquierdo, J. Esquena and C. Solans, *Adv. Colloid Interface Sci.*, 2004, **108–109**, 303.
20. C. Wagner, *Zeits. Elektrochem.*, 1961, **65**(7), 581.
21. Exova, http://www.exova.com/news/resources/tween80 (accessed on 13 May 2014).
22. Malvern Instruments Ltd., Surfactant micelle characterization using dynamic light scattering, *Application Note*, 2006.
23. M. Ochomogo, R. Garg, E. R. da Conceicao, Y. Neta, L. Yang, L. Arnt and A. Monsalve-Gonzalez, *US20130064954 A1*, 2013.
24. D. S. Bernardi, T. A. Pereira, N. R. Maciel, J. Bortoloto, G. S. Viera, G. C. Oliveira and P. A. Rocha-Filho, *J. Nanotechnol.*, 2011, **9**, 44.
25. L. Lee and N. Niknafs, *Trends Food Sci. Technol.*, 2013, **31**(1), 72.
26. Compass Foods Pte Ltd., SoluClearTM citrus flavor solubilization protocol, *Technical application sheet*, 2014.
27. N. Anarjan and C. Tan, *Molecules*, 2013, **18**, 768.
28. A. M. M. Eid, N. A. Elmarzugi and H. A. El-Enshasy, *Int. J. Pharm. Pharm. Sci.*, 2013, **5**(supplement 3), 434.
29. H. Bueschelberger, Lecithin, in *Emulsifiers in Food Technology*, ed. R. J. Whitehurst, Blackwell Publishing Ltd., 2006, vol. 2.
30. S. C. de Arau'jo, A. C. de Mattos, H. F. Teixeira, P. M. Coelho, D. L. Nelson and M. C. de Oliveira, *Int. J. Pharm.*, 2007, **37**, 307.
31. A. J. Choi, C. J. Kim, Y. J. Cho, J. K. Hwang and C. T. Kim, *Food Sci. Biotechnol.*, 2009, **18**, 1161.
32. R. H. Skiff, J. Baaklini and J. F. Vlad, Patent no. WO/2007/026271.
33. D. Amendola, F. Donsì, M. Sessa, G. Ferrari and D. M. De Faveri, *J. Food Eng.*, 2013, **114**, 207.
34. S. Zainol, M. Basri, H. Bin Basri, A. Fuad Shamsuddin, S. S. Abdul-Gani, R. A. Karjiban and E. Abdul-Malek, *Int. J. Mol. Sci.*, 2012, **13**, 13049.
35. M. Kahlweit, G. Busse and B. Faulhaber, *Langmuir*, 1995, **11**, 1576.
36. J. Zhang, *Novel Emulsion-based Delivery Systems*, PhD thesis, University of Minnesota, 2011.
37. S. Jafari, Y. He and B. Bhandari, *Int. J. Food Prop.*, 2006, **9**, 475.
38. Y. Yang, M. E. Leser, A. A. Sher and D. J. McClements, *Food Hydrocolloids*, 2013, **30**, 589.
39. C. Chen and G. Wagner, *Chem. Eng. Res. Des.*, 2004, **82**, 1432.
40. Ingredion Inc., *Personal communication through J. Li* on 04 April 2014.

41. R. Rodiguez, Novel ingredient solutions for formulating clear-type beverages, *Ingredion technical white paper*, 2011, Ingredion Inc., Bridgewater, NJ.
42. P. Tomlinson, Patent application on Soluble bioactive lipophilic compounds compositions, international application No. WO 2008/101344 A1.
43. A. Shah, O. H. Campanella and B. R. Hamaker, Conjugated linoleic acid, nano-scale soluble self-assembling complex for healthy nutrient delivery, *IFT 2009*, Annaheim, CA.
44. H. Arai and K. Shinoda, *J. Colloid Interface Sci.*, 1967, **25**, 396.
45. K. Shinoda and H. Saito, *J. Colloid Interface Sci.*, 1969, **30**, 258.
46. C. Solans, P. Izquierdo, J. Nolla, N. Azemar and M. J. Garcia-Celma, *Curr. Opin. Colloid Interface Sci.*, 2005, **10**, 102.
47. S. Friberg and C. Solans, Emulsification and the HLB temperature, *J. Colloid Interface Sci.*, 1978, **66**, 367.
48. P. Izquierdo, J. Esquena, T. F. Tadros, J. C. Dederen, M. J. Garcia and N. Azemar, *Langmuir*, 2002, **18**(1), 26.
49. P. Izquierdo, J. Esquena, T. F. Tadros, J. C. Dederen, J. Feng and M. J. Garcia, *Langmuir*, 2004, **20**, 6594.
50. P. Izquierdo, J. Feng, J. Esquena, T. F. Tadros, J. C. Dederen and M. J. Garcia, *J. Colloid Interface Sci.*, 2005, **285**, 388.
51. H. Marino, *Phase Inversion Temperature Emulsification: From Batch to Continuous Process*, PhD thesis, 2010, University of Bath, Department of Chemical Engineering, Bath, UK.
52. C. Malcolmson, D. J. Barlow and M. J. Lawrence, *J. Pharm. Sci.*, 2002, **91**(11), 2317.
53. J. Rao and D. J. McClements, *J. Agric. Food Chem.*, 2011, **59**(9), 5026.
54. C. B. Sáinz, R. J. A. Bustillos, D. F. Wood, T. G. Williams and T. H. McHugh, *J. Agric. Food Chem.*, 2010, **58**(22), 11932.
55. K. Bouchemal, S. Briancon, E. Perrier and H. Fessi, *Int. J. Pharm.*, 2004, **280**, 241.
56. B. K. Paul and S. P. Moulik, *Curr. Sci.*, 2001, **80**(25), 990.
57. Y. Yang, C. Marshall-Breton, M. E. Leser, A. A. Sher and D. J. McClements, *Food Hydrocolloids*, 2012, **29**(2), 398.
58. A. H. Saberi, Y. Fang and D. J. McClements, *Food Res. Int.*, 2013, **54**(1), 812.
59. N. Garti, A. Yaghmur, M. E. Leser, V. Clement and H. J. Watzke, *J. Agric. Food Chem.*, 2001, **49**, 2552.
60. T. Panagiotou, S. V. Mesite and R. J. Fisher, *Ind. Eng. Chem. Res.*, 2009, **48**, 1761.
61. T. Panagiotou, K. J. Chomistek and R. J. Fisher, *NanoFormulation*, Conference Proceedings, 2012, 135.

62. J. O. Wilkes, *Fluid Mechanics for Chemical Engineers*, Prentice-Hall, Inc., New Jersey, 1999, 1st edn, vol. 9, p. 426.
63. K. J. Chomistek and T. Panagiotou, *MRS Proc.*, 2010, **1209**, P03-01.
64. B. Johnson and R. Prud'homme, *AIChE J.*, 2003, **49**, 2264.
65. A. A. Mahajan and D. J. Kirwan, *J. Phys. D: Appl. Phys.*, 1993, **26**, B176.
66. M. Cortes-Munoz, D. Chevalier-Lucia and E. Dumay, *Food Hydrocoll.*, 2009, **23**, 640.
67. M. Stang, H. Schuchmann and H. Schubert, *Eng. Life Sci.*, 2001, **1**(4), 151.
68. J. S. Lim, H. J. Gang, S. W. Yoon, H. M. Kim, J. W. Suk, D. U. Kim and J. K. Lim, *Korean J. Food Sci. Tech.*, 2010, **42**(5), 565.
69. L. Lee, R. Hancocks, I. Noble and I. T. Norton, *J. Food Eng.*, 2014, **131**, 33.
70. E. Hebishy, M. Buffa, B. Guamis and A. Trujillo, *CIDIC Conf. Ser.*, 2013, **11**, 161.
71. S. Freitas, H. P. Merkle and B. Gander, *J. Control. Rel.*, 2005, **102**, 313.
72. C. Sabliov, New and emerging food applications of polymeric nanoparticles for improved health, *IFT International Food Nanoscience Conference*, June 6, 2009, Annaheim, CA.

CHAPTER 8

Imaging Nanostucture

ALEXANDRA K. SMITH

Department of Food Science, University of Guelph, Guelph, ON, Canada
Email: smitha@uoguelph.ca

8.1 INTRODUCTION

Food structure is based on a system of molecules that provide form and function. A better understanding of the arrangement of the underlying molecules that contribute to the macro-molecular structure of natural foods leads to the ability to create novel food products, including those with nutraceutical application, or new products with enhanced shelf-life and quality. Microscopes give us the opportunity to image the basic architectural molecules and are tools that contribute to research into food structure.

The main building blocks of food constituents are oxygen, nitrogen and carbon. Oxygen is the most abundant element in the human body at 65% of body weight; carbon is second at 18%.[1] However, because we cannot image oxygen we must focus on carbon as the basis of biological structure.

How large is a carbon atom?

The radius of an atom is measured in picometers (pm).

$$1 \text{ pm} = 1 \times 10^{-12} \text{ meters}$$
$$1000 \text{ pm} = 1 \text{ nm.} \tag{8.1}$$

The ionic radius of C1 has been calculated[2] to equal 29 pm.

Edible Nanostructures
Edited by Alejandro G Marangoni and David Pink
© The Royal Society of Chemistry 2015
Published by the Royal Society of Chemistry, www.rsc.org

How can we see something that is less than 1/3 of a nanometer (nm)? The answer is that we cannot, even with high power microscopes. However, biological structure is built up from the atomic level to the molecular level to create the macromolecular architecture that defines biological structure and the function of those structures. The ability to study the organization of biological structure on a nanoscale level can serve to redefine our understanding of the architecture, and by extension, the functionality of the macrostructure that is our food.

The nutritional label on our food identifies the major constituents as:

- carbohydrates (sugar, starch, gums)
- fat (saturated, unsaturated, *trans-*)
- protein
- water and air.

The fundamental structures of carbohydrate, fat and protein are shown in Figure 8.1. Carbon atoms provide the backbone for all the basic food constituents.

Creating images of nanostructure in foods is currently made possible by three types of microscope: the transmission electron microscope (TEM), the scanning electron microscope (SEM), and the

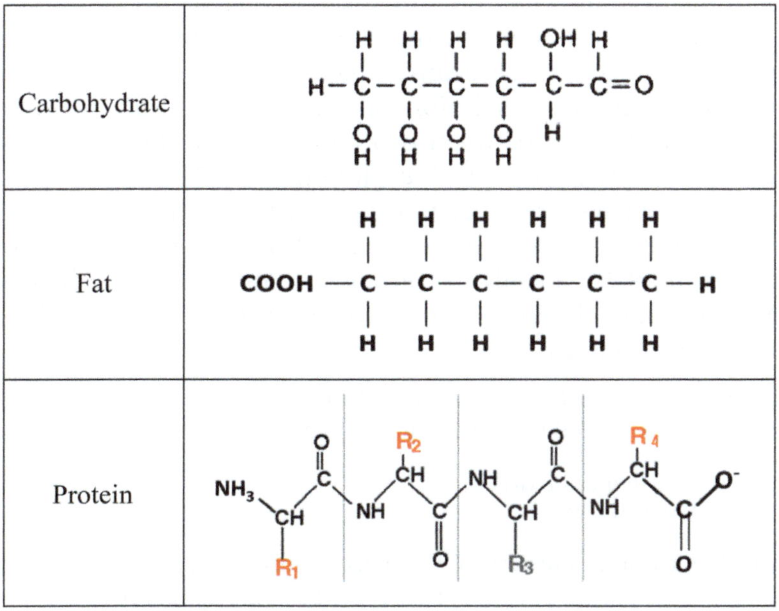

Figure 8.1 Basic molecular structures of food constituents.

Table 8.1 Microscopes capable of imaging food constituents on a nanostructure level and reported resolution capability.

Microscope	Theoretical resolution	Practical resolution	Vertical resolution
Transmission electron	0.004 nm (80 kV)[3]	0.25 nm[4]	None
Scanning electron	1.3 nm (15 kV)[6]		None
Atomic force	3.0 nm (lateral)[7]	0.15 nm[8]	0.1 nm[7]

atomic force microscope (AFM). The reported resolving power of these microscopes is shown in Table 8.1.

8.2 TRANSMISSION ELECTRON MICROSCOPY

The microscopes listed in Table 8.1 have different strengths and modes of imaging. The TEM, as shown in Figure 8.2, uses a beam of electrons generated at the electron gun, and accelerated through a condenser lens (which narrows the beam), and then through a thin sample, usually a section which is 70–90 nm thick, embedded in resin. The electrons are diffracted by the molecules in the sample; the density of the molecules determines the gray level in the image projected onto the viewing screen. If the electrons pass through areas without tissue no diffraction occurs and the image is white. The result is black and white images of ultrastructure as shown in Figure 8.3.

The transmission electron microscope (TEM) offers the best resolution in the quest to image the nanostructure of food constituents since it uses an accelerated beam of electrons in a vacuum system to interact with the tissue. Resolution, or minimum resolvable distance, is the minimum distance between two particles that allows them to be seen as separate. The theoretical resolution of the TEM is based on the wavelength, or peak-to-peak distance of the waveform, of the electron beam. Resolving power can be determined using the Abbe equation[10] [see equation (8.2)]. Resolving power will increase with increasing accelerating voltage because of the reduction in the wavelength.

Ernst Abbe equation:

Resolving power = 0.61(wavelength)/NA
           = 0.61(wavelength)/n sin alpha

NA = numerical aperture
n = refractive index

Wavelength of an electron = 1.23/accelerating voltage$^{1/2}$
           1.23 is de Broglie's constant

The resolving power of an electron microscope = 1.23/80 000$^{1/2}$

$$= 0.004 \text{ nm} \qquad (8.2)$$

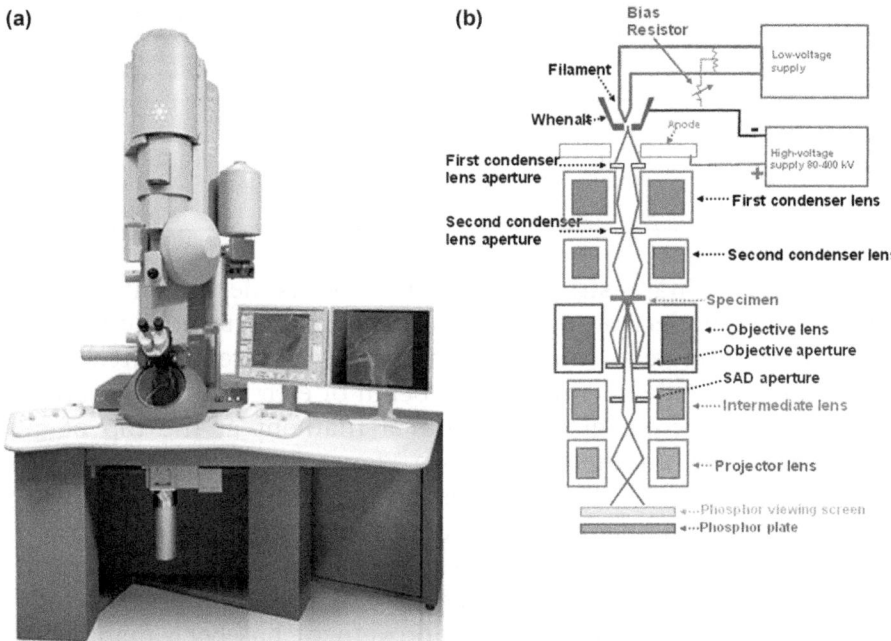

Figure 8.2 Transmission electron microscopy: (a) Technai G2 field-emission transmission electron microscope;[4] (b) electron optical column schematic.[3]

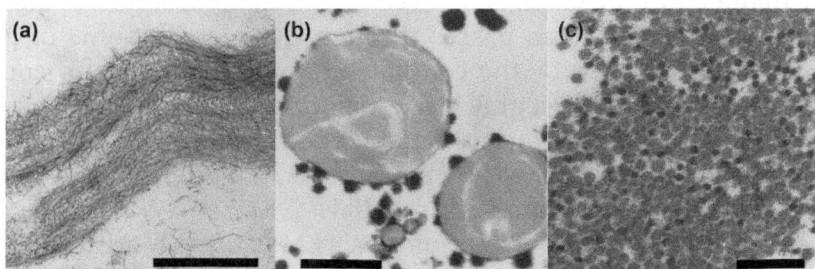

Figure 8.3 Transmission electron micrographs: (a) cellulose fibers (A. Harper, A. Smith, S. Barbut, University of Guelph), (b) fat globules in ice cream mix,[9] and (c) protein (collagen fibers in cross section). Bar a = 500 nm; Bar b = 1 μm; Bar c = 250 nm.

Therefore, at 80 kV the maximum resolving power of the electron microscope is 0.004 nm.

The theoretical resolution of the TEM is reported to be at less than 0.1 nm, but because of limitations in lens design "the Titan Tecnai TEM with a field-emission gun (FEG), operated at 200 kV, reports point-to-point resolution of 0.25 nm".[4]

Table 8.2 Comparison of electron guns for the transmission electron microscope.[11]

Source	Brightness (A cm^{-2} sr^{-1} kV)	Effective source radius (nm)	Energy spread (eV)
Tungsten hairpin	1×10^4	60 000	1.5–2.5
LaB$_6$	1×10^5	10 000	1.3–2.5
Schottky (KrW)	1×10^7	≤1000	0.35–0.7
Cold FEG	2×10^7	≤100	0.30–0.7

Resolution of nano-structure by the TEM is dependent on the elements of the electron optical column that dictate how electrons can be used to create an image. The image is generated by interaction between a prepared specimen and a beam of electrons. The beam is produced by emitting electrons from a filament at the top of an electron optical column, as shown in Figure 8.3(b), illustrating the beam path in a TEM. The beam is accelerated up to 200 keV. Increasing electron acceleration shortens the wavelength of the beam and increases resolution.[5] Historically, electrons were accelerated from heated sources, either tungsten or lanthanum hexaboride (LaB$_6$). More recently, significant improvement has been made with the advent of cold field-emission guns (FEG) and Schottky emitters. The critical improvements in electron gun design include increased brightness, smaller source size and decreased energy spread, as shown in Table 8.2.[11]

The diameter of the beam is reduced by the condenser lens to about 1 micron before the beam passes through the sample. Some of the electrons are diffracted by the components of the sample and some are transmitted to create a gray level image, depending on the degree of diffraction. After the beam has interacted with the sample it passes through an objective lens which produces a magnified image to be projected on the fluorescent screen below.

The theoretical limit to resolution in a TEM is directly attributed to the wavelength of the electrons in the beam. That being said, the electromagnetic lenses that control the beam current limit resolution because of spherical and chromatic aberration. Spherical aberration is distortion caused by the fact that some electrons in the beam are travelling a pathway that is parallel to the optic axis.[12] The divergent electrons are focused away from the center of the lens; they do not converge at the same point as those in the central part of the beam. Convergence needs to happen so that the point in the sample relates directly to the point in the image. Fortunately, in modern microscopes this can be corrected through the use of multi-pole lenses and computer control.[13] Chromatic aberration is caused by the fact that not all electrons in the beam have the same wavelength. This results in a beam of electrons with varying kinetic energy. Variations in kinetic energy can be caused by: (1) fluctuations in the voltage that heats the filament; (2) thermal spread in the electrons emitted from the source; and (3) interaction with sample elements that cause inelastic scattering "resulting in energy spread within the transmitted beam".[12]

A solution to spherical and chromatic aberration is to combine an electromagnetic lens, which focuses the beam, with the electrostatic lens that causes the divergence. The result is an achromatic lens,[12] and because the electron beam in a modern TEM has a more uniform wavelength, there is less chromatic aberration at the start. With spherical and chromatic aberration corrected, the new limit to resolution is magnetic noise. Magnetic noise is caused by thermal effects which result in currents generated from heated metal components in the column.[13]

The reported resolution of images created by the TEM, as listed in Table 8.1, is based on the assumption that the prepared sample is true to the original and is stable under the beam. The fact is that there are multiple steps involved in preparing the sample. Each step will change the native microstructure and, in turn, will change the image and the interpretation of the image. Many textbooks have been written to provide information on the best way to preserve biological structure for electron microscopy, and they include precautions regarding potential alteration to the tissue.[15,16] It is required that the sample be dry before it can be inserted into the vacuum system; high vacuum is necessary to support the beam path. The tissue must be hardened using chemical fixatives so that it will not collapse when the water is removed. Most critical to the initial preservation of structure are the pH and osmolarity of the fixative solution; these elements of fixation chemistry must match those of the biological tissue to minimize distortion. Standard preparation steps include: (1) fixation of the tissue to harden it while protecting structural arrangement; (2) solvent dehydration to remove the water; (3) substitution of solvent with resin; and (4) polymerization of the resin. The tissue, infiltrated by resin and hardened, is sectioned into 70–90 nm thick slices. This is the appropriate thickness to provide stability under the electron beam and is thick enough to provide sufficient contrast when imaged. The section is post-stained to increase the electron density of the tissue components. All of the tissues shown in Figure 8.3 were prepared using the standard protocol described. Figure 8.3(a) shows cellulose fibers; the sample was prepared as a control to help identify the cellulose used as a component in a sausage casing. Figure 8.3(b) is an image of fat in ice cream mix prepared by mixing the emulsion in low temperature setting agarose and allowing it to set. The agarose containing the mix was then cut into small cubes (2 mm^2) and fixed before embedding.[9] Figure 8.3(c) shows cross sections of collagen. Strands of muscle tissue were stretched between two pieces of foil to limit shrinkage during fixation and embedding.[17]

One must verify that the features of the tissue have been accurately preserved. How is this possible when the tissue for comparison has also been rigorously prepared for other nanostructure imaging? Fortunately, freezing, followed by low temperature processing procedures, is increasingly becoming the method of choice. Cryo-preservation methods are most effective when the water (which makes up 70–95% of biological cells) is frozen to a vitreous state. The advent of high pressure freezing has made it possible to freeze bulk tissue vitreously to a depth of 10–100 µm.[20] The application of 2100 bar of hydrostatic pressure lowers the water crystallization temperature to –90 °C.[16] Since the application of pressure lowers the crystallization temperature and limits the rate of nucleation and growth of ice crystals, distortion of the tissue is minimal.[19] The preservation of structure is almost instantaneous (potentially within 1 ms), guaranteeing that physical structures are immobilized and chemical processes are stopped.[20]

Frozen tissue can be further processed by freeze-substitution (fixation and dehydration steps) and embedded in low temperature resin at –80 °C to –90 °C.[20] Polymerization of the resin can be accomplished using ultraviolet (UV) light, while still at low temperature (–70 °C).[15] Following polymerization, the resin can be sectioned at ambient temperature. Another approach is to section the frozen material directly using a cryo-ultra-microtome. Crystalline water will shatter when cutting is attempted, and therefore it cannot be sectioned. However, if the water is vitreous it can be sectioned to provide thin sections of frozen bulk samples for cryo-TEM observation. One concern is that unstained biological sections lack sufficient contrast when imaged by the TEM. However, it has been shown that the signal-to-noise ratio is actually higher than in conventionally prepared materials and thus more information can be recovered.[18] The best approach is to image the frozen sample directly by transferring the frozen hydrated sample, by means of a transfer device, directly into the column of the cryo-TEM. Direct imaging requires that the tissue be small enough to fit into the thin layer of water (<1 µm), a category which includes, for example, fat crystals, casein micelles, viruses and

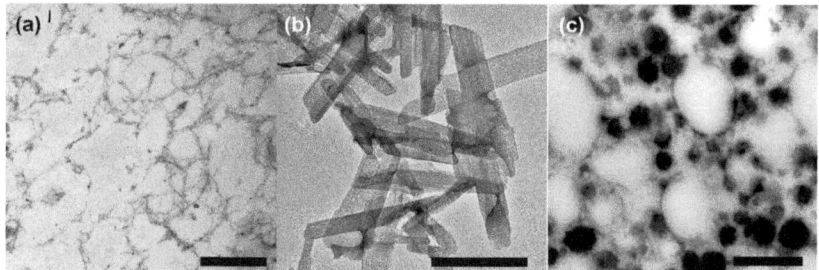

Figure 8.4 Cryo-transmission electron microscopy images: (a) agarose prepared by freeze-substitution;[21] (b) fat crystals imaged directly with cryo-TEM (A. Marangoni, University of Guelph); (c) casein micelles shown as black globules in ice cream mix prepared by freeze-substitution.[21] Bar a = 500 nm; Bar b = 250 nm; Bar c = 250 nm.

liposomes. The images shown in Figure 8.4(a) include an image of polysaccharide used to modify viscosity in a dairy emulsion in a sample prepared by freeze-substitution and low temperature embedding.[21] Figure 8.4(b) is an image of fat crystals imaged directly by cryo-TEM. The ice cream mix shown in Figure 8.4(c) was prepared by freeze substitution.[21] Casein micelles, 100 nm or less, can be seen as black spheres.

8.3 SCANNING ELECTRON MICROSCOPY

Scanning electron microscopy (SEM) is a surface imaging technique. The SEM uses a beam of primary electrons to scan the surface of a bulk sample to generate a topographic image. The image is based on gray levels (determined by the height of surface features), crystal structures and/or atomic number. Because the beam has a long focal length the image has a large depth of field and, therefore, two-dimensional (2D) images appear to be three-dimensional (3D).

The SEM uses an electron gun to generate the beam and then narrows the beam through condenser lenses, just as in a TEM column. A typical SEM is shown in Figure 8.5(a).[6] A schematic of the electron optical column of the SEM is illustrated in Figure 8.5(b).[14] The SEM column differs from that of the TEM because the beam passes through the objective aperture above the sample surface. Also, as the objective aperture is narrow, it creates a small aperture angle which is well above the sample, and the resulting image has a large depth of field.[22] The large depth of field allows for an image that appears 3D (actually 2D), as shown in Figure 8.6. Unfortunately, depth of field is enhanced at the expense of signal-to-noise, because

(a)   (b)

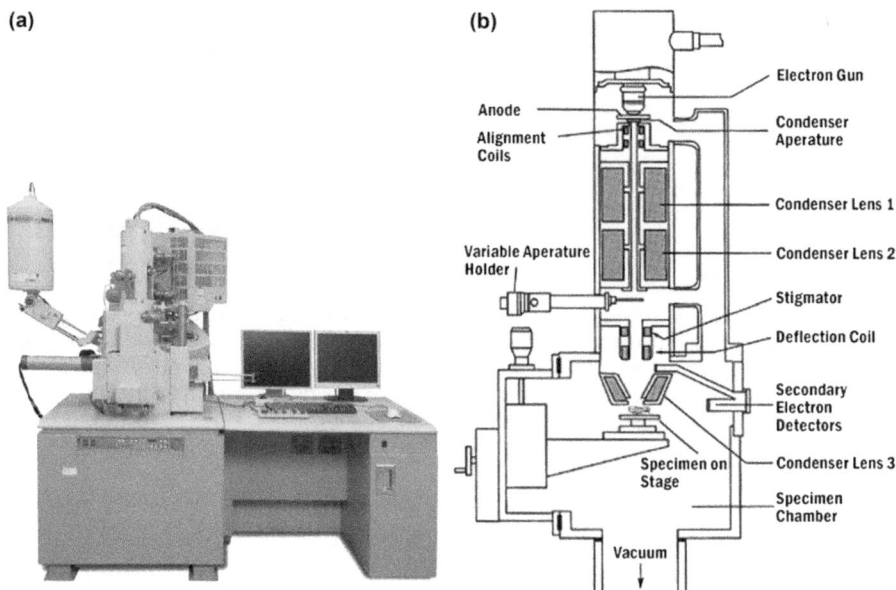

Figure 8.5   The scanning electron microscope: (a) SU6600 Analytical VP FE-SEM;[6] (b) electron optical column diagram.[14]

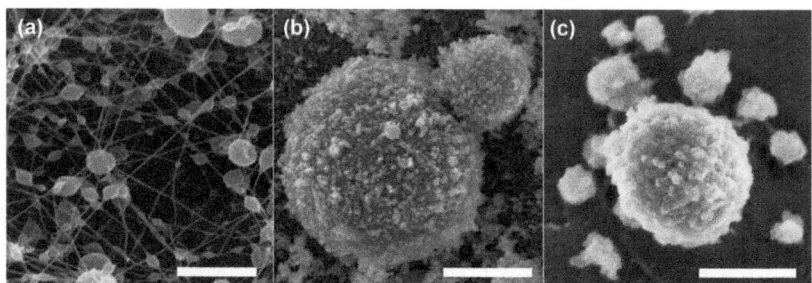

Figure 8.6   Scanning electron micrographs of: (a) electro-spun fibres (alginate/pectin/PEO);[26] (b) raw milk fat (A. K. Smith, University of Guelph); (c) FE-SEM image of casein micelles.[27] Bar a = 1.5 μm; Bar b = 3 μm; Bar c = 250 nm.

with the smaller aperture there are fewer primary electrons to scan the sample.

Primary electrons are driven by scan coils in the final lens, in a square raster pattern across the surface of the sample (hence scanning EM). The interaction of primary electrons with the surface creates several signals that can be detected and that provide information about the sample surface. The most important and widely used image signals are secondary electrons and backscattered electrons.[22] Secondary electrons are generated from inelastic scattering when primary electrons interact with the electrons in the atoms of the sample surface to a depth determined by the atomic number of the sample and the accelerating voltage of the beam. By definition, secondary electrons have energy of <50 eV and most (90%) have energy of <10 eV.[23] The number of secondary electrons generated is largely based on topography and, owing to the high points of the topography, they generate more of these areas on the image which are brighter. Secondary electrons are collected at a 90° angle into a charged aluminium scintillator where photons are produced. The photon signal is amplified in a photomultiplier tube and is converted back to electrons that drive the video boards of the viewing screen. The scintillator/photomultiplier detectors are known as Everhart–Thornley detectors from their 1960s design.[24] Backscattered electrons (BSE) are created from elastic scattering events between primary electrons and nuclei of sample atoms and come from deeper in the sample.[25] As the number of backscattered electrons depends on the atomic number of the atoms, they can provide an image that displays atomic number contrast; the higher the atomic number the brighter the image. Backscattered electrons are collected by a four-quadrant detector that is positioned above the sample, close to the pole piece on the base of the electron optical column. This detector, most commonly a Robinson detector, is also a scintillator/photomultiplier design and takes advantage of the large scattering angle of the backscattered electrons.[24] Using backscattered electrons to form an image is also a way to avoid charging artifacts caused by the absorption of primary electrons in a non-conductive specimen.

Resolution in SEM images, as in TEM images, is based on the wavelength of the electron beam and the design of the electromagnetic lenses. Because the accelerating voltage in the SEM is limited to 30 kV the resolution is less than for TEM, as demonstrated by the values reported in Table 8.1. SEM design has improved greatly

Table 8.3 Comparison of electron gun parameters for the scanning electron microscope.[28]

Source	Brightness (A cm^{-2} sr^{-1} kV)	Effective source radius	Energy spread (eV)
Tungsten hairpin	1×10^5	50 μm	3
LaB$_6$	5×10^6	10 μm	1.5
Schottky (KrW)	5×10^8	0.015–0.03 nm	0.3–10
Cold FEG	1×10^8	0.005 nm	0.3

with the introduction of the field-emission SEM (Table 8.3), largely due to the design of the field-emission electron gun (FEG) which offers increased beam brightness, decreased source size and a smaller energy spread. The dramatic increase in beam brightness values can be seen in Table 8.3. Because brightness drops as accelerating voltage is reduced, high beam brightness at the source is crucial to the success of low voltage imaging, the condition best suited to biological imaging.[28] In addition, a smaller source contributes to a narrower beam coming from the filament.

A standard tungsten hairpin has a diameter of about 50 μm, compared to the tip of a FEG tungsten wire with a diameter of 5 μm. The hairpin provides a broader beam. If the source is small to begin with, less demagnification of the beam is required before it reaches the sample. Also, the energy spread is less, which contributes to higher resolution images. The best resolution, according to the values reported in Table 8.3, would come from the cold FEG. However, with this electron source, contamination is a problem. Even with high vacuum (10^{-7} torr), the tip acquires a layer of gas molecules that must be removed by "flashing", which means that the tip is heated by the application of high voltage until it is white.[28] The solution is to use a Schottky FEG, a heated source which makes use of ZrO to flow to the tip, reducing the work function. The tip is heated and therefore remains clean and operational for up to two years of continuous use.[28] The drawback of using the Schottky FEG is the loss of resolution at low voltage, which is the ideal condition for the biological electron microscopist.

Needless to say, the advent of the FEG for the SEM has enabled imaging of biological samples at the nanometer level of resolution with this instrument. Not only is the theoretical resolution improved, but also the secondary electron signal is generated closer to the surface of the material because beam penetration is limited by the ability to use a lower accelerating voltage. The interaction volume of the primary beam in biological tissue is modeled using a Monte Carlo scattering simulation (Figure 8.7) and this is a good way of demonstrating the advantage of FE-SEM. Figure 8.7(a) shows an example of the volume of interaction in a carbon-based sample with a density of 1.5 g cm^{-1} from a single scattering event[29] when the accelerating voltage of the electron beam is set at 10 kV (tungsten hairpin filament) and each grid dimension is 500 nm. This graph can be contrasted with Figure 8.7(b), where a 2 kV beam is generated by the FEG and where the grid dimensions, illustrating the penetration of the beam, represent 50 nm. The penetration of the beam contributes

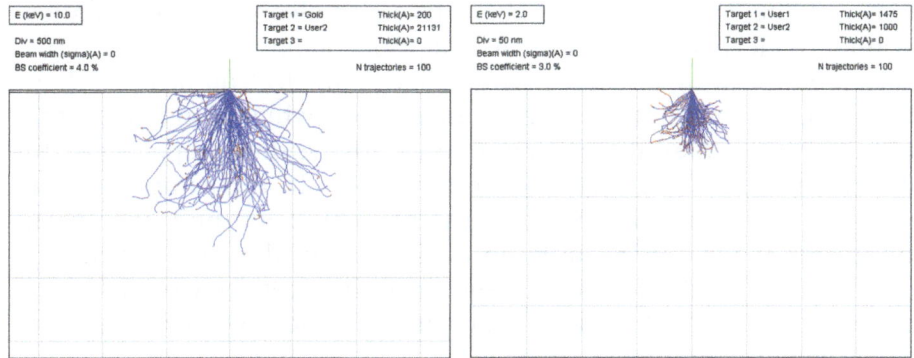

Figure 8.7 Monte Carlo electron scattering simulation of the volume of interaction from a single electron in: (a) carbon-based sample[a] coated with 20 nm of gold scanned at 10 kV; (b) carbon-based sample, not coated, scanned at 2 kV. Note that in the scale for (a) each division is 500 nm and (b) each division is 5 nm. ([a]The density of the bulk carbon sample was estimated[30] at 1.5 g cm^{-1}.)

to charging artifacts causing distortion in the image, a common problem when using a high beam current on a non-conductive sample. The lower beam current and the opportunity to control the surface charge at the sample level, by biasing the surface to prevent charge build-up within the specimen, mean that sputter coating is not required. Removing this step from sample preparation further improves resolution.

Conventional chemical methods for preparation of biological tissue for SEM require that the sample be fixed, to harden the tissue and add conductivity, dehydrated to remove the water, and critical-point dried or freeze dried.[5,15] The drying methods are designed to remove water without causing the tissue to collapse, which is a challenge because biological tissue can have anywhere from about 75 to 95% water. The effectiveness of the drying is directly dependent upon the quality of the fixation and the nature of the tissue. Some tissues which are full of water and have fewer structural elements will shrink and possibly collapse.

Critical-point drying makes use of the fact that liquid CO_2 at 1070 psi and 31.2 °C has the same vapour pressure as gaseous CO_2. Under these conditions there is no surface tension and, therefore, no distortion occurs as the liquid leaves the tissue. Freeze-drying sublimates the water from the tissue. Both drying methods will also extract tissue constituents. Once the samples are mounted they are thinly coated (<20 nm) with an element of high atomic number, usually gold or gold/palladium, and imaged.[5,15] Sputter coating increases the conductivity at the surface and also serves to enhance secondary electron formation. As mentioned, one of the problems caused by scanning biological samples with an electron beam is that, owing to the low atomic number of carbon (which constitutes the bulk of the sample), there is poor conductivity; the sample tends to absorb the beam. A negative charge build-up in the sample, called charging, will cause distortion of the primary beam and, therefore, the image. If the charge build-up is significant, the sink of primary electrons will reach the detector and result in excessively bright areas in the image (bulk charging) or bright lines and loss of focus. Fixatives such as osmium tetroxide are used to increase the atomic number of the tissue. Sputter coating is a standard preparation step for imaging with older high voltage SEMs. It is required to increase conductivity at the surface and enhance the production of secondary electrons. As mentioned earlier, the charging problem has been overcome in more recent years with the introduction of FE-SEM.

Just as artifacts are created during preparation of tissues for TEM, the same is true during preparation of samples for SEM. Again, cryo-preparation techniques are the solution. These methods are particularly important for food materials that are high in water, fat and air. Water and fat are lost during chemical preparation methods and leave behind void areas that confuse the interpretation of the images of food structures.

The general protocol to prepare samples for cryo-SEM is to use a dedicated cryo-preparation unit in order to: (1) freeze the sample in liquid nitrogen slush (−207 °C), which offers a higher freezing rate than liquid nitrogen; (2) fracture the surface to provide a fresh surface unadulterated with frost; (3) sublimate to remove surface water; and (4) sputter coat to enhance the atomic number of the surface. The frozen, partially freeze-dried sample is then moved onto the cold stage (about −150 °C) of the SEM by means of a transfer device so that the sample remains frozen and under vacuum. Bulk samples are usually the subject for cryo-SEM imaging and, therefore, the freezing rate will not be fast enough to eliminate ice crystal formation. When solutions freeze the solutes form a eutectic around the ice crystals.[30,31] Although ice crystal formation is not desirable, this consequence of freezing may provide information about the amount and distribution of the water in the sample. The sublimation process exposes a pattern of solutes in ice crystals that allows the ice to be identified. In all SEM images, morphology is the only means of identifying image components, without X-ray microanalysis. Sometimes it is necessary to prepare basic samples to become familiar with their characteristics in

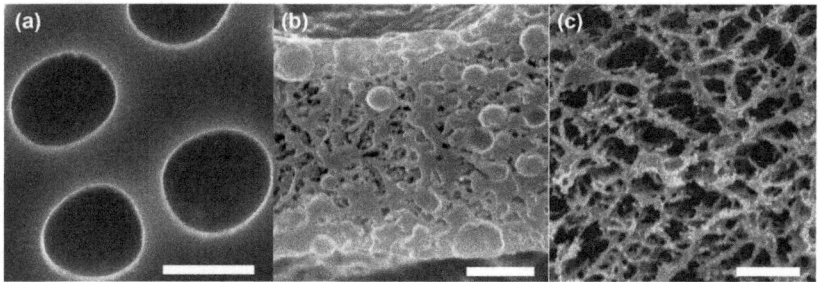

Figure 8.8 Cryo-scanning electron microscope images of: (a) carbohydrate in the form of a canola-based organogel (A. Zetzl, University of Guelph); (b) fat globules in non-dairy whipped topping (A. K. Smith, University of Guelph); (c) meat batter protein emulsion (A. Gravelle, A. Marangoni, S. Barbut, University of Guelph). Bars = 1 μm.

order to identify them as components in a food system, *e.g.* starch, fat and sugar. Cryo-SEM is the only method available to image ice cream, but as the ice cream is already frozen there is no concern over creating ice crystals during sample preparation. Cryo-SEM is the method of choice for water-in-oil emulsions such as margarine or butter; oil-in-water emulsions such as yoghurt, low fat margarine and meat emulsions; and air and oil-in-water emulsions such as whipped cream and ice cream. Figure 8.8 shows examples of research conducted using the cryo-SEM in the Department of Food Science at the University of Guelph (Ontario, Canada). Figure 8.8(a) shows a canola-based organogel from which the oil has been removed to expose the gel structure. A non-dairy whipped topping (Figure 8.8b) shows the interface between two air bubbles; the fat globules can be identified as globules of about 500 nm in the serum phase. Figure 8.8(c) shows a protein matrix in a meat batter. The protein forms a network which can be seen clearly because the ice has been removed by sublimation.

8.4 ATOMIC FORCE MICROSCOPY

The atomic force microscope (AFM) signal is derived from a flow of electrons between the probe and the surface, and therefore this is not a traditional electron microscopy technique where electrons interact with the sample to produce an image. The AFM is a scanning probe microscope. The first scanning probe microscope was a scanning tunneling microscope (STM), developed by Binnig and Rohrer in 1982, which was "the first instrument to allow surface imaging with atomic resolution in real space".[32] The STM was designed to create a topographic map of a surface by scanning a conductive sample with a very fine metallic probe at a distance of <1 nm from the surface, in a high vacuum environment. With the tip maintained at a constant distance from the sample, variations in current could be measured to provide vertical and lateral dimensions. However, the STM cannot image biological samples because they are not conductive.[33]

The AFM[34] as shown in Figure 8.9a evolved from the STM and was designed to scan the sample with a stylus mounted on the end of a cantilever[7] as illustrated (Figure 8.9b). The radius of the stylus tip is in the 0.11/10 nm range.[33] The system works under ambient conditions and is suitable for both conductive and insulating samples. A signal is generated in x, y, and z dimensions and, therefore, the AFM generates true 3D images, as represented in Figure 8.9, below.

The two basic AFM modes are contact mode and tapping mode. An AFM in either mode will raster the stylus (tip) mounted on the end of a cantilever across the surface, maintaining a constant force between the tip and the surface. A change in topography will force the tip to deflect. The tip can either be static, as in contact mode, or it can oscillate, as in tapping mode. The dimension of the tip, the applied force and the frequency of the oscillation determine the resolution capability of the system. Either the tip will scan the surface or the sample will be moved under the tip. As the tip responds to negative or positive forces generated by the sample, it will deflect proportionally.

In contact mode the stylus is positioned within a few angstroms of the surface and responds to repulsive forces which cause the tip to bend. The interaction between the tip and the surface is measured in

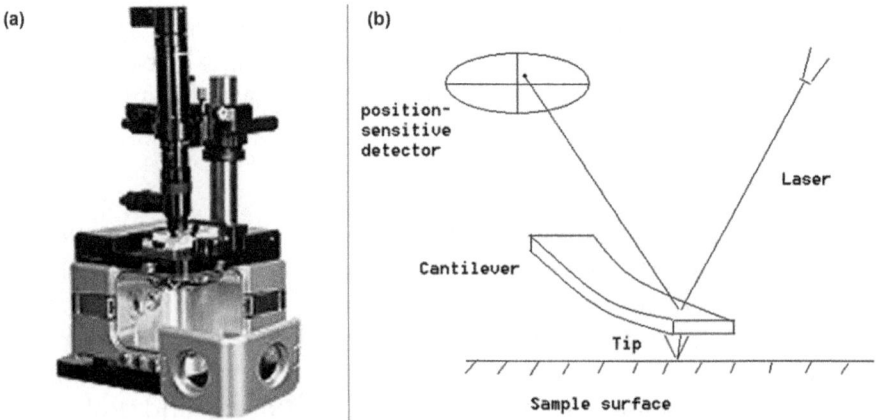

Figure 8.9 Atomic force microscopy: (a) the instrument;[34] (b) schematic of the generation of signal from interaction between the probe and the sample surface.[7]

nano-Newtons.[33] Non-contact mode AFM measures attractive forces. The deflection of the tip caused by interaction with the sample surface is detected by the change in reflection of a laser beam that is reflected back from the cantilever to a photodiode. The reflected signal changes position on the photodiode and is dependent on the strength of the response. The result is a force–distance curve.

One drawback of contact mode AFM is that the interaction between the tip and the surface causes distortion of the signal as a result of friction, thus limiting lateral resolution. As soon as the fragile tip comes in contact with the sample, it becomes dulled and particles can be displaced from the surface. Another limit to resolution is the noise of the photodiode.[33] It is this noise value that controls the displacement minimum of vertical resolution.

Tapping mode AFM (oscillating mode)[33] was designed to overcome the problem of lateral drift caused when the tip comes in contact with the surface. The tapping mode instrument uses an oscillating tip, which only comes into contact with the surface for a very short time. The response of the tip to forces generated by the sample is measured indirectly by an amplitude modulation detector with a lock-in amplifier. Tapping mode AFM operates best in a steady state environment and does not respond well to large changes in signal. It is limited to providing only topographic information because indirect measurement means that the detector can only measure average response. Thermal noise, generated by the oscillation of the cantilever, also limits vertical resolution. Thermal drift correction is available to compensate.

Resolution with AFM, as with any microscopy method, is limited by sample preparation steps which cause unavoidable changes to the sample. The basic requirements are to use a smooth ultra-clean surface upon which to immobilize the sample and to prepare the sample such that the areas of interest can be resolved. The sample must remain immobilized throughout the experiment, without slip or creep. The solution containing dispersed particles must be clean and, therefore, dialysis or numerous washing steps may be required. It could be necessary to fix the sample chemically to ensure stability throughout the experiment.

Examples of the application of AFM for imaging of food-based nanostructures are shown in Figure 8.10. Image 8.10(a) shows an ethylcellulose organogel prepared by mounting directly on cleaved mica. The trapped oil was removed by immersion of the sample in isobutlanol and then AFM was conducted in contact mode, in water. Figure 8.10(b) shows cocoa butter crystallized for two years at 22 °C

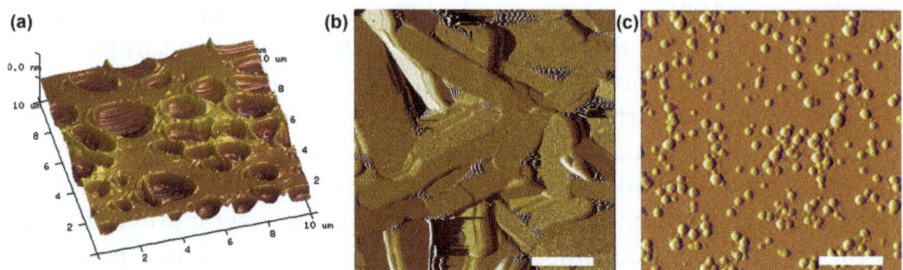

Figure 8.10   Atomic force microscopy images of: (a) organogel with height in nanometers and lateral dimension in microns (A. Marangoni, University of Guelph); (b) cocoa butter crystals in chocolate, Bar = 500 nm (A. Marangoni, University of Guelph); (c) protein assembly showing vertical dimension in nanometers (O. G. Jones, Purdue University, Department of Food Science, 745 Agriculture Mall Dr., West Lafayette, IN, 47907).

and imaged directly at room temperature using tapping mode AFM. The outline of the cocoa butter crystals can be clearly seen. Figure 8.10(c) is an amplitude image of a dried microgel assembled from a beta-lactoglobulin:pectin complex (4:1 ratio) at pH 4.5. The image was taken with a 2 N m^{-1} silicon cantilever with an aluminium reflective coating on an Asylum MFP-3D AFM.

Dry mode AFM, and/or the use of fixed samples, provides images with better resolution.[35] However, advances in AFM include the fact that oscillating mode AFM has advanced to the point of allowing biological complexes to be imaged in an aqueous environment.[35] Unfortunately, in aqueous mode, the AFM tip itself can become coated with organic material,[36] which will drastically reduce the sensitivity of the tip and limit the resolution of the images over time. This problem could limit the usefulness of wet cell AFM for some applications. There is also the opportunity to run cryo-mode AFM. By dramatically lowering the temperature of the measuring system within a vacuum system the structure of the tissue is preserved.[35] In cryo-mode traces of biological activity may be preserved and, therefore, measured.[36]

8.5   CONCLUSIONS

Nanometer resolution of food structure is possible through the use of transmission and scanning electron microscopy and atomic force microscopy, based on the fundamental design of these microscopes. There is a significant challenge presented by the fact that biological samples are non-conductive. This limitation must be conquered or accommodated without compromising the integrity of the original tissue, or the interpretation is meaningless. Equipment design does not guarantee nanometer resolution if the preparation of the sample does not provide an outcome that is true to the original. Fortunately, freezing the sample and using cryo-techniques have provided the solution for many TEM and SEM applications.

It is recommended to expand the knowledge base for any specimen by using multiple techniques. It is possible, for example, to corroborate SEM results by further processing the same sample for TEM. AFM can be conducted within the high vacuum environment of the SEM. Non-microscopy techniques can be used to correlate visual changes in microstructure with chemical or rheological changes in the tissue.

REFERENCES

1. http://www.chemistry.about.com/cs/howthingswork/f/blabundant.html. about.com chemistry accessed 07/2014.
2. https://www.webelements.com/carbon/atom_sizes.html. webelements accessed 07/2014.
3. http://www.globalsino.com/EM/. Practical Electron Microscopy and Database – An Online Book – accessed 07/2014.
4. http://www.fei.com/products/tem/tecnai-g2/?ind=MS. FEI accessed 07/2014.
5. M. Hayat, *Principles and Techniques of Electron Microscopy, Biological Applications*, Van Nostrand Reinhold Company, New York, NY, 1973, vol. 3, chapter 1, p. 3.
6. http://hitachi-htc.ca/products/electron-microscopes-and-focused-ion-beam/standard-and-variable-pressure-sem. Hitachi High Technologies Canada accessed 07/2014.
7. http://www.nanoscience.gatech.edu/zlwang/research/afm.html. Professor Zhong L. Wang's Nano Research Group accessed 07/2014.
8. http://www.chembio.uoguelph.ca/educmat/chm729/afm/resolution.htm. Hong-Qiang Li University of Guelph accessed 07/2014.
9. H. D. Goff, M. Liboff, W. K. Jordan and J. E. Kinsella, The effect of polysorbate 80 on the fat emulsion in ice cream mix: evidence from transmission electron microscopy studies, *Food Microstruct.*, 1987, **6**, 193.
10. http://en.wikipedia.org/wiki/Transmission_electron_microscopy. Wikipedia accessed 07/2014.

11. http://web.pdx.edu/~jiaoj/lect3_Contrast%20Formation-ph451-551-2012.pdf. Portland State University accessed 07/2014.
12. R. Egerton, *Physical Principles of Electron Microscopy: An Introduction to TEM, SEM, and AEM*, Springer Business and Media Inc., New York, NY, 1973.
13. S. Uhlemann, H. Müller, P. Hartel, Z. Joachim and M. Haider, Thermal magnetic field noise limits resolution in transmission electron microscopy, *Phys. Rev. Lett.*, 2013, **111**, 046101.
14. http://cmrf.research.uiowa.edu/scanning-electron-microscopy. University of Iowa accessed 07/2014.
15. J. J. Bozzola and L. D. Russell, *Electron Microscopy Principles and Techniques for Biologists*, Jones and Bartlett Publishers, Sudbury, MA, 1999.
16. M. A. Hayat, *Principles and Techniques of Electron Microscopy: Biological Applications*, Cambridge University Press, New York, NY, 4th edn, 2000.
17. K. L. Goh, D. F. Holmes, H. Y. Lu, S. Richardson, K. E. Kadler, P. P. Purslow and T. J. Wess, Ageing changes in the tensile properties of tendons: influence of collagen fibril volume fraction, *J. Biomech. Engineer.*, 2008, **13**, 021011.
18. A. Al-Amoudi, J.-J. Chang, A. Struder and J. Dubochet, Cryo-electron microscopy of vitreous sections, *Eur. Molec. Biol. J.*, 2004, **23**, 3583.
19. R. Dahl and A. Staehlin, High-pressure freezing for the preservation of biological structure: Theory and practice, *J. Electron Microsc. Techn.*, 1989, **13**, 165.
20. L. Staehelin and B. H. Kang, Nanoscale architecture of endoplasmic reticulum export sites and of Golgi membranes as determined by electron tomography, *Plant Physiol.*, 2008, **147**, 1454.
21. A. K. Smith, H. D. Goff and B. D. Sun, Freeze-substitution and low-temperature embedding of dairy products for transmission electron microscopy, *J. Microsc.*, 2004, **113**, 63.
22. http://www.emal.engin.umich.edu/courses/semlectures/focus.html. Electron Microbeam Analysis Lab accessed 07/2014.
23. http://www.emal.engin.umich.edu/courses/semlectures/se1.html. Electron Microbeam Analysis Lab accessed 07/2014.
24. http://www.ammrf.org.au/myscope/sem/background/whatissem/detectors.php. Australian Microscope & Microanalysis Research Facility accessed 07/2014.
25. http://www.emal.engin.umich.edu/courses/semlectures/bse1.html. Electron Microbeam Analysis Lab accessed 07/2014.

26. S. Alboriz, L. T. Lim and Y. Kakuda, Electrospinning of sodium alginate-pectin ultrafine fibers, *J. Food Sci.*, 2010, **75**(1), 100.

27. D. G. Dalgleish, P. A. H. Spagnuolo and D. Goff, A possible structure of the casein micelle based on high-resolution field-emission scanning electron microscopy, *Int. Dairy J.*, 2004, **14**, 1025.

28. www.physics.utah.edu/Phys5739/lecture/Lecture%202b.ppt. University of Utah accessed 07/2014.

29. Y. Yokoi and A. Kishida, On the relationship between two indices ("bulk density" and "dry-matter content") of dry-matter accumulation in plant organs, *Botanic. Mag., Tokyo*, 1985, **98**, 335.

30. P. Echlin, *Low Temperature Microscopy and Analysis*, Plenum Press, New York, NY, 1992, p. 43–48.

31. A. W. Robards and U. B. Sleytr, in *Practical Methods in Electron Microscopy*, ed. A. M. Glauert, Elsevier Science Publishing Co., New York, NY, 1985, vol. 10, pp. 11–20.

32. F. J. Guessivl, AFM's path to atomic resolution, *Mater. Today*, 2005, 32.

33. P. J. De Pablo and M. Carrión-Vazquez, *Imaging Biological Samples with Atomic Force Microscopy*, Cold Springs Harbor Laboratory Press, New York, NY, 2014, p. 167.

34. http://www.home.agilent.com. Agilent Technologies accessed 07/2014.

35. M. Gaczynska and P. A. Osmulski, AFM of biological complexes: what can we learn?, *Curr. Opin. Colloid Interface Sci.*, 2008, **13**, 351.

36. K. Tiede, A. B. A. Boxall, S. P. Tear, J. Lewis, H. David and M. Hassello, Review. Detection and characterization of engineered nanoparticles in food and the environment, *Food Addit. Contam.*, 2008, **25**, 795.

CHAPTER 9

Computer Simulation Techniques for Modelling Statics and Dynamics of Nanoscale Structures

DAVID A. PINK,*[a,b] M. SHAJAHAN G. RAZUL,[a,c] T. GORDON,[d] B. QUINN[a] AND A. J. MacDONALD[e]

[a] Physics Department, St. Francis Xavier University, Antigonish, NS, Canada; [b] Department of Food Science, University of Guelph, Guelph, ON, Canada; [c] Atlantic Computational Excellence Network (ACEnet), Antigonish, NS, Canada; [d] Physics Department, Boise State University, Boise, ID, USA; [e] OneZero Software, Fredericton, NB, Canada
*Email: dpink@stfx.ca

9.1 INTRODUCTION

Computer simulation of physical and chemical processes, and especially engineering applications, has played leading roles in creating
understanding and knowledge in the design of fabricated structures.
The use of computer simulation techniques spans all areas of physics,
and many areas of chemistry and engineering. However, there is one
area in which computer simulation does not yet play a role equivalent
to that played in those fields: food science, the rational design of food,
whether for aesthetic, economic, medical or nutritional purposes. It
might be argued that the science of designing food is essentially
satisfied by trial and error: that the accumulated wisdom of at least
10^3 centuries of cooking activities will not be significantly advanced

Edible Nanostructures
Edited by Alejandro G Marangoni and David Pink
© The Royal Society of Chemistry 2015
Published by the Royal Society of Chemistry, www.rsc.org

by any attempted mathematical modelling and computer simulation. This comment would miss the point and the cases of (i) protein folding and (ii) edible oil structures illustrate this. Finally, we cannot easily deconstruct spatial scales. Our book is about "Edible Nano-structures", but it is from those nanostructures that one obtains, *via* self-assembly, the micro- and milli-scale structures that are essential. Two examples illustrate this.

9.1.1 Protein Folding

Proteins catalyze nearly all reactions in organisms, contribute to structure, and perform many other functions necessary for the existence of life. The amino acids are linked sequentially and fold into the diversity of protein structures. The problem of predicting how a three-dimensional (3D) protein folds from a one-dimensional (1D) sequence has been called the "Protein-Folding Problem" (PFP). The role of proteins in disease processes[1–3] and drug design[4,5] is of immense interest. We now turn briefly to highlighting how such a fundamental problem is addressed. There has been considerable progress in the last 50 years, with the elucidation of about 84 000 experimental structures found in the Protein Data Bank (PDB)[6] which have identified the key molecular interactions for the diversity of structures. Experimental techniques such as X-ray crystallography,[7] nuclear magnetic resonance (NMR) spectroscopy[8] and cryo-electron microscopy[9] have provided the most information on protein structure. Experimental probes into the search for mechanisms of protein-folding have spurred the development of sophisticated Temperature-Jump methods using fluorescence of specific amino acids and even single molecules, which have been investigated on time scales of >1 μs.[10] Today, the plethora of experimental data contained in numerous bioinformatics databases[11] enables structure prediction, which is predicated on the idea that similar sequences may result in similar structures, utilizing sophisticated algorithms such as machine learning and neural networks.[11] To gain insights into the physics of the process, a variety of physical models capturing molecular interactions (*via* force fields)[12] and the development of lattice-based models[13] have been used on highly parallel computers and other novel computational techniques.[14] Despite advances in experimental and computer-based techniques, however, the PFP is still largely unresolved and has been summarized by Dill and MacCallum:[15] (i) What is the folding code? How does a 1D sequence fold into a 3D structure?; (ii) What is the mechanism and kinetics of folding?; and (iii) Can a computer algorithm be used to predict it? The literature on protein-folding is vast.[11,15] The PFP highlights and underscores the difficulty of simulating biological molecules. Applications to food science will also benefit from an improved understanding of molecular level detail of protein-folding. This brief review can only touch on the basics and the highlights of approaches to the protein folding problem.

9.1.2 Edible Oil Structures

Functionality is a term describing the suitability of an edible oil for a given culinary purpose. Functionality is related to "oil binding capacity", the ability of an edible oil to retain the liquid oil in a state appropriate to its use. It is of considerable interest to understand how hierarchies of fat crystal networks in edible oils retain liquid oil because this can lead to understanding what determines oil binding capacity and viscoelasticity and, ultimately, functionality.[16,17] The importance of understanding this arises because one would like to replace the *trans* fats, which form solids in the edible oil and assist in oil binding capacity, with "something else". But, in order to replace them, one must know what characteristics the solid fats create in the oil in order to mimic them by a different healthier component such as, for example, edible polymers. Oils and solid fats are composed of a variety of triacylglycerol (TAG) molecules. The possibility that separation of their components into many co-existing phases – possibly micro- and even nano-phases, depending upon the environment and involving interfaces of different characteristics and complexities – must be considered. Indeed it is possible that it is exactly phenomena such as this which determine functionality.[18–22] Although much work has been done using transmission electron microscopy (TEM) and rheology, a knowledge of the 3D structures of fat crystal networks is not yet clear and it is in this area, for example, that mathematical modelling and computer simulation can play an essential role on at least two scales. Small-scale modelling, with a characteristic length scale of ~ 10 nm and a time scale of ~ 5–20 ns, can provide information about nanoscale phase separation. For this it is essential to make use of atomic scale molecular dynamics. Larger structures with lengths scales of $\sim 10^2$–10^5 nm and time scales of up to 10^6 ns require the use of coarse-grained models in which the essential aspects of the structures are emphasized.

This short chapter builds on a recent review,[23] but it has developed some aspects, omitted others, and introduces examples not in the review. The intent is to outline major computer simulation techniques available for modelling static and dynamic structures, justify their use and outline how to use them, and comment on recent applications which have been used to model foods and systems similar to food. There are many very good reviews of computer simulation techniques. Those listed reflect a cross section of the many excellent works. One area in which much progress has been made in the last five years is the coarse-graining of interactions between moieties. This has essential application to any approach which seeks to replace atomic scale phenomena by a model which represents individual atoms by aggregates of atoms in order to reduce the complexity of the model and the time taken to carry out simulations. This approach has become essential if, for example, one is to further advance understanding of the dynamics of protein conformational changes. Complex systems such as solvents with proteins and other molecules in which the time scales for competitive processes can be of the order of 1 ms or longer cannot, at present, be simulated using atomic scale models and can require massive coarse-graining. It is essential therefore, that one understands how to move from the atomic to the mesoscale.

This chapter is not about models but about computer simulation techniques to investigate models or to use models to predict or explain experimental data. Nonetheless, one cannot divorce the techniques from the models because the techniques are frequently invented in order to study particular mathematical models. We have been selective and described techniques which either continue to play fundamental roles in computer simulation of matter or have shown an explosive use in the last decade. Thus, this chapter does not explicitly address models such as effective field, mean field or self-consistent field approaches, but it does describe models which cannot be utilized without computer simulation and have shown greatly increased use, such as Dissipative Particle Dynamics (DPD) and Lattice Boltzmann (L-B) theory. This chapter is intended for researchers who want to model food structures and related systems mathematically, and to carry out the computer simulation required by the unavoidably complex models. We have attempted to tell the story "the way it is" and have not held back in describing the "technology of computer simulation". Nonetheless, we have included a number of applications, the descriptions of which vary in the amount of detail. Not all of them come directly from edible nanostructures, but they are all concerned with modelling soft biological materials on a nanoscale. We see these as "problems" that can be used to test one's knowledge because the "answers" are also given. At the very worst, if the results of the reader's simulation do not agree with the "answers", they will have an opportunity to write a dissenting note to the journal in question. This chapter begins with an outline of the background to statistical mechanics, intended to ease one into a thermodynamic and statistical mechanical frame of mind. One need not study this section before moving on to the parts that describe "how to do it", but it should be here, somewhere, and we have decided that the beginning is best. This is followed by descriptions, with selected examples, of Atomic Scale Molecular Dynamics. Subsequently, the coarse-graining of atom–atom and atom–molecule interactions is treated, and the chapter ends with descriptions and applications of Monte Carlo approaches and the use of Dissipative Particle Dynamics.

9.2   THEORY I: SOME BACKGROUND

9.2.1   Statistical Mechanics and Thermodynamics

There are many books on this subject and this Background is intended only to give an overview of aspects relevant to carrying out computer simulation. For this section, books on a range of levels are by Yourgrau, van der Merwe and Raw,[24] Chandler,[25] Mandl,[26] Evans and Morris,[27] Hecht,[28] Hoover,[29] Pathria,[30] Greiner, Niese and Stöcker,[31] Baierlein[32] and Reichl.[33]

9.2.2   Ensembles

All thermodynamic potentials are expressed in terms of conjugate dynamical variables and modelling involves choosing which of those variables to keep constant. Those chosen define the ensemble. Thermodynamics tells us that we can keep three such variables fixed, but none of them may be mutually conjugate. Examples of mutually conjugate variables are absolute temperature (T) and entropy (S), volume (V) and pressure (p), chemical potential (μ) and particle number (N), and stress (σ_{jk}) and volume-weighted strain $(V\varepsilon_{kj})$. The NpT and the NVT ensembles are popular because they reflect conditions closely associated with many experiments: those in which the amount of substance and the temperature are kept fixed as well as either the applied pressure or a fixed volume.

   One can choose to use the Microcanonical (NVE) where the energy, E, is fixed, the Canonical (NVT), and the Grand Canonical (μVT), ensembles. The physical meanings of these ensembles are easily understood: the Microcanonical ensemble describes a system of fixed volume and fixed number of objects which is isolated from the external world and is unaware of the surrounding temperature – there is no thermal contact and no energy transfer. The Canonical ensemble relates to systems where the number of objects is fixed but which is in thermal contact with an environment (a "heat bath") characterized by temperature, T. The Grand Canonical ensemble is useful when there is not simply thermal contact, and thus energy exchange between two environments, but also the exchange of objects (particle exchange); it makes use of the fundamental relationship that two systems in thermal contact (energy exchange) and particle exchange must possess the same value of the chemical potential, μ.

9.2.3 Ergodicity

An important concept in statistical mechanics, and one that underpins successful implementations of computer simulation techniques, is ergodicity. Consider a system with a set of generalized coordinates – variables which define the state of a system. The set of all the variables defines the coordinates of the system in phase space. At any time, the values of the variables identify the location of the system in phase space. As the values of these variables change, the system moves on a trajectory through phase space. The Ergodic hypothesis states that for certain systems the time average of their properties is equal to the average over all of phase space. A system is ergodic if the average values of its variables have been obtained by sampling all of space. That is, the computer simulation has been run for sufficiently long that all of phase space has been adequately sampled. The power of the Monte Carlo method is that it is based upon "importance" sampling so that it is unnecessary to sample all of phase space.

9.2.4 Boundary Conditions (BCs)

These determine the structures or flows that emerge from simulations and here we mention the BC common to all methods: Periodic BCs. All computer simulations are carried out in finite-size spaces and are accordingly bounded by walls or enclosing surfaces. Periodic BCs were designed to eliminate their effect. If a simulation volume is bounded by two parallel surfaces separated by a mutually perpendicular vector $\vec{L}$, then, if a component of the system possessing velocity $\vec{v}(s)$ and position $\vec{r}(s)$ between the surfaces at simulation step (time or Monte Carle step), s, were to move to position $\vec{r}(s+1)$ lying outside the surfaces with velocity $\vec{v}(s+1)$, the position is reset to

$\vec{r}(s+1) - \vec{L}$ with velocity $\vec{v}(s+1)$. Below we shall mention BCs specific to particular approaches. Other BCs appropriate to physical surfaces involve surfaces that cannot be penetrated by any part of an object of the simulation. This implies an infinite repulsive force at the interface and so cannot be treated by Atomic Scale Molecular Dynamics (below). Surfaces can have a variety of soft algebraic repulsive potentials such as $1/d^{\alpha}$ depending upon their origin.

A BC of major importance at solid–fluid interfaces is the no-slip BC: the requirement that the local speed of the fluid parallel to a stationary surface is zero.[34] In solving the Navier Stokes equation, this BC is imposed and determines the flow. In the case of models which involve a particle- or object-based theory, such as Dissipated Particle Dynamics, or one in which distributions are propagated from one lattice site to another, such as Lattice Boltzmann theory, one must devise kinetics to realize this BC. For both theories, Bounce-Back BCs have been devised. These are described below under the sub-section Theory IV.

9.3 THEORY II. MOLECULAR DYNAMICS

9.3.1 General Equations

In the following paragraphs we will outline the theoretical basis for molecular dynamics. First, we will simply provide the connection between molecular dynamics (*via* statistical mechanics) and thermodynamics. Consider a system of N atoms,[35] with the positions and momenta of the atoms given by $\vec{r^{N}} \equiv (\vec{r_1}, \ldots, \vec{r_N})$ and $\vec{p^{N}} \equiv (\vec{p_N}, \ldots, \vec{p_N})$ respectively. These can be thought of as coordinates in a $6N$ dimensional space, more commonly termed phase space. A particular point in phase space can be defined by the abbreviation, $\Gamma = (\vec{r^{N}}, \vec{p^{N}})$. Therefore, for a particular instantaneous property, A_{int}, the value can be written as the function $A_{\text{int}}(\Gamma)$. If the system evolves in time, so that Γ changes, $A_{\text{int}}(\Gamma)$ will also change as a result. The experimentally observable "macroscopic" property, A_{obs}, is then given by the time average of $A_{\text{int}}(\Gamma)$ taken over a long time interval, t_{obs}:

$$A_{\text{obs}} = \lim_{t_{\text{obs}} \to \infty} \frac{1}{t_{\text{obs}}} \int_0^{t_{\text{obs}}} A_{\text{int}}(\Gamma(t)) \mathrm{d}t. \qquad (9.1)$$

The time evolution of $A_{\text{int}}(\Gamma(t))$ for a large number of molecules or atoms is rather complex, thus the time average given above can be replaced by the ensemble average for a closed system; this is the essence of the ergodicity mentioned earlier.[26] The ensemble average can be regarded as a collection of points, Γ, where, in phase space, these points are associated with a probability density distribution, ρ_{ens}, given by

$$\rho_{ens}(\Gamma) = \frac{\exp[-\beta H(\Gamma)]}{\iint \exp[-\beta H(\Gamma)]\mathrm{d}\overrightarrow{r^N}\,\mathrm{d}\overrightarrow{p^N}}, \tag{9.2}$$

where the Hamiltonian of the system is given by

$$H(\Gamma) = \frac{1}{2m}\sum_{i=1}^{N}|\overrightarrow{p_i}|^2 + U(\overrightarrow{r^N}) \tag{9.3}$$

where $\beta = 1/k_B T$ for a closed system at constant temperature, T, k_B is Boltzmann's constant, and U the potential. A_{obs} can then be expressed as

$$A_{obs} = \sum_{\Gamma} A(\Gamma)\rho_{ens}(\Gamma). \tag{9.4}$$

The above expression may be re-expressed by replacing ρ_{ens} by a "weight" function, $w_{ens}(\Gamma)$, that satisfies

$$\rho_{ens}(\Gamma) = Q_{ens}^{-1}w_{ens}(\Gamma), \tag{9.5}$$

with

$$Q_{ens} = \sum_{\Gamma} w_{ens}(\Gamma). \tag{9.6}$$

Q_{ens} is called the partition function, which acts to normalize the probability density.[30] Therefore, in the Canonical ensemble (defined as constant N, V and T), the probability density is proportional to

$$\exp[-\beta H(\Gamma)], \tag{9.7}$$

with the partition function expressed in a quasi-classical form given by

$$Q_{NVT} = \frac{1}{N!\,h^{3N}}\iint \exp[-\beta H(\Gamma)]\mathrm{d}\overrightarrow{r^N}\,\mathrm{d}\overrightarrow{p^N} \tag{9.8}$$

Since the partition function is a dimensionless quantity, the integral in equation (9.8) must be divided by a parameter that contains the dimensions of phase space.[24] Planck's constant, h, is chosen for two important and fundamental reasons. The first is that it has dimensions of length multiplied by momentum, therefore a division by h^{3N} formally satisfies the requirement for Q to be dimensionless. Second, the uncertainty principle is satisfied (where $h = dpdr$ is the element within which one is unable to be sure that the energy has varied). The origin of the 1/N! term stems from the indistinguishability of particles.[31] The appropriate connection to a thermodynamic function is through the Helmholtz free energy,[29]

$$f = -\frac{1}{\beta} \ln Q_{NVT}. \tag{9.9}$$

Relationships to other thermodynamic functions are easily obtained with the appropriate differentiation.

Now, with the connection between statistical mechanics and thermodynamics established, the link between statistical mechanics and molecular dynamics is made through the equations of motion, particularly Hamilton's equation of motion. Formally this link is established by Liouville's theorem, which states that the rate of change of phase space density, ρ, experienced by a fixed point in phase space is given by

$$\frac{\partial \rho}{\partial t} = -\sum_{i=1}^{N} \left(\frac{d\vec{r_i}}{dt} \frac{\partial \rho}{\partial \vec{r_i}} + \frac{d\vec{p_i}}{dt} \frac{\partial \rho}{\partial \vec{p_i}} \right) \tag{9.10a}$$

which depends explicitly on time. However, an alternative formulation can be derived from a phase space point travelling in space; the rate of change of phase space density, ρ, in this case is now 0, after substituting equation (9.10a) in equation (9.10b) for $\partial\rho/\partial t$,

$$\frac{d\rho}{dt} = \frac{\partial \rho}{\partial t} + \sum_{i=1}^{N} \left(\frac{d\vec{r_i}}{dt} \frac{\partial \rho}{\partial \vec{r_i}} + \frac{d\vec{p_i}}{dt} \frac{\partial \rho}{\partial \vec{p_i}} \right) = 0, \tag{9.10b}$$

where $d\vec{r_i}/dt$ is the velocity and $d\vec{p_i}/dt$ is the force. This result also satisfies the condition that at thermodynamic equilibrium the ensemble average is time independent, hence the phase space density does not explicitly depend on time.[24]

9.3.1.1 Molecular Dynamical Equations. A collection of interacting atoms or molecules is followed by solving the equations of motion for each particle, through time.[36] This motion can be described by the Newtonian equation,

$$\vec{F_i} = m_i \frac{\mathrm{d}^2 \vec{r_i}}{\mathrm{d}t^2},$$ (9.11)

where $\vec{F_i} \equiv -\overrightarrow{\nabla_r} U(\overrightarrow{r^N})$ is the force exerted on atom i, m_i the mass, $\vec{r_i}$ is the position vector of particle i, and $U(\overrightarrow{r^N})$ is the interaction potential function of atom i that includes all particles. The Newtonian equation can be written in Hamiltonian form,

$$\frac{\partial H}{\partial \vec{p_i}} = \frac{\vec{p_i}}{m} = \frac{\mathrm{d}\vec{r_i}}{\mathrm{d}t}$$ (9.12)

and

$$\frac{\partial H}{\partial \vec{r_i}} = \frac{\partial U(\overrightarrow{r^N})}{\partial \vec{r_i}} = -\frac{\mathrm{d}\vec{p_i}}{\mathrm{d}t}.$$ (9.13)

Hamiltonian equations give $6N$ first order differential equations as opposed to Newton's $3N$ second order equations. In principle the latter enables more efficient algorithms to be developed to solve for the motion of individual particle trajectories. Therefore, in a typical molecular dynamics simulation, the motion of a phase space point (or the trajectories of the particles) is followed in time. For numerical treatment, time is discretized and the equations of motion are solved (numerically) at each step in time, to provide positions and momenta for all the particles. The number of time steps can typically vary from thousands to millions, depending on the system of interest.

There are two principal techniques utilized to solve the equations of motion: the predictor–corrector and the Taylor series methods. Modifications of these two principal methods are also used.[37] All these methods allow one to obtain numerical solutions to differential equations. As an example, a general predictor–corrector algorithm is outlined below:[38]

a) at a time t, use the current positions and current velocities and their time derivatives to predict new positions and velocities at a new time $(t + \delta t)$
b) at the new positions, forces are evaluated
c) utilizing these forces, correct the positions, velocities and their derivatives

d) finally, variables of interest are calculated and time averages are performed; the process now iterates by returning to step a).

In this way, the dynamics in the system evolves, one time step at a time. Phase space is adequately sampled in a typical molecular dynamics simulation after a sufficient amount of time has elapsed depending on the nature of the system and the lifetimes inherent to the properties under investigation.

9.3.2 Analysis of Data from Molecular Dynamics

A molecular dynamics (MD) simulation produces trajectories of its particles, and the particles composing the system of interest are typically contained within a simulation cell. Periodic boundary conditions, where a simulation box is surrounded by images of itself, are applied to mimic bulk behaviour.[36] This is done so every particle in the simulation cell is unaffected by boundary constraints, which would be problematic if only bulk behaviour is desired. Properties of interest are then calculated in various ways to generate useful data that may be compared to experimental results. In the following paragraphs, a few select properties will be described to show the range of properties that can be obtained from MD.

The temperature of the system is compared to the thermodynamic temperature through the equipartition theorem,[38]

$$\sum_{i=1}^{N} \frac{1}{2} m v_i^2 = \frac{3}{2} N k_B T, \tag{9.14}$$

where the sum is over all particles in the system, m is the mass, v_i is the velocity, N is the number of particles and T is the temperature. The pressure is calculated from the virial theorem,

$$P = \frac{N k_B T}{V} - \frac{1}{3} \sum_{\text{all pairs}} \left(\frac{-\vec{r_{ij}} \cdot \vec{F_{ij}}}{V} \right) \tag{9.15}$$

where P is the pressure, V is the volume, $\vec{r_{ij}}$ is the distance vector between particles i and j, and $\vec{F_{ij}}$ is the force between the particles.

Another useful property obtained in liquid state simulations is the diffusion constant, which can be calculated from the mean square displacements of the particles in the system. The diffusion constant, D, is then obtained through the Einstein relation,

$$6Dt = \langle R^2 \rangle, \tag{9.16}$$

where t is the time elapsed in moving. In principle, profile functions of any other system property, such as the energy or pressure, can be similarly extracted, but are always recorded in 2D slabs of the simulation box.[38]

One simple structural function commonly obtained is the radial distribution function, $g(r)$. This is defined as the average number of molecules (or atoms) found at a particular distance from a specific reference position. The radial distribution function provides information on near neighbour distances and coordination numbers. For a model water system, this approach can yield g_{OO}, g_{HH} and g_{OH} distribution functions. The correlation functions for O–O, O–H and H–H can be experimentally determined from X-ray diffraction (g_{OO}) and neutron diffraction (g_{OH} and g_{HH}). Other functions that can be used to probe the local orientational order are the dipole moments of the water molecules, the O–H and H–H vectors.

Molecular dynamics enables the calculation of many thermodynamic properties including heat capacities, isothermal compressibilities and dielectric constants, to name a few.[36–38] In most simulation studies, key thermodynamic quantities such as the Gibbs free energy and enthalpy are of more practical use. However, the free energy is difficult to estimate for biomolecular systems and liquids because a standard MD simulation does not adequately sample phase space in regions that would make contributions to the free energy. The vast majority of free energy calculations are generally formulated in terms of estimating the relative free energy differences, ΔG, between two equilibrium states, since the thermodynamic properties between two such states are of practical importance. Methods such as thermodynamic perturbation and thermodynamic integration are used to calculate such free energy differences, these methods are described elsewhere.[39]

9.3.3 Empirical Force Fields

No discussion of MD would be complete without a discussion of force fields use in biomolecular simulations. Many force fields for atomistic and biomolecular simulations have been developed over the past decades.[12] No attempt is made in the present review to provide an exhaustive review of all such systematic development of force fields. Instead, the primary focus taken here is on the form of a force field as currently used in biomolecular simulations and limitations of its use.[40]

The most commonly used form of a force field, called a "Class I force field", is given by

$$U(\vec{r}) = \sum_{\text{bonds}} k_b(b - b_0)^2 + \sum_{\text{angles}} k_\theta(\theta - \theta_0)^2 + \sum_{\text{torsions}} k_\varphi[\cos((n\phi + \delta) + 1)]^2$$

$$+ \sum_{\text{non-bonded pairs}} \left\{ \frac{A_{ij}}{r_{ij}^{12}} - \frac{C_{ij}}{r_{ij}^6} \right\}_{LJ} + \left\{ \frac{q_i q_j}{r_{ij}} \right\}_{\text{Coulomb}}$$

(9.17)

where $U(\vec{r})$ denotes the potential energy of a molecule which is a function of the position $\vec{r}$ of N atoms. The bonded interactions are given by the first three summations: bonds (1–2 interactions), angles (2–4 interactions), and torsions (1–4 interactions) respectively, as illustrated in Figure 9.1.

The first term in equation (9.17) takes into account the bond length, b, between atoms 1 and 2 and is given by a harmonic potential between it and the reference value, b_0. Likewise, the bond angle (atoms 2–3–4) is given by a harmonic potential. The third term models the energy changes as the torsional angle (or dihedral angle, atoms 1–2–3–4), ϕ, changes. The torsion term usually includes "improper" dihedral angles; this term is included to ensure planarity around planar rings and sp^2 atoms, for example. The atoms (usually four) need not be connected to the atom that has such planar characteristics. In addition, other functional forms also exist to ensure rigidity and planarity. The non-bonded interactions are typically represented by the short ranged Lennard-Jones (LJ) 6-12 potential and a coulombic term. The former term captures a combination of the dispersion and exchange repulsion forces and is often called the "van der Waals" term. The latter term describes electrostatics that use partial charges, q_i on each atom, that interact *via* Coulomb's law; the final sum (over pairs of atoms i and j) excludes 1–2 and 1–3

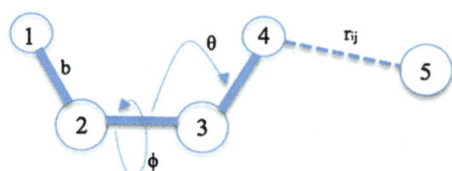

Figure 9.1 Force field parameters illustrated with a schematic representation. Atoms are represented by circles; covalent bonds are represented by thick solid lines, b; non-bonded interactions by a dashed line, r_{ij}.

interactions and often uses separate parameters for 1–4 interactions as compared with those used for atoms separated by more than three covalent bonds. The parametrization that describes the bonded interactions is quite accurately obtained from quantum chemical calculations or spectroscopic data.[41–43] It is difficult and computationally expensive to obtain accurate parameters describing the non-bonded van der Waals interactions from quantum mechanics. Therefore they are usually obtained by fitting molecular simulation results to experimental data. Vapor–liquid equilibrium (VLE) properties are relatively sensitive to the non-bonded interactions, and often force-field parameters are adjusted to describe these data accurately. It is well established in the literature that the van der Waals parametrization assigned to an atom in one molecule is fairly well transferable to the parametrization of the same atom in a different molecule.[40,44–47] Equation (9.17) is thus the simplest potential energy function that can reproduce the basic features of a biomolecular energy landscape at an atomic level of detail, and it has proven to provide significant insights into a broad range of properties. Such a potential is called a force field; all the initial parameters and constants of the potential energy function as described have been derived from experiments and quantum chemical calculations from the appropriate molecule or fragments of the different chemical environments. The force field is also further tested and tuned to reproduce structural and thermodynamic properties before it is used in production simulations.[48]

The common force fields use in biomolecular simulations are CHARMM,[12,49] AMBER,[50] GROMOS[51,52] and OPLS[53] among others, all belonging to the so called "Class I force fields". Class II force fields are developed to contain more details, such as higher order terms (for bonds, angles and dihedrals) and more accurate descriptions of spectra (*e.g.* vibrational properties).[40] These force fields have provided reasonable agreement with an array of experimental data, often within the accuracy of the varied experimental methods. The use and acceptance of these force fields have provided detailed insights into structure–function relationships in biomolecules. Most of these simulations are undertaken to provide detailed knowledge on systems inaccessible by current experiments.

The vast majority of force fields are tuned to suit the temperature and pressure conditions, thus going outside the prescribed tuned range could be problematic. Even the choice of the available water models typically used could bias the results towards a more or less structured equilibrated structure of the biomolecule and must be chosen with care.[54–56] Another key limitation is the effective inclusion of polarizability into force fields.[57] It is clear that no force field can accurately reproduce the myriad complex interactions and properties of real systems. However, they can produce qualitative and quantitative data that address static properties (*e.g.* structure and energetics) and dynamic properties (*e.g.* diffusion constants and correlations). All simulation studies should ultimately be checked against experiments to confirm their validity.

9.3.4 The Potential of Mean Force (PMF)

The potential of mean force (PMF) is an important concept in the statistical mechanical theories of liquids and complex molecular systems.[58] The PMF can be simply defined as that potential that provides the average force over all configurations of all the particles acting on a particle *i*. The idea can be extended and illustrated with two particles, *i* and *j*, separated by a distance, *x*, in a solvent mixture. If the coordinate chosen is *x*, then the potential that gives the average force over all the configurations of the solvent (during an MD run) along the coordinate *x* of particles *i* and *j* is the PMF. The PMF can be considered here to be the (Helmholtz) free energy. Therefore, along some appropriately selected coordinate, ξ, the free energy (*i.e.* the PMF) may be defined by the following expression,

$$A(\xi) = -k_B T \ln \left[\frac{\langle p(\xi) \rangle}{\langle p(\xi^*) \rangle} \right] + const, \tag{9.18}$$

where $\langle p(\xi) \rangle$ is the average distribution function and $\langle p(\xi) \rangle$ is an arbitrary constant. The average distribution function can be defined along the selected coordinate, ξ, defined from a Boltzmann weighted average,

$$\langle p(\xi) \rangle = \int d\vec{R} \delta(\xi'[\vec{R}] - \xi) e^{-\frac{U(\vec{R})}{k_B T}} \times \left(\int d\vec{R} e^{-\frac{U(\vec{R})}{k_B T}} \right)^{-1} \tag{9.19}$$

where $U(\vec{R})$ is the total potential energy of the system in some co-ordinate system $\vec{R}$ and $\xi'[\vec{R}]$ is a function dependent on an appropriate degree of freedom in a dynamical system, which may be the position of a single particle or a fixed distance as given earlier. The expression in square brackets as given in equation (9.18) can be replaced by the radial distribution function, $g(r)$, which provides a convenient route to calculating the PMF.[59] However, for MD simulations that involve processes with large free energy barriers, or if the timescales are long or too computationally demanding to be simulated directly, the required PMF may not be calculated by regular MD.[40] In other words, the adequate sampling of phase space to obtain appropriate averages required for the calculation of a free energy profile may not be possible. To ensure that proper sampling can be achieved, the umbrella sampling method may be used.[60] Both MD and MC simulation techniques may employ the umbrella sampling method. The method involves first selecting an appropriate reaction coordinate, and restraining potentials are used over a series of sampling windows (to obtain the ensemble averages) that span the range of the reaction coordinate. The restraining potentials in the system bias the sampling regions of conformational space that would not otherwise be accessible during direct sampling, enabling unfavourable states to be sampled. The resulting simulations then contain a series of histograms which show the biased distribution of the reaction coordinate from each sampling window. These histograms are then unbiased and combined,[61,62] usually using the weighted histogram analysis method (WHAM), among others,[63] to enable the PMF to be calculated.

9.3.5 Carrying out an AMD Simulation

In what follows, we are concerned primarily, but not exclusively, with describing the procedure followed by the simulation package GROMACS (GROningen MAchine for Chemical Simulations).[64] GROMACS incorporates many analysis packages, one of which is the ability to compute PMFs. The general approach to running an AMD simulation is as follows:

[1] Define the questions about a particular phenomenon or event for which you wish answers.
[2] Decide whether AMD can provide these answers or elucidate some parts of the question so that you can formulate another method of answering your question.
[3] If AMD is applicable, determine what tools are needed to perform your AMD simulation:
    (a) What software is available for you to use and how easy is it to get help if you run into problems?
    (b) Do you have a coordinate file to describe your molecule of interest? Have you identified and specified the location of all the atoms of the molecules that you want to model? Do you know which pairs of atoms are chemically bonded to each other *via* covalent bonds? If not, do you have the

ability or resources to answer these questions? Can you create the files in your computer that adequately describe the molecules of interest?

(c) What force fields are appropriate for the phenomenon under study?

[4] Generate topologies for the molecules under study by identifying and specifying the relative locations of all the atoms of the molecules that you want to model and specifying which pairs of atoms are bonded to each other.

[5] Define the simulation box by specifying its dimensions and whether you are using periodic boundary conditions or not.

[6] Insert the appropriate number of molecules in the appropriate orientations into the simulation box.

[7] Fill the simulation with "solvent". This can be TAGs in the case of oils or, more commonly, water. It is likely that you will have to create the TAGs, but libraries of water molecules exist. Simulations can also be carried out in vacuum.

[8] Determine the appropriate simulation parameters for the equilibrium simulation that are consistent with the chosen force field and ensemble. For example:

(a) Choose the cut-off length for the van der Waals and electrostatic interactions. These are the distances beyond which the interaction is set equal to zero because it is very small, or it is handled by a procedure that is not a force field. Thus, electrostatic interactions beyond their cut-off can be handled by the Ewald summation procedure.[65–68]

(b) Temperature.

(c) Pressure.

(d) Magnitude of the simulation time step. This must be small enough so that the system satisfies the requirement of ergodicity.

[9] Run the simulation such that the system relaxes sufficiently, as shown by the energy coming to equilibrium and not changing further.

[10] Run production simulations for a length of time such that it sufficiently samples phase space, *i.e.* it satisfies the requirement of ergodicity. This journey through phase space – the space defined by the dependent variables of the simulation – is called the "trajectory". It is at this step that we carry out averaging.

[11] Analyse the trajectory to obtain information about the observed phenomenon or event.

Example 1: *Nanoscale phase separation in edible oils*

Recently, simulations were carried out to investigate the validity of the assumption that a two-component TAG liquid oil was homogeneous in the neighbourhood of a solid fat particle. The particle was assumed to be a single nanocrystal (below) of tristearin and the oil was a mix of triolein with a low concentration of another molecule, either with one oleic moiety replaced by an elaidic chain, or two oleic chains replaced by two elaidic chains. These two minority components can be called OOE and EEO, respectively, with triolein denoted OOO, and are shown schematically in Figure 9.2(A) where the position of the *cis* or *trans* double bond can be seen. The radial interactions between non-bonded TAG atomic moieties were

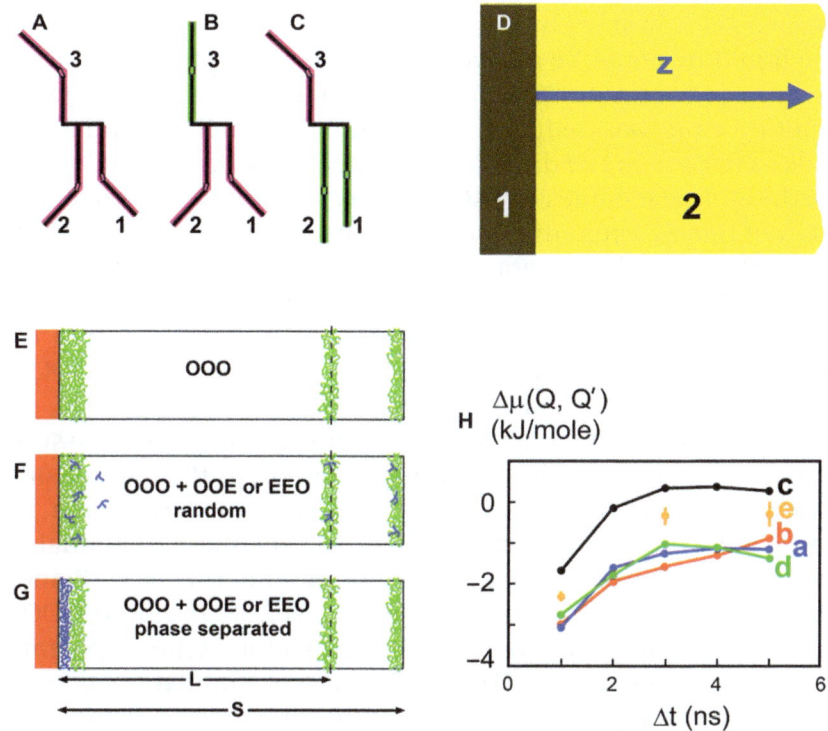

Figure 9.2 Nanoscale phase separation in edible oils. (A, B, C) Schematic models of OOO, EOO and EEO showing the *trans* and *cis* double bonds. (D) The model continuum tristearin solid (1) and the liquid oil (2). (E, F, G) The simulations run. Pure OOO (E), OOO + randomly-mixed EOO or EEO (F), and OOO + phase separated EOO or EEO. The total simulation length is *S*, while the length for averaging is *L*. (H) Chemical potentials for the four cases (F and G) with that of OOO as reference state. Error bars are shown in orange.

modelled using GROMACS.[69,70] TAGs possess only CH, CH_2, CH_3, O
and C=O moieties, and the force fields given by Berger *et al.*[71] were
used together with the v-rescaled thermostat that keeps the temperature constant.[64]

Although the *NpT* ensemble is the "natural" ensemble to use, the
NVT ensemble was employed in order to make use of the experimentally known density of the oils. Because the tristearin molecules
in the solid vibrate at frequencies much higher than the molecules in
the liquid oil and do not undergo diffusion on the short time scales of
the simulation, the liquid oil and the tristearin nanocrystals were
modelled differently. Tristearin nanocrystals were modelled as continuum solids possessing a known density while the liquid oil components were represented by atomic scale models. This is shown
schematically in Figure 9.2(D–G). The van der Waals interactions
between a static average infinitely large solid continuum and an atom
in the liquid oil could be analytically computed. By treating the solid
as an average continuum and not as a crystal composed of atoms, the
computing time was reduced significantly. Periodic boundary conditions were used in the directions parallel to the nanocrystal surface.
Since the atomic moieties of TAGs are very similar, the coefficients, C_6^{ij}
and C_{12}^{ij} of the 6-12 potential interaction between atoms labelled i and
j were replaced by their averages over the atomic moieties making up
the solids. This was not done for the TAG molecules of the liquid oils.

Three cases were considered for each of the two minority components of the liquid oil (Figure 9.2E–G). These were pure OOO (E),
OOO + EOO or OOO + EEO randomly mixed together (F), and
OOO + EOO and OOO + EEO with the minority components assumed
to have phase separated by becoming "bound" to the surface of the
tristearin slab (G). The free energy was computed from the PMF
(above) and the results are shown as chemical potentials in
Figure 9.2(H). There the parameter Q represents the cases F and G,
while Q' indicates the reference system, taken as the pure OOO oil
without the minority components (E). Results for chemical potentials
can be seen in Figure 9.2(H) as functions of simulation times (in
nsec), where the approach to equilibrium can be seen. Orange points
(e) show the error bars at different times. Blue (a) or red (b) curves
show the chemical potentials for the cases when EOO molecules are
randomly mixed or are phase separated onto the surface of the tristearin slab. It can be seen that although the chemical potentials for
the cases OOO + EOO are essentially the same, the chemical potentials for phase separated OOO + EEO (c, green) and OOO + EEO (d,
black) for the randomly mixed oil are significantly different. This was

taken to indicate that, if minority oil component molecules possess sufficient *trans* content then they will phase separate from OOO onto a nanoscale crystalline surface.[72]

9.3.6 Molecular Dynamics in Food Science Research

In the last section, we discussed the synergy between experiment and MD simulations and the benefit of simulations in research into physicochemical properties. A review by Limbach and Kremer[73] contains a limited amount of MD work restricted to polymers. More recent reviews by van der Sman[74] and Ho *et al.*[75] have only included a few works on MD. Ho *et al.* stated that, in food science, MD is "hardly applied". The latter point presents the difficulty in ascribing MD to food science topics directly, because any research that improves the application of force fields to biomolecules is a potential development in food science. This chapter, because of space restrictions, cannot undertake a comprehensive investigation of carbohydrate, lipid and protein force field developments which have been occurred in the last few decades. However, the following discussion is presented to illustrate the diversity of food science issues directly tackled by MD; the discussion is not a comprehensive overview of all MD work undertaken in food research, instead some relatively interesting and recent examples have been chosen.

In a recent study,[76] heat induced changes to the secondary and tertiary structures of β-lactoglobulin were carried out utilizing the AMBER force field. The analysis that followed from simulation data supported the experimental data obtained from fluorescence spectroscopy. The authors remarked that such structure–function relationships would aid in allowing the functionality of the protein to be understood for its use as a supplement for special food products. Protein structure and function in food chemistry, particularly in surface adsorption processes, gelation and aggregation, have also been simulated with MD.[77,78] Even complicated processes like protein folding of *Bacillus circulans* xylanase have been simulated recently[79] to model the optimal pH of this acidic xylase, important in food and animal feeds. Modelling of larger protein structures has been recently undertaken with a combination of homology modelling and MD.[79] Clearly, modelling of larger systems and longer time scales for important processes in food science is required. Such developments will be addressed in Section 9.4. The requisite protein is usually built using X-ray crystallographic structures and is easily accessible from the Protein Data Bank (www.pdb.org).[80]

For the preservation of food, cryoprotectorant agents such as carbohydrate aqueous solutions (disaccharide α,α-trehalose) have been routinely investigated with MD for more than a decade and a half now.[81,82] Even complex fundamental trehalose–lysozyme interactions in water, jointly studied with MD and inelastic neutron scattering, have been used to confirm microscopic mechanisms of hydrogen bonding.[83] Other carbohydrates, including the complex cyclodextrin–peptide structures, routinely the subject of experimental investigations, show more potential applications in food systems because of the diversity of host–guest complexes they form. The range of structures can be readily investigated with MD using model systems, because these complicated structural interactions and structures are typically not accessible to experiments. The basic building of carbohydrates has attracted much interest in terms of testing and tuning the sugar forcefields.[84–86] Moreover, validation with NMR experiments also continues with modifications to existing carbohydrate forcefields.[87] The effect of complex formation in terms of peptide structure is not well known yet and MD (and *ab-initio* MD) have been recently used to investigate the structure, dynamics and energetics of such a system.[88] Lipid structures have been routinely investigated by MD for over two decades now,[89] owing to the important lipid bilayer of cells. Today, state of the art simulations such as the incorporation of omega 3/omega 6 polyunsaturated fatty acids by spring dextrin encapsulation for delivery to the small intestine are routinely modelled, among other such systems.[90] Simulations involving edible lipids are also growing in importance.[91] A recent simulation investigating crystalline nanoparticles (CNPs) and "oil binding capacity", an important concept regarding the ability of fat particles to retain oil, and the ability of these CNPs to bind oil, is important in designing healthy foods. Atomic scale molecular dynamics computer simulations were carried out to understand the behaviour of a TAG oil (triolein) in nanoscale confinements between tristearin CNPs.[72] The force field used in this study is well characterized, and has been tuned to reproduce experimental lipid properties such as surface tension and surface bilayers (and monolayers). This enables such lipids to be used in unique simulation investigations and the results can be compared against experiments. Earlier work on triacylglycerols by Sum *et al.*[92] used a NERD force field and Gaussian quantum mechanical computations for the polar groups of the glycerol core, to compute thermodynamic quantities which were found to be in good agreement with those measured. Hall *et al.*[93] modelled tripalmitin and triolein and attempted to observe the liquid-to-crystalline phase transition in the former. Although the results were encouraging and suggested the structure of the crystalline phase, they did not compute the X-ray scattering to show they had achieved a sufficient approximation to crystallinity. Hsu and Violi[94] were concerned with the thermal stability of various triacylglycerols when used as high-strength lubricant films. They defined stability as "the ability of triacylglycerols to stay in a disordered phase during the cooling process" and found that this was related to the order–disorder transition temperature.

9.4 THEORY III. COARSE-GRAINED MESOSCALE MODELS

9.4.1 Coarse-grained Interactions: Nano- to Mesoscale[95–102]

How we model a system depends upon what spatial scale interests us. As far as the Coulomb interaction in atomic systems is concerned, we could model a molecule by treating the nuclei as point objects possessing charge and spin, and explicitly describe the electronic states, most of which exhibit non-spherical distributions, as well as electronic polarizabilities and spin and orbital angular momentum states. However, many of these degrees of freedom are irrelevant to, for example, interactions between oil-in-water colloids or casein micelles. One objective of modelling must be to eliminate those coordinates which undergo motions on time scales much faster than the time scales in which we are interested. We then replace their detailed dynamics by their average values. Such a procedure is called "coarse-graining". Although the references above describe coarse-graining as applied to phospholipid bilayer membranes, they are adequate for illustrating the principle of the procedure. In studying protein dynamics, one can consider the section in which one is interested in detail, while representing the remainder of the protein by an average static and dynamic structure which reflect the detailed segments that have been coarse-grained.[103] Coarse-grained models have also been used to model the outer leaflet of the outer membranes of Gram-negative bacteria.[104,105]

The topic of coarse-graining in computational biology was recently reviewed by Saunders and Voth.[102] Many useful coarse-grained models such as the elastic network model (ENM)[106,107] and the MARTINI model (MARTINI CG),[96,108] exist, among others.[102] To overcome the limitations of the MD technique, great strides have been made in coarse-graining development,[102] thereby allowing larger time and length scales to be simulated by incorporating fewer interaction sites. An important issue raised in the Saunders and Voth[102] review was the important issue of connecting all-atom simulation to coarse-grained models. The foundation of the MD technique is firmly established in statistical mechanics, presented in Section 9.3. Therefore, a natural extension of the idea of coarse-graining is to connect it to all-atom MD *via* statistical mechanics principles. A connection can be established from

$$\exp(-\beta f) \propto \int d\vec{r} \exp[-\beta U(\vec{r})] \qquad (9.20)$$

which is a variant of equation (9.9), where f is the Helmholtz free energy and $U(\vec{r})$ denotes the potential energy of the whole system. The importance of this equation is that it links the atoms/molecules to the potential energy function. The structural form of equation (9.18) enables distribution functions, equilibrium averages and system properties to be calculated.[59] For coarse graining, Voth *et al.*[102,109] transform equation (9.18) by replacing the atomic coordinates with coarse-grained ones:

$$\int d\vec{r} \exp[-\beta U(\vec{r})] \equiv \int d\overrightarrow{R_{GG}} \exp\left[-\beta U_{GG}(\overrightarrow{R_{GG}})\right] \quad (N_{\overrightarrow{R_{GG}}} \ll N_{\vec{r}}) \qquad (9.21)$$

where the free energy expression has been replaced with a set of coarse-grained particles, $\overrightarrow{R_{GG}}$. Clearly, the coarse-graining has resulted in fewer interaction sites, $\overrightarrow{R_{GG}}$, compared with the all-atom sites, $N_{\vec{r}}$. The potential is also now expressed as an effective coarse-grained potential, U_{GG}. Voth has defined mapping operators, $M_{\overrightarrow{R_{GG}}}$, that map the atomic positions onto coarse-grained sites. After some approximations and rearranging, this leads to an expression connecting equations (9.18) and (9.19),

$$\exp[-\beta U(\overrightarrow{R_{GG}})] \equiv \int d\vec{r}\, \delta(M_{\overrightarrow{R_{GG}}}(\vec{r}) - \overrightarrow{R_{GG}})[-\beta U(\vec{r})] \qquad (9.22)$$

with the mapping operator defined in this case as the centre of mass of a group of atoms. The set of integrals has been taken over delta functions of all the possible CG sites and is a product of $N_{\overrightarrow{R_{GG}}}$ sites. The important quantity required here is the CG potential, $U_{GG}(\overrightarrow{R_{GG}})$. It is a multi-dimensional free energy surface for the CG variables. Another way to define this quantity is that it is the many-body potential of mean force (PMF), defined as the potential that yields the same force as that resulting from the ensemble average over all the appropriate and relevant configurations of a given system. Voth has devised methods and procedures for calculating such a multi-body PMF,[109] and this can be characterized as a Force Matching scheme.

The CG models such as the United Residue (UNRES) model[110] include limited PMF interactions together with empirical fitting (knowledge based) for CG, and have proven to be successful. Other CG force fields are typically optimized to reproduce bulk thermodynamic and/or structural properties of the system *via* the Iterative Boltzmann Inversion[111] and related techniques.[112,113] Other rigorous schemes including the Relative Entropy method[114] have also been successful for investigating small proteins.[115] The motivation for developing such rigorous CG models is to connect to realistic atomic-scale interactions rooted in fundamental physics. Such a connection to statistical mechanics, *via* entropy ideas (Relative Entropy method) and a multi-body PMF (Force Matching) can lead to the possibility of a true bottom-up CG method that could have a significant impact on MD coarse-graining schemes in food science.

9.5 THEORY IV. STOCHASTIC PROCESSES

9.5.1 The "Metropolis" Monte Carlo (MMC) Method[116–119]

In addition to the references listed here, one could also read the review and reassessment by Kastner,[120] who sets out to provide a unifying foundation for Monte Carlo (MC) simulations. The paper reviews the different approaches that have been taken, identifies the basic strategies of the variety of realizations of the method and brings order in unifying the diverse techniques which have been employed. Once one has traversed the mathematical fundamentals, this article is a pleasing addition to the fundamental literature. Here we present the technique in a less rigorous way.

The Metropolis Monte Carlo (MMC) method is designed to compute averages of characteristic quantities for **systems in equilibrium**. Such quantities might be the **energy**, the number of objects in a given physical state, the **average length of a polymer chain**, and the **local density** of a given type of object. It involves only the **interaction**, and **internal, energies** of the model being simulated and is not concerned with kinetic energies and so does not involve forces, accelerations or velocities. What this implies is that, since we have no forces in this approach, **we cannot, in general, use the Metropolis Monte Carlo**

techniques to model the dynamics of a system – the path through
phase space generated by the method is not necessarily one which
would be generated in response to the forces of the system if one were
to use a method of MD (however, see the Kinetic Monte Carlo method
below). Kikuchi *et al.*[121] argued that, under certain conditions, the
MMC method can establish a relationship between the time elapsed
and the number of MC steps. At the end of this section, we briefly
describe the Kinetic MC (KMC) technique. It is not implausible that, if
one were to use only physically realizable transitions such as very
small displacements (small compared to other length scales) at each
MC step, one might establish a connection between the MMC step
and the elapsed time of an experiment. However, since acceleration
and velocity are not involved in the MMC method, we could observe
seemingly unphysical events involved with the attempt by one object
to impinge on part of the space occupied by another. In the MMC
method such an event would be discarded even though we know that
a larger object can displace a smaller one. The effect of this is that,
without introducing additional rules for elastic scattering, a massive
object will not bring about the displacement of a much less massive
object *via* a seeming collision. This unphysical effect arises because
we are attempting to interpret the MMC approach beyond its validity.
Our error is in thinking that, somehow, the MMC method can say
something about the dynamics of a system being simulated. Again,
because inertial terms are not explicitly present in the MMC method,
any attempt to model flow can be valid only for small Reynolds
numbers. At best, the MMC method should be used for cases in which
random motion dominates inertial terms in the energies: the surrounding heat bath provides the impetus for changes of state, and the
inertial terms involving velocities and accelerations (*e.g.* $mv^2/2$) are
irrelevant.

The Inverse (or Reverse) MC method involves the same technique as
the standard MMC method but tunes the various parameters so that
the outcomes of the simulation agree with those of a particular set of
data. We are not concerned with this approach here.

A standard MMC method utilizes the following procedure:

[1] The system is set up in any initial state that is convenient. This
   is defined by the values of a set of variables, $a_1, a_2, \ldots a_K$. One
   MC step involves trying to change each of these values to a
   different value. At any point in the simulation, let the state of
   the system be labelled A, possessing energy E_A, and let us try to
   change the system to state B, possessing energy E_B, by

attempting to change one or more of the variables, $a_1, \ldots, a_K$.
Define $\Delta E = E_B - E_A$. The following example is intended to make
this description clear.

Assume that we have a system comprising N spheres at
temperature, T. Each sphere can be in two states: a ground
state (g) with energy E_g, and an excited state (e) with energy
$E_e > E_g$. When in the unique state g, the sphere possesses a
radius, R_g, and because the state is unique, possesses degeneracy, $D_g = 1$. The excited state, e, possesses a radius
$R_e > R_g$ and a degeneracy, $D_e > D_g$. Two spheres, labelled j and
k, interact *via* a (possibly complicated) interaction, $V(s_j, s_k, r_{jk})$,
where s_j and s_k are the states (g or e) of the two spheres and r_{jk}
is the centre-to-centre distance between the two spheres.
We shall visit each sphere in a random sequence and attempt
to carry out the following changes: change of state and
change of location. Note that because there are no forces,
accelerations or velocities involved, the simulation is strictly
not dynamic; it is a technique for discovering equilibrium
properties of the system. It is not necessary to attempt
to carry out all possible changes at once: the energy differences involved might require that we carry out very long
simulation runs.

However, let us attempt to move the sphere labelled j and
change its state. Let it be initially in its state e with its centre
located at r_j. We attempt to change its state to g. In doing so,
the internal free energy would change by the amount

$$\Delta E_1 = E_g - E_e - k_B T l n D_g + k_B T l n D_e \qquad (9.23)$$

where k_B is Boltzmann's constant. The energy difference is
negative $(E_e > E_g)$ while the entropy term is positive. Let the
sphere attempt to relocate at r_j'. Compute the interaction energy
difference,

$$\Delta E_2 = \sum_k{}' V(g, s_k, r_{jk}') - \sum_k V(e, s_k, r_{jk}) \qquad (9.24)$$

where the first sum would be over all spheres which interact
with sphere j after it has carried out the attempted move, while
the second sum is over all spheres which interact with sphere j
before its attempted move. The distance r_{jk}' is the distance
between the sphere j in its new location and another sphere k.
The sums are not necessarily over the same sets of spheres.
The total energy change is then $\Delta E = \Delta E_1 + \Delta E_2$.

[2] If $\Delta E \leq 0$ then the change is accepted and the state of the system becomes B; that is, the sphere j will now be in its state g with its centre located at a new position r_j'.

[3] If $\Delta E > 0$ then choose a random number u with $0 \leq u \leq 1$.

[4] If $u \leq \exp(-\beta \Delta E)$, where $\beta = 1/k_B T$, then the state of the system becomes B. Otherwise the system remains in state A.

[5] When we have tried to change all variables (*i.e.* tried to change the states and positions of all spheres in the example above) then one MC step has been carried out.

[6] There are two aspects to a simulation: (a) initializing the system, which entails permitting the system to come to equilibrium after the start of the simulation; and (b) calculating average values while the system samples equilibrium states.

    (a)   After the simulation has begun, let us keep track of the total energies at each MC step, 0, 1, 2, ... Let these energies be $E(0)$, $E(1)$, $E(2)$, ... We will notice that these change in a regular way, apart from fluctuations, and that they approach a value, $\langle E \rangle$, which does not change (apart from fluctuations) as the simulation proceeds further. This is the average equilibrium value of the energy and tells us when the process of initialization has been completed.

    (b)   The simulation is continued and the average values of other variables are calculated by carrying out a further S MC step and computing the average of the set $\{a_n, n = 1, \ldots K\}$. Let the sequence of values obtained from these S MC steps be $a_n(1)$, $a_n(2)$, ..., $a_n(S)$. Then the average over the S MC steps is,

$$\langle a_n \rangle = \frac{1}{S} \sum_{r=1}^{S} a_n(r) \qquad (9.25)$$

[7] It has been shown that, if a simulation is carried out for a sufficient number of MC steps, then the system being simulated will come to thermal equilibrium at the temperature selected for the simulation, and will exhibit average values and fluctuations characteristic of thermal equilibrium.

    We can use the methods described in the section on atomic scale molecular dynamics to compute the free energy *via* the PMF (see above). The latter can be computed from knowledge of the pair correlation functions (above) which can be computed from MMC simulations. Structure functions can give information about aggregation and can assist in the interpretation of X-ray, neutron and light scattering.

9.5.2 Structure Functions

Consider an assembly of N points, $\{1, 2, \ldots j, \ldots\}$ identified by their position vectors $\{\vec{r_j}\}$. Define the structure function,

$$S(\vec{q}) = \frac{1}{N}\sum_j \sum_k \exp(i\vec{q} \cdot (\vec{r_j} - \vec{r_k})) \tag{9.26}$$

If we are relating this to a scattering experiment, then $\vec{q} = \vec{k} - \vec{k}'$ where $\vec{k}$ is the wave vector of the incident particles (photons, neutrons) and the wave vector, $\vec{k}'$, is that of the elastically scattered ($|\vec{k}'| = |\vec{k}|$) particles. Otherwise it is simply a Fourier analysis of the distribution of points carried out to discover whether there are characteristic lengths in the assembly. As usual, one can relate spatial lengths, λ, to q *via* $\lambda = 2\pi/q$. If the experiment is believed to possess no axis that breaks rotational symmetry on the average, so that the structure function depends only upon the magnitude of $\vec{q}$, $|\vec{q}| = q = |\vec{k} - \vec{k}'|$, then one can average over the directions of $\vec{q}$ to obtain $S(q)$,

$$S(q) = \langle S(\vec{q})\rangle_{|\vec{q}|=q} \tag{9.27}$$

One can identify the fractal dimension, D, of the assembly of points, for some range of q, *via* the relation,[122–124]

$$S(q) \sim q^{-D} \quad \text{or} \quad \log[S(q)] \sim -D\log[q] \tag{9.28}$$

The range of q should be sufficiently large and in practice, when relating a simulation to experiments, should be at least one decade.

In what follows, we give examples from the peer-reviewed literature of implementations of the MMC approach. We shall specify the system, outline one way of modelling it *via* coarse-graining, describe how the MMC is carried out, show some of the results and, finally, summarize the important conclusions. The intent is that a reader could then model the system themselves and see whether they get similar results.

Example 2: *Coarse-grained models of aggregates of triacylglycerol crystalline nanoplatelets (CNPs) on different spatial scales*[125–127]

Work by Nuria Acevedo identified the most stable form of solid fat structures in triacylglycerol (TAG) edible oils at room temperature to be crystalline nanoplatelets (CNPs). Much work had been done on studying the structures which arose in these systems on spatial scales ranging from ~ 10 nm to millimetre lengths, *via*, for example, polarized light microscopy, transmission electron microscopy (TEM), rheology and X-ray diffraction. No model had been proposed which could account for the range and morphology of the spatial structures

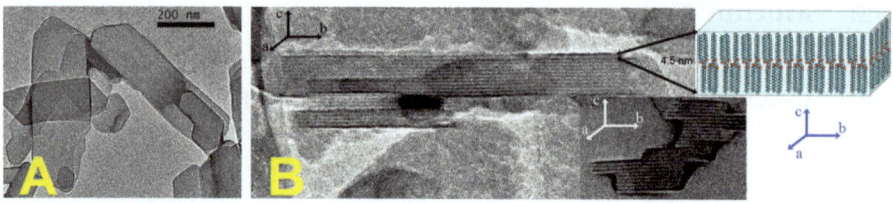

Figure 9.3 Crystalline nanoplatelets (CNPs). (A) Plan view. The bar indicates
200 nm. (B) Side view. Each layer exhibits the thickness (∼5 nm) of a
TAG bilayer.
(Reproduced by kind permission of Dr Nuria Acevedo.)

observed. The samples made by Acevedo and Marangoni involved
cooling an oil, removing the liquid oil, shearing the system at speeds
up to 3×10^4 rpm, coating the resulting solid mass with uranyl acetate
and observing the result using cryoTEM. Some of the more striking
images are shown in Figure 9.3.

In order to model the aggregation characteristics of these highly
anisotropic structures, one must represent them so that the interactions between different parts of the CNPs are correct. To represent
them by slabs, examples of which can be seen in Figure 9.3A, would
lead to complexities in accounting for their orientation and the instantaneous distances between any two parts of two such CNPs. We
ignore the solvent oil because it will not contain structures possessing
correlation lengths on the spatial scales exhibited by the CNPs and
their aggregates. The interactions are predominantly van der Waals,
though some electrostatic effects due to the glycerol core with three
carbonyl groups cannot be discounted. The model which is likely to
be easiest to implement is one in which the structures shown in
Figure 9.3A are represented by rigid arrays of close-packed spheres.
Such an array can be described by the number of spheres along the
two axes of a close-packed hexagonal or equilateral triangular lattice
together with the number of such layers perpendicular to the lattice
plane, {k, m, n}. The van der Waals interaction between two spheres is
known and, if one assumes that the total van der Waals interaction is
simply the sum of all such pair-wise interactions, then such a model
lends itself to easily coded computer simulation.

We write the interaction between two identical homogeneous
spheres, each of radius R, a centre-to-centre distance r apart as

$$V(r) = V_d(R,r) \quad r > 2(R + \Delta) \tag{9.29}$$

$$V(r) = V_B \quad r \leq 2(R + \Delta) \tag{9.30}$$

where Δ is the thickness of a "boundary layer" region into which another sphere, belonging to a different CNP, cannot penetrate, and $V_B < 0$ is a "binding energy". The distance 2Δ is approximately the nearest distance to which a pair of parallel TAGs can approach and is approximately 0.2 nm. One obvious choice for V_B is $V_B = V_d(R, 2R + 2\Delta)$ while other choices can explore aspects of very tight binding. The attractive interaction free energy, $V_d(R, r)$, was given by Hamaker[128] and by Parsegian[129] as

$$V_d(R, r) = \frac{-A}{6}\left[2R^2\left(\frac{1}{r^2 - 4R^2} + \frac{1}{r^2}\right) + \ln\left(\frac{r^2 - 4R^2}{r^2}\right)\right] \qquad (9.31)$$

where A is the Hamaker coefficient. Here $A \approx 10^{-20}$ J because the two spheres are immersed in liquid triolein oil. In air or vacuum,[130] $A \approx 5 \times 10^{-20}$ J.

Figure 9.4 shows the layer of thickness Δ around two spheres, an example of a model CNP composed of one layer of spheres, and the

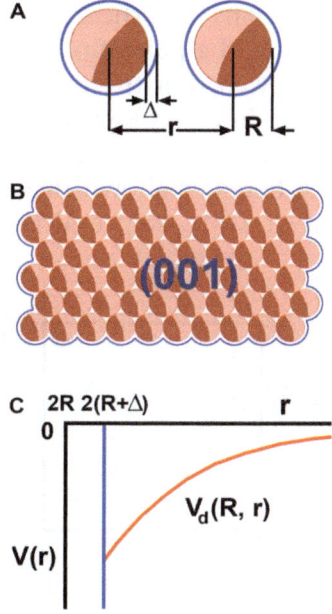

Figure 9.4 Model CNP composed of a rigid array of spheres. (A) Two such spheres, each of radius R. They possess a surface coating of thickness, Δ. (B) A model CNP defined by the integers $\{k, m, n\}$. A CNP with $k = 10$, $m = 6$ and $n = 1$ is shown. (C) The van derWaals interaction between spheres belonging to different CNPs.
(Reproduced with permission from Pink *et al. J. Applied Physics*, **114**, 234901 (2013). Copyright 2013, AIP Publishing LLC.)

attractive van der Waals interaction between a pair of spheres belonging to different model CNPs.

The MC code must define the position of the centre of each sphere and specify to which CNP it belongs. One can represent the radius of the spheres by a "soft" repulsive potential acting between any pair of spheres belonging to different CNPs. Such a potential could have the form of a phenomenological power law such as s^{-n}, where s is the shortest distance between the surfaces of two spheres belonging to different CNPs, $s = r - 2R$, and n is a sufficiently large integer such as $n = 8$.

One must also define an axis passing through the centre of mass of each model CNP in a randomly selected direction and attempt to carry out a rotation of the entire CNP around this axis by a random angle. In addition to these movements, one must attempt to move (translate) the entire CNP in a random direction by a random distance. Since there are no boundaries constraining these systems, periodic boundary conditions along the x-, y- and z-axes can be employed.

The results of MC simulations with the surface coating, $\Delta \approx 0.12$ nm, as a typical closest approach of a pair of parallel tristearin molecules in their ground states, showed that the model CNPs formed 1D aggregates ("TAGwoods"), as shown in Figure 9.5(Aa). In order to study TAGwood aggregation, these linear structures were represented as rigid linear arrays of L spheres as shown in Figure 9.5(Ab). $S(q)$ was computed as a function of MMC step and the results for $L = 5$ are

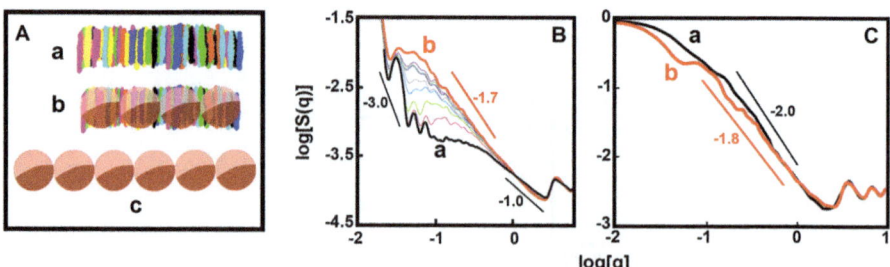

Figure 9.5 (A) A (one-dimensional) TAGwood (a), its representation by a rigid linear array of L spheres ($L = 4$) (b), and an example of an $L = 6$ TAGwood (c). Log[$S(q)$] *vs.* log[q] for $L = 5$. (B) Initial $S(q)$ (a) and after a large number of MMC steps (b). (C) Relaxation from DLCA aggregation (b) to RLCA aggregation (a).
(Reproduced with permission from Pink *et al. J. Applied Physics*, **114**, 234901 (2013). Copyright 2013, AIP Publishing LLC.)

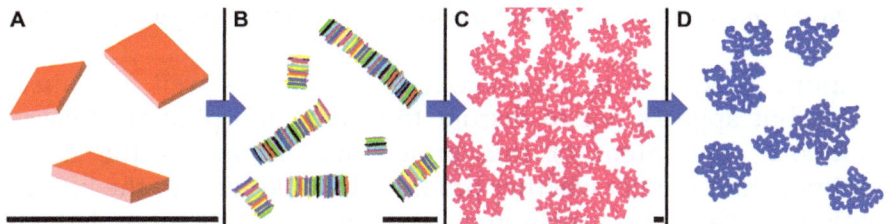

Figure 9.6 The Heirarchy of Being for aggregating CNPs with uniformly smooth surfaces in a single-component liquid oil. (A) Individual CNPs. (B) CNPs aggregate into TAGwoods. (C) TAGwoods aggregate into DLCA/RLCA clusters. (D) TAGwood clusters uniformly distributed in space. (Reproduced with permission from Pink *et al. J. Applied Physics*, **114**, 234901 (2013). Copyright 2013, AIP Publishing LLC.)

shown in Figure 9.5(B and C). The first shows the aggregation of TAGwoods beginning with a slope of -1 at higher q (a) and a slope consistent with -3 at lower q. This shows that the 1D TAGwoods are initially unaggregated and uniformly distributed in space. As the simulation proceeds, the curve changes to one with a slope of about -1.7, characteristic of diffusion limited cluster–cluster aggregation (DLCA) (b).[124] As the simulation continues to run (Figure 9.5C), we see that the slope changes from -1.8 (b) to about -2 (a), indicating that the structure is relaxing and appears as reaction limited cluster–cluster aggregation (RLCA).

The results of MC simulations showed that: (i) model CNPs formed 1D aggregates (TAGwoods), (ii) such TAGwoods aggregated *via* DLCA and relaxed *via* RLCA to form clusters exhibiting fractal dimensions $D = 1.7$–2.1, and (iii) these clusters were, on average, uniformly distributed in space $(D = 3)$, as shown in Figure 9.6.[131]

Example 3: *Coarse-grained modelling of the interaction of the antimicrobial peptide protamine with membrane surfaces, showing the effects of divalent cations*[132–136]

Protamine (Ptm; clupeine) is a highly charged cationic antimicrobial peptide that can be obtained from herring milt, where it compacts the DNA. Ptm is used as a carrier for injectable insulin and is an antagonist to heparin. Ionized clupeine possesses 31 peptide groups of which 20 are arginine, each of which possess an approximately proton positive charge at 20 °C and over a wide pH range. It exhibits antimicrobial characteristics against a range of bacterial components.

The inner membrane of Gram-negative bacteria – a phospholipid bilayer:[137]

Protamine is known to internalize in bacteria as well as spheroplasts, but appears not to disrupt the inner phospholipid bilayer membrane. The intent of this project was to model the inner membrane and predict whether protamine would bind to the bilayer membrane. The inner membrane comprises ~75% phosphatidylethanolamine (PE), ~20% phosphatidylglycerol (PG) and ~5% cardiolipin but, for all practical purposes, each cardiolipin can be replaced by 2 PG molecules.

Figure 9.7 shows the coarse-grained models used to represent PE, PG and Ptm. There we see that atomic aggregates (pairs of CH_2 groups, carbonyl, PO_4^-, NH_3^+, arginine residues at the ends of the hydrocarbon chains) were represented by spherical coarse-grained "atoms". The charges associated with each of these "atoms" were located at the centres of the respective spheres. Blue represents negatively charged and red represents positively charged moieties. Ions in the water were also represented by spheres with the appropriate charges in the centres of the spheres. The lengths of all hydrocarbon chains were taken to be equal. The advantage of this is that, instead of moving model molecules, one needed only move head groups.

The spatial resolution required rendered Derjaguin-Landau-Verwey-Overbeek (DLVO) theory inadequate.[138] DLVO theory involves representing the surface by its average charge density. In this case, where one is investigating whether the highly charged cationic Ptm

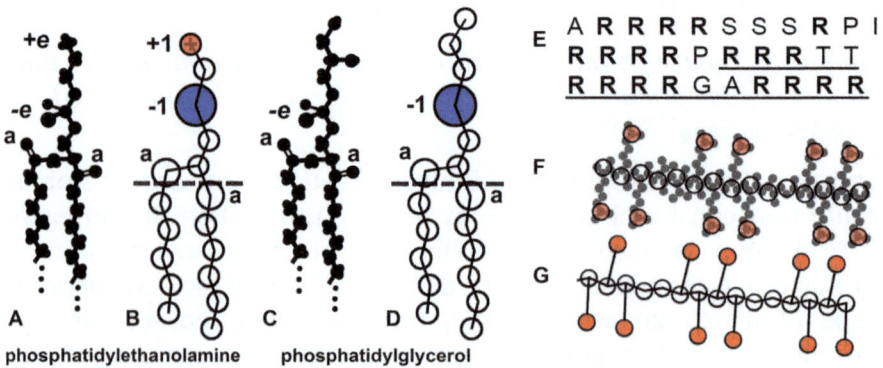

Figure 9.7 Coarse-grained models of phosphatidylethanolamine (A, B), phosphatidylglycerol (C, D) and protamine (E, F, G). Red (blue) indicates positively (negatively) charged moieties.

molecule will bind to the surface, it is clear that one must be prepared to consider membrane polarization *via* the movement of PG head groups from one region to another. In addition, such polarization can change as the Ptm molecule moves above the surface. Such an effect is not allowed for in DLVO theory. Although monovalent ions in solution might be represented by a continuum characterized by a Debye screening length, κ^{-1}, this approach cannot be used to describe multivalent ions because the latter can take part in bridging between negatively charged molecular moieties, which can exhibit a high degree of three-body correlation.

Electrostatic interactions: Unit vectors $\hat{x}$ and $\hat{y}$ defined the $x-y$ plane, the plane of the membrane, with $\hat{z}$ along the z-axis. The aqueous solution $(z< -z_{HC}/2)$ and $(z>z_{HC}/2)$, and the hydrocarbon-chain region $(-z_{HC}/2<z<z_{HC}/2)$ were represented as two dielectric continua with relative permitivities $\varepsilon_w = 81$ and $\varepsilon_{HC} = 5$, respectively.[138] Part of the monovalent ion screening in the aqueous solution was represented by linearized Poisson–Boltzmann theory.[138–140] The remainder of the monovalent ions were represented by explicit ions. All divalent ions were represented by explicit ions since these can take part in three-body bridging interactions. Charge-induced membrane polarization was represented by image charges, and the electric potential at $\vec{R}=x\hat{x}+y\hat{y}+z\hat{z}$ due to an electric charge Q at $\vec{R}_0 = x_0\hat{x}+y_0\hat{y}+z_0\hat{z}$ in the aqueous solution was calculated. Netz and co-workers[141–143] derived expressions for the electrical potential at $\vec{R}$. Since $\varepsilon_{HC} \ll \varepsilon_w$ (giving $\eta = \varepsilon_{HC}/\varepsilon_w \approx 0.06$), the expression for the potential simplifies. Defining $\chi = (1-\eta)/(1+\eta)$, the electrical potential at $\vec{R}$ becomes

$$V(Q,\vec{R}_0,\vec{R}) = (1/4\pi\varepsilon_w\varepsilon_0) \sum_j Q_j f(|\vec{R}_j|), \qquad (9.32)$$

$$f(|\vec{R}_j|) = e^{-\kappa|\vec{R}_j|}/\vec{R}_j, \qquad (9.33)$$

where the sum is over $j=1$ to 2, $\vec{R}_1 = \vec{R}-\vec{R}_0$, $\vec{R}_2 = \vec{R}-\vec{R}_0+2z_0\hat{z}$, $Q_1 = Q$, $Q_2 = \chi Q$, κ^{-1} is the Debye screening length and ε_0 is the permittivity of free space. $V(Q,\vec{R}_0,\vec{R})$ goes to the correct limiting values as $\varepsilon_{HC}\to 0$ and as $\kappa\to 0$.

MMC simulation was used together with periodic boundary conditions along the x- and y-axes. At each MC step an attempt was made to change the position of each lipid headgroup sphere, each Ptm sphere, and each ion. The z-component of the simulation volume ranged from $-22.5 \leq z \leq 22.5$ nm with the aqueous solution occupying the spaces $-22.5 \leq z \leq -2.5$ and $2.5 \leq z \leq 22.5$ nm. In addition, movement of the spheres was restricted so that no centre of the sphere was permitted to leave the region in which they were initially placed. The temperature was $T = 300$ K. The system was equilibrated for $10^5 - 10^6$ MC steps, and the properties of the system were measured for a further $10^4 - 10^5$ steps.

The models shown in Figure 9.7 are essentially linear polymers and use can be made of standard methods for moving such objects. Thus, in attempting to move the headgroup spheres or the

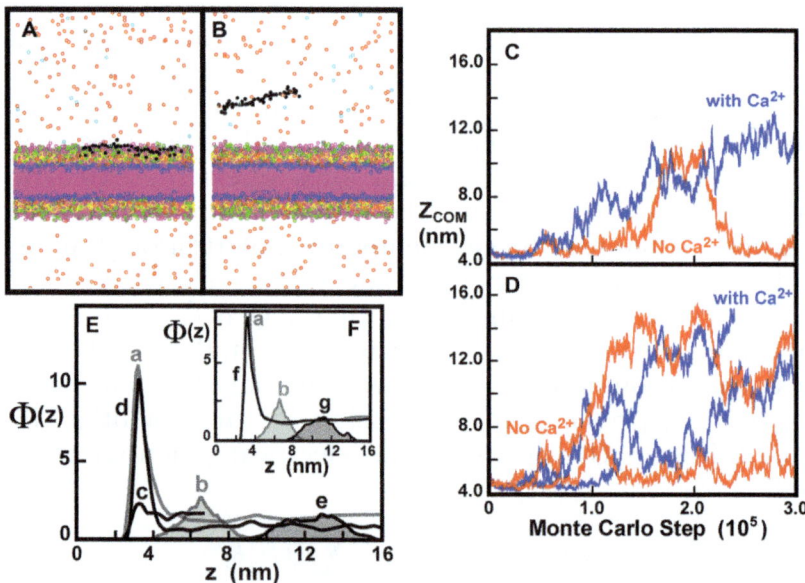

Figure 9.8 Protamine (black) location near a PE–PG bilayer membrane in the presence of Ca^{2+} ions (red). (A) 5×10^4 MC steps. (B) 15×10^4 MC steps. Instantaneous protamine centre of mass location, with (blue) and without (red) Ca^{2+} ions for different PE:PG ratios. (C) PE:PG ratio [65:35]. (D) PE:PG ratio [75:25]. Number density, $\Phi(z)$, as a function of the distance from the centre of the bilayer at $z = 0$ nm with the oil–water interface at $z = 2.5$ nm. (E) Effect of 40 mM Ca^{2+} ions on a [65:35] PE–PG bilayer membrane. Na^+ ion (grey **a**) and single Ptm (grey **b**) distributions in the absence of $CaCl_2$. Na^+ (black **c**), Ca^{2+} (black **d**) and Ptm (black **e**) distributions in the presence of $CaCl_2$. (F) Comparison of [65:35] and [75:25] PE–PG bilayer in the absence of $CaCl_2$; **a** and **b** (grey) as in (E). Na^+ (black **f**) and single Ptm (black **g**) distributions in the absence of $CaCl_2$ for [75:25] PE–PG membrane.

Ptm spheres, the bond, connecting it to an adjacent sphere, was permitted to change its length thereby permitting the moiety to escape from being trapped.[144]

The Ptm molecule was initialized within ∼5 nm of the head-group region and the system permitted to explore its phase space. Some results are shown in Figure 9.8. Figure 9.8(A and B) show instantaneous snapshots of the Ptm as it drifts away from an initially favourable binding position near the membrane, while Figure 9.8(C) shows the movement of the Ptm centre of mass if one has a membrane comprising 65% PE and 35% PG (PE:PG ratio = [65:35]), with and without 40 mM Ca^{2+} ions present. Figure 9.8(D) shows the same cases for the membrane containing PE:PG = [75:25], the more realistic case. It can be seen that, even with the more strongly attractive, more unrealistic PG concentration, Ptm does not bind to the membrane surface even in the absence of Ca^{2+} ions. Figure 9.8(E and F) compare the average spatial distributions of Ptm from the oil–water interface at 2.5 nm as functions of the PE:PG ratio and Ca^{2+} concentration. Figure 9.8(E) shows the effect of 40 mM Ca^{2+} ions on a [65:35] PE–PG bilayer membrane. There we see distributions of Na^+ ions (grey **a**) and that of a single Ptm molecule (grey **b**) in the absence of $CaCl_2$. In the presence of $CaCl_2$ we see that the Na^+ ions are displaced by the Ca^{2+} ions. This is seen in the distributions of Na^+ (black **c**), Ca^{2+} (black **d**) and Ptm (black **e**). Figure 9.8(F) compares a [65:35] bilayer with a [75:25] bilayer in the absence of $CaCl_2$. We compare the results of Figure 9.8(E) for Na^+ and Ptm [**a** and **b** (grey)] with the corresponding distributions for a [75:25] PE–PG membrane. Shown are the distributions for Na^+ (black **f**) and a single Ptm (black **g**) in the absence of $CaCl_2$. Even in the absence of Ca^{2+} ions, a single Ptm will not bind to an unrealistically highly charged membrane.[145]

***The outer cell wall of* Campylobacter fetus *(C. fetus). Lipopoly-saccharides (LPS) of the outer membrane (OM), surface layer proteins (SLP), and the effect of Ca^{2+} upon membrane stability.*[146–151]**

The *C. fetus* cell envelope is typical of Gram-negative bacteria and includes an inner membrane, a peptidoglycan layer and an outer membrane containing lipopolysaccharide (LPS). The LPS of the outer layer of the outer membrane (OM) of Gram-negative bacteria possesses, (i) on the average about five to seven hydrocarbon chains which, together with two saccharide groups to which they are attached, form Lipid A; (ii) a polysaccharide core comprising the charged inner core and the, generally uncharged, outer core; and (iii) an O-antigen or O-side chain comprising sequences of saccharide units. The *C. fetus* LPS molecules are homogeneous in length (smooth-type LPS) and are expressed as one of two unique O-antigen types which distinguish the bacterium as belonging to serotype A, B or AB. The bacterium that was modelled, *C. fetus* subsp. *fetus* strains, may be of serotype A, B or AB; serotype B for which the O-antigen is composed of charged trimers was the one selected to model. External to the LPS is an additional cell envelope component, a proteinaceous surface layer. The surface-layer proteins adhere to the O-antigen component of the LPS, forming a 2D paracrystalline array on the external side of the outer membrane. This array exhibits hexagonal or, less frequently, tetragonal lattice symmetry depending upon the size of the protein subunits.

Accordingly, even if one ignores other proteins associated with the OM, the structure of the outer layers of the cell wall requires that at least three LPS regions and the SLP layer be modelled. One must also consider the effects of both monovalent and divalent ions and the roles they play in stabilizing the outer layers of the cell wall. The reason for this is that there are two scenarios which have led to the advance of different hypotheses to account for the stability of the hexagonal SLP lattice. In addition, one must also consider the interaction of protamine (Ptm) with the components of the cell surface. Figure 9.9 shows models of the *C. fetus* Lipid A, core and part of the O-antigen (A and B) and an atomic model of an SLP (C) (Protein Data Bank) where its orientation with the average plane of the membrane (brown band) is shown. Each sphere (shown as circles) of (A and B) (except for the two large spheres of Lipid A) represents a charged or uncharged saccharide moiety. In (A and B), the red discs indicate positively charged moieties, grey discs represent uncharged moieties and blue discs represent negatively charged moieties. It can be seen that the representation of the six hydrocarbon chains involves only two large spheres. The reason for this is that there is interest only in processes in the aqueous regions, and the hydrocarbon chain region provides simply a base to stabilize the core and O-antigen. An SLP was assumed to be a rigid structure, the centre of mass of which could either translate randomly in any direction or rotate around a randomly chosen axis, passing through the centre of mass, by a random angle restricted by a requirement that approximately half of the moves attempted should be successful. Again the

Carmesin–Kremer bond stretching algorithm was used so that all polymers could adequately explore phase space. All charges were located in the centre of the charged spheres. The interactions were the van der Waals interaction together with electrostatics.

MMC techniques with periodic boundary conditions along the *x*- and *y*-axes were used to study the equilibrium properties of the system of a monolayer of LPS model molecules, representing the outer leaflet of the OM. The *z*-axis was perpendicular to the plane of the membrane. The oil–water boundary was taken so that all sugar moieties lay in the water, consistent with the assumption that the maximum volume of the two large hydrocarbon chains should not be in the aqueous solution. Model SLP molecules were then arranged in a hexagonal array and oriented as shown in Figure 9.9(C) with as close a packing as feasible. This case is labelled SLP+. Cases where the SLPs are absent were labelled SLP−. Monovalent and divalent ions were then randomly

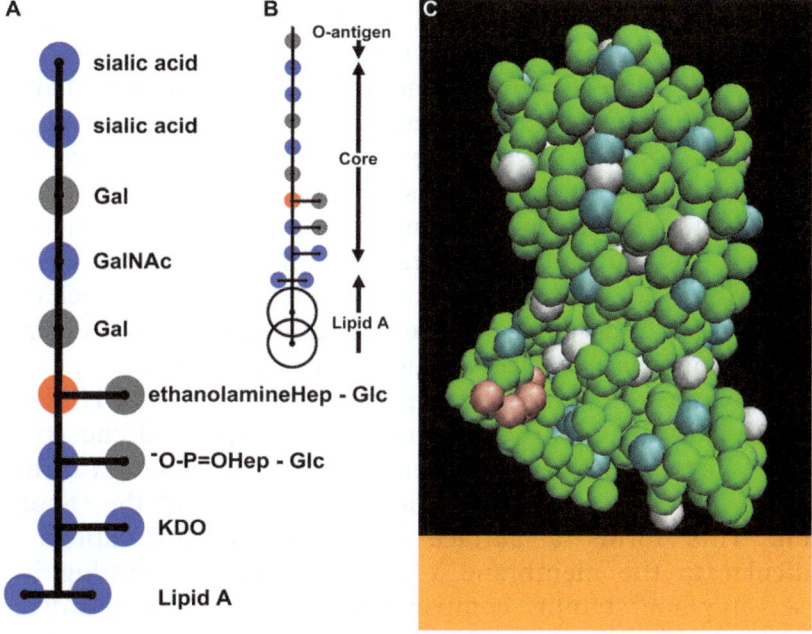

Figure 9.9 (A) Model of *C. fetus* core. Red indicates a positive charge, blue a negative charge and grey uncharged. (B) Model of *C. fetus* Lipid A, core and O-antigen using the same colour coding. (C) Atomic representation of an SLP from the Protein Data Bank. The horizontal brown band indicates the average plane of the outer leaflet of the OM.

distributed in the aqueous region consistent with total charge neutrality. Finally a single Ptm molecule together with its counterions was placed randomly some distance away from the OM and the effects of ionic concentration were studied. Figure 9.10 summarizes the results.

Figure 9.10(A and B) confirmed[152] that $CaCl_2$ is essential for the appearance of a hexagonal lattice arrangement of SLP distribution in the *x–y* plane, but there are two hypotheses proposed to explain how this self-assembly is brought about.

Dworkin *et al.*[147] proposed that divalent cations neutralized the net negative charge on SLP subunits, thus reducing their electrostatic repulsion and facilitating aggregation. Yang *et al.*[146] proposed instead that divalent cations form "bridges" between the negatively charged core LPS molecules, thereby enabling stronger interactions between SLPs and the LPS region. By reducing the repulsive forces between individual LPS molecules, a larger number of LPS molecules are able to interact with the N-terminal region of the surface-layer protein subunits.

A comparison of Figure 9.10(C) and (D) shows that, in S+ simulations, a decrease in Cl^- concentration is manifested in the region of the SLPs when compared to S– simulations. This decrease in anion concentration occurs as a result of the net negative (–4*e*) charge of surface-layer protein atomic moieties. Figure 9.10(C and D) also show that the Na^+ distribution increases slightly in the S+ case compared to the S– simulations. The regions where this takes place are indicated by pairs of arrowheads. It can be seen, however, that the Ca^{2+} distribution remains essentially unchanged. This indicates that no aggregation of calcium ions with surface-layer proteins was observed in computer simulations.

These results show that the hypothesis of Yang *et al.* is likely to be correct: divalent ions bring about a "collapse" of the core region and do not associate preferentially with the SLPs. In turn, the collapse of the LPS cores brings about a collapse of the O-antigen region. This forms a surface with reduced fluctuations perpendicular to the membrane *x–y* plane, thereby resulting in increased in-plane stability compared to the case in which divalent ions are absent. These conclusions are supported by Figure 9.11, where we see Ca^{2+} ions bringing about a collapse of B band (charged) O-antigens, accompanied by an enhanced concentration (curve b) of SLP atoms close to the outer section of the LPS layer.

Figure 9.10(E–H) shows the effect of $CaCl_2$ upon the Ptm distribution. It can be seen that it is not the SLP layer that prevents

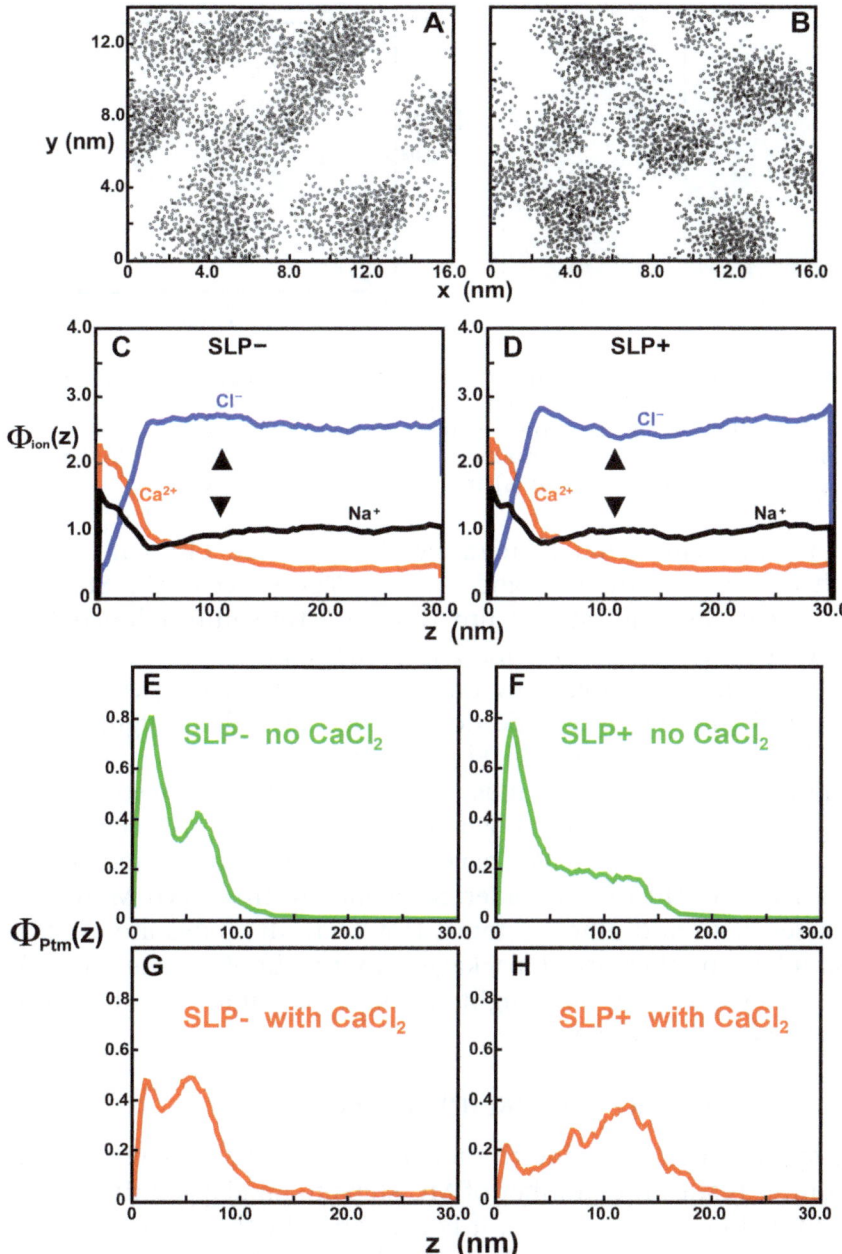

Figure 9.10 (A) Plan view of the distribution of six SLPs in the *x*–*y* plane with no $CaCl_2$ present. (B) Plan view of the distribution of six SLPs in the *x*–*y* plane with ~50 mM $CaCl_2$ present. (C) Ion distributions perpendicular to the OM in the absence of an SLP layer. (D) Ion distributions perpendicular to the OM in the presence of an SLP layer. Ptm distributions perpendicular to the OM. (E) SLP layer absent and no $CaCl_2$ present. (F) SLP layer present and no $CaCl_2$ present. (G) SLP layer absent with ~50 mM $CaCl_2$ present. (H) SLP layer present with ~50 mM $CaCl_2$ present.

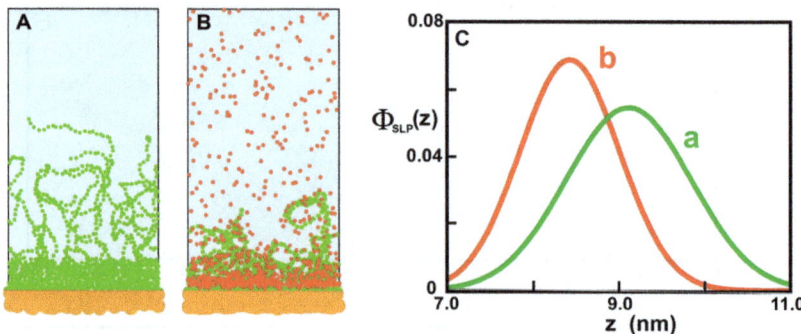

Figure 9.11   Effect of Ca^{2+} ions upon hydrophilic regions of LPS (green silhouettes). (A) Typical equilibrium conformations in the absence of Ca^{2+}. (B) Typical equilibrium conformations in the presence of excess Ca^{2+} (red discs). (C) Distributions of SLP atoms perpendicular to the membrane plane without (a) and with (b) $\sim$50 mM $CaCl_2$.

the Ptm molecule from penetrating down to the LPS core region, which is located between 0 and $\sim$2 nm. The effect of the SLP layer is to smear the peak at $\sim$7 nm (Figure 9.10E) into a broad band located between 5 and 13 nm. The penetration of Ptm to the core region is actually inhibited by Ca^{2+} ions (Figure 9.10G) even in the absence of the SLP layer. Addition of that layer then augments the Ca^{2+} inhibition, as shown in Figure 9.10(H).[152]

These examples serve to illustrate the use of MMC techniques to discover equilibrium thermodynamic distributions associated with soft biological interfaces. We have described the simulation process in sufficient detail that a reader can generate their own MMC code to model these and other systems. However, MC codes are available in standard mathematical packages. In the brief section that follows, we outline how one might introduce a "time" variable into a Monte Carlo simulation.

9.5.3   Kinetic Monte Carlo (KMC) Method[153–158]

This approach enables one to model events of systems not in equilibrium. It relates the number of MC steps to an elapsed time that is determined by rate constants. It is described by the Master equation that relates the rate at which the probability of being in a state s, $P(s, t)$, changes depending upon transition rates, $W(s' \rightarrow s)$ and $W(s \rightarrow s')$, into and out of that state,

$$\frac{\mathrm{d}P(s,t)}{\mathrm{d}t} = \sum_{s'} [W(s' \rightarrow s)P(s',t) - W(s \rightarrow s')P(s,t)] \qquad (9.34)$$

This Master equation is actually a set of coupled first-order differential equations which are, in general, difficult to solve analytically without simplifying assumptions regarding the transition rates. The transition rate coefficients, $W(s' \to s)$ and $W(s \to s')$ cannot be determined by the KMC method, but must be determined in some other way and used as input parameters to equation (9.2). Let us assume that the system dynamics is described by a set of rates $\{W_1, W_2, \ldots, W_K\}$ and let the instantaneous time be denoted by t. To implement the KMC procedure, that is, **to solve the Master equation numerically using computer simulation**, one proceeds as follows:

Define

$$w_k = \sum_{j=1}^{k} W_j \qquad (9.35)$$

[1] Select a transition event randomly, by choosing a random number, u, $0 \leq u < 1$ and select the rate, $W_l \equiv u w_K$ that satisfies $w_{l-1} < u w_K \leq w_l$.
[2] Carry out the transition event labelled l.
[3] Choose another random number, v, $0 \leq v < 1$, and update the time, $t \to t + \Delta t$, where

$$\Delta t \equiv -w_K^{-1} \ln(1/v) \qquad (9.36)$$

In the case in which there is only one transition, $W(s \to s') = W$, then

$$\frac{dP(s, t)}{dt} = -WP(s, t) \qquad (9.37)$$

so that $P(s, t) = \exp(-Wt)$. In this case, $w_K = w_1 = W$ so that $\Delta t \equiv -w^{-1} \ln(1/v)$.

The KMC method has been most commonly used to simulate "rare events" – events which take place individually and do not involve cooperative mechanisms and for which the occurrence is on time scales much longer than other dynamical processes of the system.

9.5.4 Dynamic Monte Carlo (DMC) Method

Although this method is similar to the KMC method, it is not concerned with transition rates and rare events. Instead, its intent is to map the number of MC steps onto a time variable. Within the context of lateral movement leading to the computation of a diffusion co-efficient, this has been done in the following way: (i) select a maximum displacement, δr, and try to achieve a move for a randomly chosen displacement, $u\delta r$, where $0 < u \leq 1$. (ii) Define the average acceptance probability of a move $u\delta r$ to be a. Now, naively, one might define the diffusion coefficient to be $D \sim \langle r^2 \rangle / M$ where M is the average number of MC steps and $\langle r^2 \rangle$ is the average distance moved. However, Sanz and Marenduzzo[157] showed that the average time elapsed in M Monte Carlo steps is $\Delta t = aM$. This approach has not been extended to accommodate other transitions such as rotations or changes of state.

9.5.5 General Comments

Above we commented on apparent "collisions" in which one massive object (object 1) attempts to occupy part of the space already occupied by another much less massive object (object 2). In the MMC, such a move might be rejected or else admitted with a high energy. If, in the next step, object 1 seeks to move away, then it will do so because of the high energy state it is in. It will then appear as though object 2 displaced the much more massive object 1. However, we would have expected that, physically, it is object 1 which should displace object 2. We can accommodate a more physical response by defining a rule that a massive object will displace a much less massive one. This can be realized by scaling the distances moved per MC step by a factor of $m^{-1/2}$ where m is the mass of the object being moved.

From the previous sections, it should be clear that other MMCs can be defined. One requires that a functional which defines a characteristic of a state of the system, $\mathscr{F}(\{\alpha\})$, exists where the set $\{\alpha\}$ descibes the state of the system and the functional, $\mathscr{F}$, is a scalar identifying the state. Then, if at an MC step, M, the system is in state $\{\alpha\}$ with $\mathscr{F}$ and, if the system tests whether it makes a transition, at MC step $M + 1$, to state $\{\alpha'\}$ with $\mathscr{F}'$, then we define $\Delta\mathscr{F} = \mathscr{F}' - \mathscr{F}$ and use the procedures described above with $\mathscr{F}$ playing the role of E. MC methods are widely used in evaluating integrals, in finance, and in finding the statistically best solution to problems as in simulated annealing. This article, however, is not the place to discuss these. An example of this in simulated annealing, in which an optimum solution to a problem is defined in terms of the functional $\mathscr{F}$ may be seen at http://mathworld.wolfram.com/SimulatedAnnealing.html and http://www.mathworks.com/discovery/simulated-annealing.html.

9.5.6 Applications

Mezzenga and Fisher[159] reviewed models of protein aggregation of importance to food science, and discussed the forces driving aggregation, including irreversible aggregation, in one, two and three dimensions. They discussed acid-induced aggregation as well as aggregation of casein micelles, and interfacial rheology. Although this article does not describe the use of MC methods, it is wide-ranging and essential reading for anyone who wants to begin simulating aggregation, effects of conformational change and the mechanical properties of proteins.

Proteins are essentially linear folded peptide polymers and here we touch on recent work on simulating polymers where we restrict ourselves to the last decade and a representative selection. Polyions, ionized polymers, complexing with whey proteins at their isoelectric point were studied by de Vries.[164] His result that some binding occurs *via* large charge patches, while others occur *via* multiple charge areas, has been found in similar studies. Binder and Paul[160] is an excellent review by experts on developments in MC methods of simulating polymers. Although it is written for models on lattices, the techniques can be used for off-lattice simulations. The authors address important questions concerning the escape of systems which have become trapped: extensions to the Carmesin–Kremer bond stretching algorithm and escape from tubes. Other recent work and reviews are on critical phenomena, the inverse MC method, the pruned-enriched Rosenbluth method (PERM) applied to polymers, and the technique of Computational Steering, resulting in a non-stochastic process, and which directs the course of the simulation, by manipulating the selection probability function, the function that enables the choice of successive MC steps.[161–164] Techniques for computing the free energy of polymers or of particles in polymers[165,166] are potentially useful for delivery systems. The free energy is conveniently computed using the method of Potential of Mean Force (PMF; see above). An example of modelling polymer adsorption is by Elli *et al.*[167] An older paper by Binder, Baschnagel and Paul[168] is worth studying.

Casein micelle gelation *via* acidification of fat-free milk was modelled by van Heijkamp *et al.*[169] They modelled cluster–cluster aggregation on a 3D lattice in the diffusion-limited (DLCA) and reaction-limited (RLCA) regimes.[124] They modelled a system in which the initial reactivity was low so that the initial state was RLCA, which evolved to DLCA as the reaction switched on *via* increased acidity.

They compared their results to measurements using spin echo small angle neutron scattering (SESANS) and ultra-small angle neutron scattering (USANS). Modelling and simulation of food colloids were described in an excellent paper by Ettelaie,[170] which addressed colloidal gels, viscoelasticity, phase separation, networks, protein displacement from interfaces, foams and bubble dispersions. Work on modelling colloids has been published by Auer and Frenkel,[171] by de Vries and Cohen Stuart,[172] and by Adamczyk.[173] Wetting is an important physical effect, and there is a review by Binder.[174] Other work on colloids of relevance to food researchers studied the effects of external parameters upon colloidal aggregate structure, particle-stabilized emulsions and wetting, colloidal particle deposition using KMC, macromolecules at solid–liquid interfaces, oppositely charged colloidal particles and nano- and microscale fat structures in oils.[175–180] Other applications involved the Reverse Monte Carlo method.[181–183] The applications are all directly related to the use of MC methods in food technology.[184–199]

9.6 THEORY V. SIMULATING FLUID DYNAMICS

This section deals with techniques which *simulate* fluid flow, and not with numerical solutions to the Navier–Stokes (N-S) equations. If one can represent a fluid by continuous distributions of velocities or momenta such that the length scales of the components, or their mean free paths, are much smaller than the dimensions of the containers, or the characteristics of the boundaries are not determined by the flow, then N-S equations can be used. This is generally not the case in complex fluids such as edible oils, colloids or polymeric fluids. N-S equations are non-linear differential equations for velocity distributions for which software codes ("solvers") are available. Their ranges of applicability depend upon the Knudsen number, which is the ratio of the mean free path length of the objects making up the fluid and the characteristic dimensions of the fluid container. In the last 80 years, and especially since ∼1960, the use of numerical methods to solve the N-S equations has expanded, driven by the need for understanding fluid flow over complex hard surfaces and the development of high performance computers. This is the field of computational fluid dynamics (CFD) and a good descriptive summary of these techniques may be found in, for example, http://en.wikipedia.org/wiki/Computational_fluid_dynamics. The solutions to differential equations are determined by their boundary conditions. One that is ubiquitous is the "no-slip boundary condition", that is, the velocity component of the fluid, parallel to the local tangent plane at the boundary is $v_{surf\parallel} = 0$. One can also set the normal component $v_{surf\perp} = 0$. Accordingly, it can be summarized as: the boundary conditions determine the flow. One can ask, however, how should one model a system where the flow erodes, distorts or destroys a boundary? Such systems probably belong to the class known as "soft matter". Examples are water flow eroding/distorting a sandbank, or oil flow which carves out easier pathways through chocolate. The problem can be summarized: boundary conditions determine the flow which, in turn, determines the boundaries. One can envisage accommodating such boundary conditions in CFD but it would require, at best, solving highly non-linear equations self-consistently. Accordingly, other methods have been developed which reduce to the N-S equations in appropriate limits but which can model systems that cannot be easily treated by use of the N-S equations. There are two approaches which are widely used: (1) methods which restrict the position of objects to be on a lattice, and (2) those which treat the position and velocity variables as continuous (off-lattice). Both of these make use of coarse-grained models. Thus, for example, water would be modelled as objects, representing many individual water molecules which flow as a whole.

In this section we describe two classical (non-quantum) approaches to modelling fluids: Dissipative Particle Dynamics (DPD) and Lattice–Boltzmann (L-B) theory. In contrast to Monte Carlo **techniques**, which can be applied to a variety of different systems, both DPD and L-B theory are **models**, and not techniques, for simulating a dynamical system or for obtaining equilibrium values of thermodynamic quantities. However, in general one cannot obtain thermodynamic quantities without using computer simulation, and they are sufficiently general to have been used to model the hydrodynamics of flows or fluid distribution in a variety of systems. These involve polymer brushes, entangled polymers, di- and tri-block co-polymers, membranes, vesicles, nanopores and nanochannels. Since the introduction of these models, their fields of application have expanded enormously. Their great advantages are their ease of applicability and, in appropriate cases, they correspond to the N-S equations. Here we describe the fundamentals of the approaches, briefly describe two applications, and give an approximately representative sample of the advances and applications of the last decade.

9.7   DISSIPATIVE PARTICLE DYNAMICS (DPD)

9.7.1   Fundamentals[200–204]

DPD is a method for **modelling** and **computing** properties of complex **many-component fluids**. Examples of these might be solid objects of various highly anisotropic shapes or complex polymers such as branched polysaccharides. Of course, if it can handle complex objects such as these, it can also handle simple regularly shaped solids or unbranched polymers. DPD can also model such fluids moving through a space, **the walls of which could be eroded by the fluid**. This is an important property of this powerful method since it is not easily achieved by other methods such as CFD. This is very important if one wants to model, for example, the flow of a complex oil through pathways in solid chocolate and allow the possibility that the oil could dissolve, and carry with it, components in the walls of the pathways, or open up new pathways.

    In DPD a physical object is represented by a point at its centre which identifies the position, $\vec{r}(t)$, velocity, $\vec{v}(t)$, and acceleration, $\vec{a}(t)$, of the object at time t. These are shown in Figure 9.12 where,

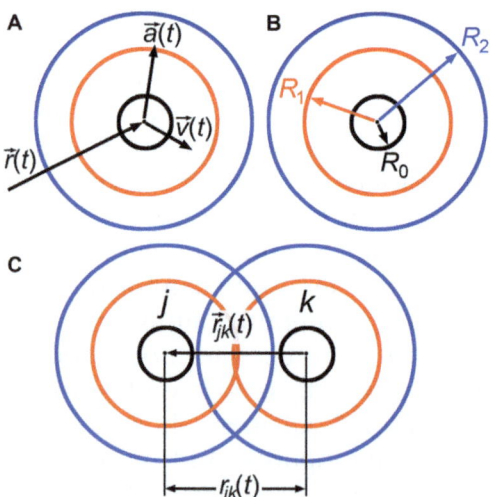

Figure 9.12   (A) A physical object is represented by a point possessing position, $\vec{r}(t)$, velocity, $\vec{v}(t)$, and acceleration, $\vec{a}(t)$, at time t. The concentric spheres define the ranges of various interactions associated with the object. (B) The ranges, R_0, R_1 and R_2, of the three interactions. (C) Two objects, j and k, interact *via* the pair of blue interactions and the pair of red interactions because they overlap, respectively, but not *via* the black interactions.

in addition to the black point which identifies the object, a set of concentric spheres surround the point. It is these concentric spheres which give the object its size and which specify its interactions with its surroundings. Thus the size of the object could be defined by the black sphere of range R_0, the red sphere could represent the range of the van der Waals interaction while the blue sphere might represent the effective range of electrostatics. If we have K such identical objects labelled by an integer, $j = 1, \ldots , K$, with the possibility that each object might be in one of several (internal) states, $\{\alpha, \beta, \delta \ldots\}$, then a more complete description of the object would be the set $\{\vec{r_j}(\alpha, t), \vec{v_j}(\alpha, t), \vec{a_j}(\alpha, t)\}$. If we define the relative positions and velocities of two such objects, j and k, as (Figure 9.12C),

$$\vec{r_{jk}}(\alpha, \delta, t) = \vec{r_j}(\alpha, t) - \vec{r_k}(\delta, t) \tag{9.38}$$

$$r_{jk}(\alpha, \delta, t) = \left|\vec{r_{jk}}(\alpha, \delta, t)\right| = \left|\vec{r_j}(\alpha, t) - \vec{r_k}(\delta, t)\right| \tag{9.39}$$

$$\widehat{u_{jk}} = \vec{r_{jk}}(\alpha, \delta, t)/r_{jk}(\alpha, \delta, t) \tag{9.40}$$

$$\vec{v_{jk}}(\alpha, \delta, t) = \vec{v_j}(\alpha, t) - \vec{v_k}(\delta, t) \tag{9.41}$$

where object j is in state α and object k is in state δ, and $\widehat{u_{jk}}$ is a unit vector pointing from object k to object j. To avoid non-essential complications, we shall assume that each object possesses only a single internal state so that we can dispense with the labels $\{\alpha, \beta, \delta \ldots\}$. The motions of the two objects are governed by their interactions with their surroundings *via* three pair-wise forces. These forces due to the interaction defined by the radii between objects j and k are: (i) a conserved force, $\vec{F^C_{jkn}}(r_{jk}(t), t)$, which represents elastic scattering and involves zero energy transfer between two interacting objects, (ii) a dissipative force, $\vec{F^D_{jkn}}(\vec{r_{jk}}(t), \vec{v_{jk}}(t), t)$, which results in energy loss on the average and represents friction ("molecular" or otherwise) and (iii) a random force, $\vec{F^R_{jkn}}(r_{jk}(t), t)$, which represents their interactions with the surrounding heat bath at temperature, T, which transfers energy, on the average, to objects. The dissipative and random forces are related *via* the Fluctuation–Dissipation theorem and serve as a thermostat, keeping the temperature constant. The conserved force is derivable from a potential which can be made up of different potentials representing van der

Waals, electrostatic or other interactions. In early work, these forces took the form,

$$\overrightarrow{F_{jkn}^C}(r_{jk}(t), t) = A_n[2R_n - r_{jk}(t)]\widehat{u_{jk}} \tag{9.42}$$

$$\overrightarrow{F_{jk0}^D}(\overrightarrow{r_{jk}}(t), \overrightarrow{v_{jk}}(t), t) = -\gamma[2R_0 - r_{jk}(t)]^2[\overrightarrow{r_{jk}}(t) \cdot \overrightarrow{v_{jk}}(t)]\widehat{u_{jk}} \tag{9.43}$$

$$\overrightarrow{F_{jk0}^R}(r_{jk}(t), t) = \sigma[2R_0 - r_{jk}(t)]\beta_{jk}\Delta t^{-1/2}\widehat{u_{jk}} \tag{9.44}$$

where we have assumed that the "size" of the objects is R_0. The quantity Δt is the magnitude of the time step of the simulation, and β_{jk} is a randomly fluctuating variable with Gaussian statistics (appropriate for a random force due to fluctuations driven by a surrounding heat bath at temperature T) and average equal to zero.

If we consider a system in which there are different objects, possessing radii R'_0, R'_1 and R'_2, where R'_0 represents the size of the second type of object, then, in equations (42) to (44), $2R_n$ is replaced by $R_n + R'_n$. The dissipative and random forces are not independent but are related *via* the Fluctuation–Dissipation theorem, *e.g.* Kubo,[205] which yields,

$$\sigma^2 = 2k_BT\gamma \tag{9.45}$$

where, again, k_B is Boltzmann's constant.

The conserved force of equation (9.42) corresponds to the assumption that the two objects interact *via* an harmonic potential. This can be replaced by any differentiable potential, such as an attractive van der Waals potential, of ranges R_1 and R'_1 appropriate to a pair of spheres, with the size of the system represented by a phenomenological soft-core repulsion.

We have said nothing about the spatial scales of the system: what range of values can the set $\{R_n\}$ assume? The answer is that DPD is not constrained to a particular spatial scale, as long as one can plausibly represent the components of a system by the model described here, so that it does not matter whether we are representing bacteria, sand particles, "chunks" of water or ice, or planets. That the interactions and properties of the particles must reflect the physical problem is the only requirement.

The total force on object j is then the vector sum of all the forces due to its environment comprising other objects and the surrounding heat bath,

$$\overrightarrow{F_j}(t) = \sum_{kn}[\overrightarrow{F_{jkn}^C}(r_{jk}(t), t) + \overrightarrow{F_{jk0}^D}(\overrightarrow{r_{jk}}(t), \overrightarrow{v_{jk}}(t), t) + \overrightarrow{F_{jk0}^R}(r_{jk}(t), t)] \tag{9.46}$$

When all the forces have been computed, accelerations are computed in the usual classical way, $\vec{a} = \vec{F}/m$, where m is the mass of the object and $\vec{F}$ is the total force acting on it. These are then integrated, using standard molecular dynamics routines, to yield new velocities which, in turn, yield new positions. A well-behaved integrating routine is the Velocity–Verlet algorithm. DPD ensures local conservation of linear and angular momentum, necessary in order that this approach be in accord with correct hydrodynamic characteristics as the system becomes large. In principle, the conservative forces are free to possess multiple characteristics so that they can exhibit angular dependent forces. Thus, for example, a pair of objects could represent a spherical object (*via* the repulsive potential), that interacts with other objects *via* the angular-independent attractive van der Waals force together with angular-dependent forces *via* "hotspots" at the north and south poles. Non-spherical structures can be represented by appropriate aggregates of spheres, though one must take care that Newton's laws are satisfied.

One must also ensure that the simulations are in accord with known physical characteristics of the systems being modelled. For example, correct densities and compressibilities should be used. In particular, if one is modelling water, then one must ensure that the model water is, in fact, essentially incompressible. Thus, in representing water, it has been found that the particle density had to take on a value of 6; that is, 6 objects had to occupy a space the size of the soft-core excluded volume, in order to ensure sufficient incompressibility.

Fluid flow is, in part, dependent upon the boundaries: the N-S differential equations yield solutions appropriate to a given system by satisfying the boundary conditions. A strength of DPD is that it can accommodate boundaries that interact with the flowing objects and, as a result, change dynamically depending upon the flow. This self-consistent modelling is not as easy to accommodate *via* the N-S differential equations and the treatment of complex boundary conditions remains a challenge to some implementations of the Lattice Boltzmann theory (below). Boundaries make two demands: (i) how to realize the no-slip boundary condition and (ii) how to go about defining a boundary. A choice for (i) is to reverse all velocities at the boundary. This ensures that the velocity components parallel and perpendicular to the boundary are zero. Regarding (ii), one can impose a layer of the fluid onto the boundary surface and, with the interactions already specified, one need not have the burden of working out the interactions between, say, water and a metallic or polymeric boundary. Adaptive boundary conditions have been studied by Pivkin and Karniadakis.[206]

Many people prefer to write their own code for simulations. However, there appear to be good codes available. Here we describe how to use DPD in the software package ESPResSo© which can be found, for example, on ACEnet, the Atlantic Computational Excellence Network, a consortium of Atlantic Canadian Universities.

9.7.2 Using ESPResSo DPD

ESPResSo is a versatile engine for running particle dynamics simulations. Users of ESPResSo specify system parameters and control the simulation by writing a script using the ***Tcl scripting language***. These scripts can make use all of the Tcl language commands as well as commands specific to ESPResSo, which can be found in the ESPResSo documentation. There are several tutorials, with sample scripts and explanations, which can be found either in the docs/tutorials directory of the ESPResSo source code, or online at the ESPResSo website. A powerpoint by Smiatek[207] is good.

Generally, the Tcl script for an ESPResSo simulation will begin by defining the parameters of the system. System parameters are set using the ***setmd*** command. For example, the dimensions of the simulation box can be specified with ***setmd box_l size_x size_y size_z*** where size_x, size_y, and size_z are the dimensions of the box. The user must also specify the boundary conditions using ***setmd periodic x y z***, where x, y, and z are either 1, indicating that the coordinate is periodic, or 0, indicating that the coordinate is not periodic. For example, to set the boundary conditions such that the x and y coordinates are periodic, but the z coordinate is not, the command would be setmd periodic 1 1 0. ESPResSo also supports constraints (*via* the ***constraint*** command), including walls, which can be used to keep particles from leaving the box in directions where the boundary conditions are not periodic.

Next, the user specifies the parameters of the simulation. The command ***thermostat*** is used to define the thermostat and its associated parameters. In ESPResSo, DPD is implemented as a thermostat so it is at this point that the user chooses to run the simulation using DPD. The temperature that the user sets for the DPD thermostat is used to calculate the strength of the random and dissipative forces.

ESPResSo gives the user complete control over the positions of all particles in the simulation and the interactions between them. In the Tcl script, the user specifies properties of individual particles (position, velocities, particle type, *etc.*) using the ***part*** command. Interactions are defined between particles of different types using the ***inter*** command. When running a simulation with the DPD thermostat, any of the available interaction types listed in the ESPResSo documentation can be used. ESPResSo then computes the conservative force used by the DPD thermostat from the potential associated with the specified interaction.

Once the interactions and particle properties have been set the user chooses a time step for the integration by setting the variable time_step. The ***integrate*** command carries out a Velocity–Verlet integration. It takes as input the number of steps to integrate over. Thus the simulation moves forward in time by the number of steps multiplied by the value of the time step variable each time the ***integrate*** command is called. During and after the simulation, the user can write data to file using Tcl's ***puts*** command.

ESPResSo is constructed modularly, so that most of its non-essential features are contained in separate modules which are not compiled into the program by default. The purpose of this is to increase the speed of simulations. ESPResSo is compiled separately on each ACEnet cluster so modules compiled on one cluster might not be compiled on another cluster. *Before running a simulation the user should make sure that all of the modules required for their simulation are compiled.* For instance, the DPD thermostat requires the DPD module to be compiled. Several of the interaction potentials require that their corresponding modules be compiled.

A list of compiled modules can be found in the file **src/myconfig-default.h** in the directory used to build ESPResSo. When a user requires modules which are not compiled by default, they may compile ESPResSo in their home directory with the required modules. To do this the user should copy the file **src/myconfig-default.h** to a file called **myconfig.h** in the same directory and add to it lines of the form:

#define MODULE

where MODULE is the name of the module needed. Then ESPResSo can be re-compiled, at which point the features will be available. Detailed instructions on compiling ESPResSo can be found on the website. Questions about building ESPResSo on ACEnet can be directed to ACEnet support.

Once the user has a Tcl script for the simulation and ESPResSo has been compiled with all the required modules all that remains is to start the simulation. To use ESPResSo 3.2.1 on ACEnet the modules

openmpi/pgi/1.4 and ***espressomd*** must be loaded using the following commands:

module load pgi openmpi/pgi/1.4

module load espressomd

An example submission script for an ESPReSSo simulation is at http://www.ace-net.ca/wiki/ESPReSSo. The submission script can be submitted to ACEnet's job scheduler using the ***qsub*** command.

Because it is most useful when treating mesoscale systems, one should obtain the strength of the dissipative forces from theories which can carry out first-principles or equivalent calculations. There is also the important question: what is the physical basis for the analytical form of the dissipative and random forces? Depending upon what one takes to be the "objects" (nanoscale clusters of lipid molecules, microscale crystallites or milliscale aggregates), the strength of DPD is that one can model objects of any size for times which depend upon the characteristic velocities of the system. Thus simulations have been carried out on microscale models for tens of microseconds.

Representative papers on the theory and applications of DPD are by Revenga, Zúñiga and Español,[208] Whittle and Dickinson,[209] Groot,[210] Pivkin and Karniadakis,[206] Ter-Oganessian,[211] Trofimov, Nies and Michels,[212] González-Melchor et al.,[213] Jakobsen, Besold and Mouritsen,[214] Pivkin and Karniadakis,[204] Symeonidis, Karniadakis and Caswell,[215] Jiang et al.,[216] Fedosov, Pivkin and Karniadakis,[217] Eriksson et al.,[218] Füchslin et al.,[219] Litvinov et al.,[220] Huang et al.,[221] Vázquez-Quesada, Ellero and Español,[222] Fedosov, Karniadakis and Caswell,[223] Zhang and Guo,[224] Binder, Kreer and Milchev,[225] Lísal, Brennan and Avalos,[226] Yamanoi, Pozo and Maia,[227] Khani, Yamanoi and Maia,[228] Kulkarni et al.,[229] Langeloth et al.,[230] and Warren et al.[231] They encompass surfactants, vesicles, rheology of polymer fluids, colloids, self-diffusion, flexible rods, reptation, hydrodynamics of polymer solutions, electrostatics, entangled polymers and polymer brushes.

9.7.3 Simulations Using DPD

Here we show two examples to illustrate the power of DPD.

Example 1: *Two-component system* (Adam J. MacDonald, adam. macdonald@onezero.ca). This system consists of massless green spheres which interact with each other and with massive interacting

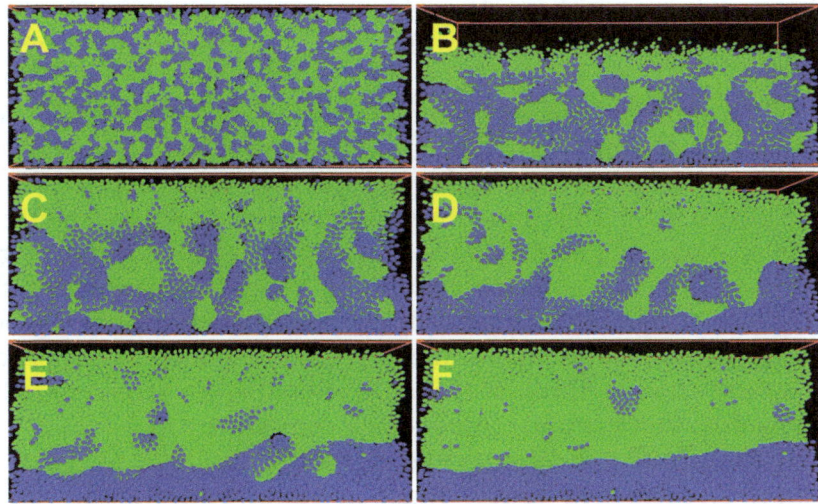

Figure 9.13 A compressible two-component system comprising massless green spheres and massive blue spheres. Gravitational attraction is vertically down. Shown is a slice through the centre of the simulation. (A) Initial state with velocity of green spheres $= 0$. (B–F) Time increases. The fluid bounces as it is released and the blue spheres begin to sink to the bottom. At the same time the green fluid begins to flow to the right. Simulation developed using Fluidix© (OneZero Software, onezero.ca). (Images used with the kind permission of Adam J. MacDonald.)

blue spheres. Because of their mass, the blue spheres interact with the gravitational field, which is oriented down.

Figure 9.13 shows instantaneous images, taken from a video, of a slice through the centre of the simulation box which occupies the space $0 \leq x \leq X$, $0 \leq y \leq Y$ and $0 \leq z \leq Z$. The images show slices of the x–z plane with the x-axis horizontal and the z-axis vertical. Each slice is centred at $y = Y/2$ and is approximately two sphere diameters thick. The system is held in the initial state (time, $t = 0$) of the simulation (A) comprising an approximately random distribution of small (3D) aggregates of blue spheres. At $t = 0$, the system is released and, because it is compressible, it bounces and quickly settles into an approximately uniform density. As t increases (B–F), the green spheres are given a velocity to the right as they enter the simulation volume at the $x = 0$ plane. Driven by gravity, the blue spheres fall towards the bottom of the box but are washed along by the motion of the green spheres. The end result (F) is that the blue spheres have formed a layer at the bottom that has been deformed by the flow of the green current. The most striking aspect is that the green flow helps to bring about aggregation of blue spheres, that it distorts the aggregates, tearing some apart and causing other aggregates to form. The complexity of the surfaces formed in this way is interesting to watch on the video and shows the power of DPD in being able to handle this system of complex, changing boundaries at the green–blue interfaces.

Example 2: *Bead pulled with a constant force through polymers in an aqueous solution.*[232] This simulation was developed to understand the motion of a paramagnetic bead as it was pulled by a constant force through an aqueous solution containing actin polymers using the principle of "magnetic tweezers" developed by the Sackmann group at the Technische Universität München.[234] This simulation involved a rigid sphere moving through a fluid comprising an incompressible component (water) and a compressible component (polymers). The rectangular simulation box, similar to that of Example 1, at a temperature, T, contained model water spheres, semi-flexible model polymers and a hard sphere.[232,233] As the sphere was driven by a constant force, its speed exhibited three regimes: (i) an initial regime where the sphere accelerated inside a viscous fluid so that, eventually, the acceleration went to zero. The speed was then at a maximum. If the fluid comprised a non-polymeric substance, then this would occur as the time, $t \to \infty$. However, because of the presence of the second, compressible, polymeric substance, this maximum occurred at finite t. The sphere achieved its maximum speed in this way because of the build-up of a polymer "bow-wave". The sphere was then moving at a speed that was too great for the polymers to be able to move away from its direction of motion and the simulation entered the second regime. (ii) In this regime, the speed of the sphere decreased in accord with a power law so that the sphere's displacement was $x \sim t^{\alpha}$. (iii) Finally, the third regime was when the sphere achieved a long-time constant-speed state. One of the goals was to compute the exponent, α. The simulation used periodic boundary conditions and consideration was given to disallow polymers moving across the boundaries to convey information from one side of the simulation to the other. Figure 9.14 shows three images from the simulation: the initial state (A), a state after some elapsed time t_1 (B), and a state after an elapsed time $t_2 \approx 2t_1$ (C). Water spheres are not shown. Polymer monomers belonging to different polymers possess different colours. The key aspect to understanding the results is that the diffusion rate of the polymers away from the (projected) path of the sphere is less than the speed of the sphere. In B and C one sees the build-up of the "bow-wave" which is a direct consequence of this. In image C, one can see an indication of the decrease in speed as the sphere gets its

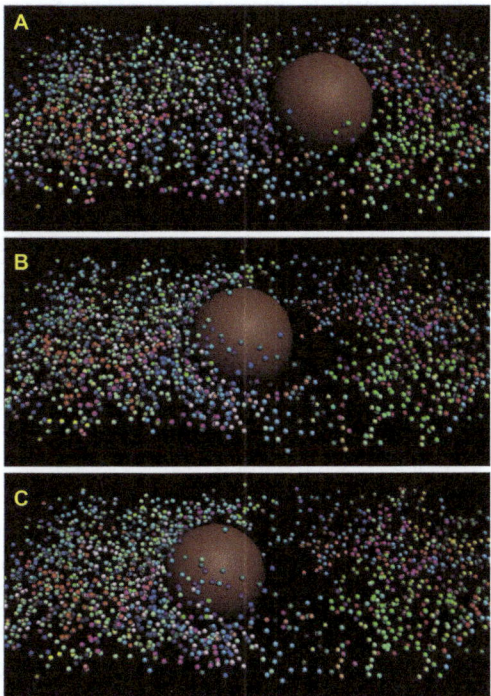

Figure 9.14 A solid sphere moving, under a constant force, through an aqueous solution containing linear polymers and simulated using Dissipative Particle Dynamics. Polymer monomers are spheres linked by harmonic spring bonds. Neither the water molecules nor the bonds are shown for clarity. Monomers with the same colour belong to the same polymer. (A) Initial state with speed $\approx$ 0. (B) state when the speed is approximately maximum. (C) State when the seed has decreased.
(Published with the agreement of Dr Nikita Ter-Oganessian and Dr Alexei Boulbitch.)

speed reduced by the high local density of polymers ahead of it. It was found that the exponent $\alpha \approx 0.5$, in complete agreement with the experimental data of Uhde *et al.*[234]

9.7.4 Lattice Boltzmann (L-B) Theory[75,235–238]

Two other unpublished but very readable contributions are by Chirila[239] and by Peng.[240] As the name suggests, L-B theory represents the flow as objects moving on a lattice. There is no physical meaning to locations that do not coincide with the sites of the lattice being used. L-B theory is one of a class of Cellular Automata and was developed to address deficiencies of the Lattice Gas Cellular Automata

(LGCA), *e.g.* the lack of Gallilean invariance, and the statistical noise. Further, L-B theory does not require the use of a coarse-grained representation of the system, as does DPD, in order to determine average values of macroscopic dynamical variables such as viscosity. Set against this is the observation that a lattice, in general, does not possess rotational invariance. Techniques have been developed to address this but one should be aware that, in complex geometries, predictions made using some models may be affected by the orientation of the lattice.[241] L-B theory is concerned with solving the Boltzmann equation, shown in equation (9.47) for thermal dilute gases, for the single-particle position and linear momentum distribution function, $f(\vec{r}, \vec{p}, t)$,

$$\frac{\partial f}{\partial t} + \vec{v} \cdot \vec{\nabla} f = C_2 \qquad (9.47)$$

where C_2 represents the local uncorrelated two-body collision operator for particles. L-B theory considers the system to be composed of fictitious particles moving between lattice sites and requires specification of: (i) the definition of the lattice to be used, (ii) the definition of dynamical variables associated with each site of the lattice, and (iii) a rule to define the dynamics at each site (updating rule) that takes into account the interactions between sites and which is divided into (a) streaming (propagation or flow of fictitious particles from one site to another) and (b) the effects of collisions between the fictitious particles upon position and velocity distributions. The use of a lattice and the restriction in the number of velocities that describe the propagation of the particles between lattice sites makes L-B theory an efficient computational solver of the Boltzmann equation for complex fluids.

Representative papers on the theory and applications of L-B theory are on adsorbed systems, fluid–fluid interfaces, microfluidics, baked bread, thixotropic systems, shear thinning, droplets, stirred systems, heat transfer, Sago, Pickering and other emulsions, moving boundaries, suspensions, viscous mixing, Bingham fluids and surfactants.[176,242–267]

9.8 CONCLUSIONS

This chapter has attempted to describe key aspects of three of the major areas of computer simulation approaches in use: Molecular Dynamics, Monte Carlo techniques, and Dissipative Particle Dynamics, with brief descriptions of other models or techniques. The intent has been to provide a foundation from which one can begin to become a practitioner, together with a representative selection of sources to give a survey of contemporary questions related to the computer simulation of models of food structures and some examples which readers can simulate since, seemingly, the "answer is known". To those whose works have been omitted, we can only say that *force majeure* required us to make selections.

ACKNOWLEDGEMENTS

It is a pleasure to thank our many colleagues, especially Alejandro Marangoni, who have contributed to our understanding of this important field. We also thank NSERC of Canada for funding.

REFERENCES

1. D. J. Selkoe, *J. Chem. Phys.*, 2001, **129**, 144108.
2. S. Murakami, R. Nakashima, E. Yamashita and A. Yamaguchi, *Nature*, 2002, **419**, 587.
3. C. Fabrizio and C. M. Dobson, *Annu. Rev. Biochem.*, 2006, **75**, 333.
4. M. E. M. Noble, J. A. Endicott and L. N. Johnson, *Science*, 2004, **303**, 1800.
5. E. Yuriev and P. A. Ramsland, *J. Mol. Recognit.*, 2013, **26**, 215.
6. F. C. Bernstein, T. F. Koetzle, G. J. Williams, E. E. Meyer Jr., M. D. Brice, J. R. Rogers, O. Kennard, T. Shimanouchi and M. Tasumi, *J. Mol. Biol.*, 1977, **112**, 535.
7. P. Bernado, L. Blanchard, P. Timmins, D. Marion, R. W. Ruigrok and M. Blackledge, *Proc. Natl. Acad. Sci. USA*, 2005, **102**, 17002.
8. H. J. Dyson and P. E. Wright, *Methods Enzymol.*, 2005, **394**, 299.
9. J. L. Jimenez, E. J. Nettleton, M. Bouchard, C. V. Robinson, C. M. Dobson and H. R. Saibil, *Proc. Natl. Acad. Sci. USA*, 2002, **99**, 9196.
10. R. H. Callender, R. B. Dyer, R. Gilmanshin and W. H. Woodruff, *Annu. Rev. Phys. Chem.*, 1998, **49**, 202.
11. M. M. Gromiha, *Protein Bioinformatics from Sequence to Function*, Elsevier, New Delhi, 2010.
12. A. D. MacKerell Jr., B. Brooks, C. L. Brooks III, L. Nilsson, B. Roux, Y. Won and M. Karplus, in *The Encyclopedia of Computational Chemistry*, 1, 271–277, ed. P. v. R. Schleyer, John Wiley & Sons, Chichester, 1998.
13. J. D. Bryngelson and J. D. Wolynes, *Proc. Natl. Acad. Sci. USA*, 1987, **84**, 7524.
14. K. A. Dill, S. B. Ozkan, M. S. Shell and T. R. Weikl, *Annu. Rev. Biophys.*, 2008, **37**, 289.

15. K. A. Dill and J. L. MacCallum, *Science*, 2012, **338**, 1042.
16. J. Gonzalez-Gutierrez and M. G. Scanlon, in *Rheology and Mechanical Properties of Fats*, ed. A. G. Marangoni, AOCS Press, Urbana, 2012, pp. 127–172.
17. F. Maleky, in *Structure–Function Analysis of Edible Fats*, ed. A. G. Marangoni, AOCS Press, Urbana, 2012, pp. 207–230.
18. K. Sato, *Chem. Eng. Sci.*, 2001, **56**, 2255.
19. A. G. Marangoni, *Trends Food Sci. Technol.*, 2002, **13**, 37.
20. A. G. Marangoni, *Fat Crystal Networks*, Marcel Dekker, New York, 2005.
21. C. Himawan, V. M. Starov and A. G. F. Stapley, *Adv. Colloid Interface Sci.*, 2006, **122**, 3.
22. K. Sato and S. Ueno, *Cur. Opin. Colloid Interface Sci.*, 2011, **16**, 384.
23. D. A. Pink and M. S. G. Razul, *Food Struct.*, 2013, **1**, 71.
24. W. Yourgrau, A. van der Merwe and G. Raw, *Treatise on Irreversible and Statistical Thermophysics: An Introduction to Nonclassical Thermodynamics*, The MacMillan Company, New York, 1966.
25. D. Chandler, *Introduction to Modern Statistical Mechanics*, Oxford University Press, New York, 1987.
26. F. Mandl, *Statistical Physics*, The Manchester Physics Series, John Wiley and Sons Ltd., 2nd edn, 1988.
27. D. J. Evans and G. P. Morris, *Statistical Mechanics of Nonequilibrium Liquids*, Academic Press, London, 1990.
28. C. E. Hecht, *Statistical Thermodynamics and Kinetic Theory*, Dover Publications Inc., New York, 1990.
29. W. G. Hoover, *Studies in Modern Thermodynamics II*, Elsevier, Amsterdam, 1991.
30. R. K. Pathria, *Statistical Mechanics*, Butterworth-Heinemann, London, 2nd edn, 1996.
31. W. Greiner, L. Niese and H. Stöcker, *Thermodynamics and Statistical Mechanics*, Springer Verlag, New York, 1997.
32. R. Baierlein, *Thermal Physics*, Cambridge, The University Press, 1999.
33. L. E. Reichl, *A Modern Course in Statistical Physics*, Wiley-VCH Verlag, Weinheim, Germany, 3rd edn, 2009.
34. T. Sochi, *Polymer Rev.*, 2011, **51**, 309.
35. D. C. Rapaport, *The Art of Molecular Dynamics Simulation*, Cambridge University Press, Cambridge, UK, 1995.
36. J. M. Haile, *Molecular Dynamics Simulation: Elementary Methods*, John Wiley and Sons, New York, 1997.

37. R. J. Sadus, *Molecular Simulation of Fluids: Theory, Algorithms and Object-Orientation*, Elsevier, Amsterdam, 1999.
38. M. P. Allen and D. J. Tildesley, *Computer Simulation of Liquids*, Clarendon Press, Oxford, 1989.
39. D. Frenkel and B. Smit, *Understanding Molecular Simulation*, Academic Press, San Diego, 2nd edn, 2002.
40. A. R. Leach, *Molecular Modelling: Principles and Applications*, Addison Wesley Longman Ltd., Harlow, 1996.
41. N. L. Allinger, Y. H. Yuh and J. M. Lii, *J. Am. Chem. Soc.*, 1989, **111**, 8551.
42. M. J. Martin and J. I. Siepmann, *J. Phys. Chem. B*, 1998, **102**, 2569.
43. D. Dubbeldam, S. Calero, T. J. H. Vlugt, R. Krishna, T. L. M. Maesen and B. Smit, *J. Phys. Chem. B*, 2004, **108**, 12301.
44. S. K. Nath, F. A. Escobedo and J. J. de Pablo, *Chem. Phys.*, 1998, **108**, 9905.
45. R. Faller, H. Schmitz, O. Biermann and F. J. Mueller-Plathe, *Computat. Chem.*, 1999, **20**, 1009.
46. P. Ungerer, C. Beauvais, J. Delhommelle, A. Boutin, B. Rousseau and A. Fuchs, *J. Chem. Phys.*, 2000, **112**, 5499.
47. E. Bourasseau, M. Haboudou, A. Boutin, A. H. Fuchs and P. J. Ungerer, *J. Chem. Phys.*, 2003, **118**, 3020.
48. T. Schlick, R. Collepardo-Guevara, L. A. Halvorsen, S. Jung and X. Xiao, *Quart. Rev. Biophys.*, 2011, **44**, 191.
49. B. R. Brooks, R. E. Bruccoleri, B. D. Olafson, D. J. States, S. Swaminathan and M. Karplus, *J. Computat. Chem.*, 1983, **4**, 187.
50. W. D. Cornell, P. Cieplak, C. I. Bayly, I. R. Gould, K. M. Merz, D. M. Ferguson, D. C. Spellmeyer, T. Fox, J. W. Caldwell and P. A. Kollman, *J. Am. Chem. Soc.*, 1995, **117**, 5179.
51. W. F. van Gunsteren and H. J. C. Berendsen, *Groningen Molecular Simulation (GROMOS) Library Manual*, Biomos, Groningen, 1987, pp. 1–121.
52. W. F. van Gunsteren, S. R. Billeter, A. A. Eising, P. H. Hünenberger, P. Krüger, A. E. Mark, W. R. P. Scott and I. G. Tironi, *Biomolecular Simulation: The GROMOS96 Manual and User Guide*, Vdf Hochschulverlag A. G. van der ETH, Zürich, 1996.
53. W. L. Jorgensen and J. Tirado-Rives, *J. Am. Chem. Soc.*, 1988, **110**, 1657.
54. M. C. Stumpe, N. Blinov, D. Wishart, A. Kovalenko and V. S. Pande, *J. Phys. Chem. B*, 2010, **115**, 319.
55. A. Kovalenko and N. Blinov, *J. Molec. Liq.*, 2011, **164**, 101.

56. P. J. Dyer, H. Docherty and P. T. Cummings, *J. Chem. Phys.*, 2008, **129**, 02458.

57. A. D. MacKerell, Jr., *J. Computat. Chem.*, 2004, **25**, 1584.

58. J. G. Kirkwood, *J. Chem. Phys.*, 1935, **3**, 300.

59. J. P. Hansen and I. R. McDonald, *Theory of Simple Liquids*, Academic Press, London, 1976.

60. G. M. Torrie and J. P. Valleau, *Chem. Phys. Lett.*, 1974, **28**, 578.

61. A. M. Ferrenberg, *J. Phys. Rev. Lett.*, 1989, **63**, 1658.

62. A. M. Ferrenberg and R. H. Swendsen, *J. Phys. Rev. Lett.*, 1989, **63**, 1195.

63. S. Kumar, D. Bouzida, R. H. Swendsen, P. A. Kollman and J. M. Rosenberg, *J. Computat. Chem.*, 1992, **13**, 1011.

64. B. Hess, C. Kutzner, D. van der Spoel and E. Lindahl, *J. Chem. Theory Comput.*, 2008, **4**, 435.

65. D. M. York, T. A. Darden and L. G. Pedersen, *J. Chem. Phys.*, 1993, **99**, 8345.

66. T. Darden, D. York and L. Pedersen, *J. Chem. Phys.*, 1993, **98**, 10089.

67. H. G. Petersen, *J. Chem. Phys.*, 1995, **103**, 3668.

68. U. Essmann, L. Perera, M. L. Berkowitz, T. Darden, H. Lee and L. G. Pedersen, *J. Chem. Phys.*, 1995, **103**, 8577.

69. H. J. C. Berendsen, D. Van der Spoel and R. Vandrunen, *Comput. Phys. Commun.*, 1995, **91**, 43.

70. C. Kutzner, D. Van der Spoel, M. Fechner, E. Lindahl, U. W. Schmitt, B. L. De Groot and H. Grubmuller, *J. Comput. Chem.*, 2007, **28**, 2075.

71. J. Berger, O. Edholm and F. Jähnig, *Biophys. J.*, 1997, **72**, 2002.

72. C. J. MacDougall, M. S. G. Razul, E. Papp-Szabo, A. G. Marangoni, F. Peyronel and D. A. Pink, *Faraday Discuss.*, 2012, **158**, 425(discussion 493).

73. H. J. Limach and K. Kremer, *Trends Food Sci. Tech.*, 2006, **17**, 215.

74. R. G. M. van der Sman, *Adv. Colloid Interface Sci.*, 2012, **18**, 176.

75. Q. T. Ho, J. Carmeliet, A. K. Datta, T. Defraeye, M. A. Delele, E. Herremans, L. Opara, H. Ramon, E. Tijskens, R. van der Sman, P. Van Liedekerke, P. Verboven and B. M. Nicolai, *J. Food Eng.*, 2013, **114**, 279.

76. N. Stănciuc, I. Aprodu, G. Râpeanu and G. Bahrim, *Innov. Food Sci. Emerg. Technol.*, 2012, **15**, 50.

77. M. Rabe, D. Verdes and S. Seeger, *Adv. Colloid Interface Sci.*, 2011, **162**, 87.

78. S. R. Euston, *Curr. Opin. Colloid Interface Sci.*, 2004, **9**, 321.

79. J. Sorensen, D. S. Palmer, K. B. Qvist and B. Schiott, *J. Agric. Food Chem.*, 2011, **59**, 5636.
80. H. M. Berman, J. Westbrook, Z. Feng, G. Gilliland, T. N. Bhat, H. Weissig, I. N. Shindyalov and P. E. Bourne, *Nucl. Acids Res.*, 2000, **28**, 235.
81. L. Lupi, L. Comez, M. Paolantoni, P. Sassi, A. Morresi, B. M. Ladanyi and D. Fioretti, *J. Phys. Chem. B*, 2012, **116**, 14760.
82. P. B. Conrad and J. J. de Pablo, *J. Phys. Chem. A*, 1999, **103**, 4049.
83. A. Lerbret, F. Affouard, A. Hedoux, S. Krenzlin, J. Siepmann, M.-C. Bellissent-Funel and M. Descamps, *J. Phys. Chem. B*, 2012, **116**, 11103.
84. V. Krautler, M. Muller and P. H. Hunenberger, *Carbohydr. Res.*, 2007, **342**, 2097.
85. L. Peric-Hassler, H. S. Hansen, R. Baron and P. H. Hunenberger, *Carbohydr. Res.*, 2010, **345**, 1781.
86. D. Q. Wang, M. L. Amundadottir, W. F. van Gunsteren and P. H. Hunenberger, *Eur. Biophys. J. Biophy.*, 2013, **42**, 521.
87. E. Autieri, M. Sega, F. Pederiva and G. Guella, *J. Chem. Phys.*, 2010, **133**, 095104.
88. V. Yeguas, M. Altarsha, G. Monard, R. Lopez and M. F. Ruiz-Lopez, *J. Phys. Chem. A*, 2011, **115**, 11810.
89. K. Raghavan, M. R. Reddy and M. L. Berkowitz, *Langmuir*, 1992, **8**, 233.
90. J. Xu, W. X. Zhao, Y. W. Ning, M. Bashari, F. F. Wu, H. Y. Chen, N. Yang, Z. Y. Jin, B. C. Xu, L. Zhang and X. M. Xu, *Carbohyd. Polymers*, 2013, **92**, 1633.
91. A. Brasiello, L. Russo, C. Siettos, G. Milano and S. Crescitelli, Multi-scale modelling and coarse-grained analysis of triglycerides dynamics, in *20th European Symposium on Computer Aided Process Engineering Book Series: Computer-Aided Chemical Engineering*, ed. S. Pierucci and B. G. Ferraris, Amsterdam, Elsevier, 2010, vol. 28, p. 625.
92. A. K. Sum, M. J. Biddy, J. J. dePablo and M. J. Tupy, *J. Phys. Chem. B*, 2003, **107**, 14443.
93. A. Hall, J. Repakova and I. Vattulainen, *J. Phys. Chem. B*, **112**, 13772.
94. W. D. Hsu and A. Violi, *J. Phys. Chem. B*, 2009, **113**, 887.
95. M. Müller, K. Katsov and M. Schick, *Phys. Rep.*, 2006, **434**, 113.
96. S. J. Marrink, H. J. Risselada, S. Yefimov, D. P. Tieleman and A. H. de Vries, *J. Phys. Chem. B*, 2007, **111**, 7812.
97. S. V. Bennnun, M. I. Hoopes, C. Xing and R. Faller, *Chem. Phys. Lipids*, 2009, **159**, 59.

98. C. Peter and K. Kremer, *Soft Matter*, 2009, **5**, 4357.
99. J. J. de Pablo, *Annu. Rev. Phys.*, 2011, **62**, 555.
100. M. Laradji and P. B. S. Kumar, *Adv. Plan. Lipid Bilayers Liposomes*, 2011, **14**, 201.
101. E. Brini, E. A. Algaer, P. Ganguly, C. Li, F. Rodriguez-Ropero and N. F. A. van der Vegt, *Soft Matter*, 2013, **9**, 2108.
102. M. G. Saunders and G. A. Voth, *Annu. Rev. Biophys.*, 2013, **42**, 73.
103. L. Monticelli, S. K. Kandasamy, X. Periole, R. G. Larson, D. P. Tieleman and S.-J. Marrink, *J. Chem. Theory Comput.*, 2008, **4**, 819.
104. R. G. Oliveira, E. Schneck, B. E. Quinn, O. V. Konovalov, K. Brandenburg, T. Gutsmann, T. A. Gill, C. B. Hanna, D. A. Pink and M. Tanaka, *Phys. Rev. E*, 2010, **81**, 41901.
105. E. Schneck, T. Schubert, O. V. Konovalov, B. E. Quinn, T. Gutsmann, K. Brandenburg, D. A. Pink and M. Tanaka, *Proc. Natl. Acad. Sci. USA*, 2010, **107**, 9147.
106. M. M. Tirion, *Phys. Rev. Lett.*, 1996, **77**, 1905.
107. I. Bahar, A. R. Atilgan and B. Erman, *Fold. Des.*, 1997, **2**, 173.
108. C. A. Lopez, A. J. Rzepiela, A. H. de Vries, L. Dijkhuizen, P. H. Hunenberger and S. J. Marrink, *J. Chem. Theory Comput.*, 2009, **5**, 3195.
109. W. G. Noid, J. Chu, G. S. Ayton, V. Krishna, S. Izvekov, G. A. Voth, A. Das and H. C. Andersen, *J. Chem. Phys.*, 2008, **128**, 244114.
110. J. C. Shelley, M. Y. Shelley, R. Reeder, S. Bandyopadhyay and M. L. Klein, *J. Phys. Chem. B*, 2001, **105**, 4464.
111. H. A. Karimi-Varzaneh, H.-J. Qian, X. Chen, P. Carbone and F. Müller-Plathe, *J. Comp. Chem.*, 2011, **32**, 1475.
112. A. P. Lyubartsev and A. Laaksonen, *Phys. Rev. E*, 1995, **52**, 3730.
113. W. Tschöp, K. Kremer, O. Hahn, J. Batoulis and J. Bürger, *Acta Polymer*, 1998, **49**, 61.
114. M. S. Shell, *J. Chem. Phys.*, 2008, **129**, 144108.
115. S. P. Carmichael and M. S. Shell, *J. Phys. Chem. B*, 2012, **116**, 8383.
116. N. Metropolis, A. W. Rosenbluth, M. N. Rosenbluth, A. H. Teller and E. Teller, *J. Chem. Phys.*, 1953, **21**, 1087.
117. K. Binder, *Rep. Prog. Phys.*, 1997, **60**, 487.
118. K. Binder and D. Heermann, *Monte Carlo Simulation in Statistical Physics: An Introduction*, Springer, Berlin, 2010.
119. M. Feig, *Computational Modelling in Lignocellulosic Biofuel Production*, ACS Symposium Series, American Chemical Society, Washington DC, 2010, vol. 1052, Chapter 8, pp. 155–178.
120. M. Kastner, *Commun. Nonlinear Sci. Numer. Simulat*, 2010, **15**, 1589.

121. K. Kikuchi, M. Yoshida, T. Maekawa and H. Watanabe, *Chem. Phys. Lett.*, 1991, **185**, 335.
122. R. Jullien, *J. Phys. I France*, 1992, **2**, 759.
123. M. Lach-hab, A. E. Gonzalez and E. Blaisten-Barojas, *Phys. Rev. E.*, 1998, **57**, 4520.
124. T. Viscek, *Fractal Growth Phenomena*, World Scientific, Singapore, 2nd edn, 1999.
125. N. C. Acevedo and A. G. Marangoni, *Crystal Growth Des.*, 2010a, **10**, 3327.
126. N. C. Acevedo and A. G. Marangoni, *Crystal Growth Des.*, 2010b, **10**, 3334.
127. F. Peyronel, J. Ilavsky, G. Mazzanti, A. G. Marangoni and D. A. Pink, *J. Applied Phys.*, 2013, **114**, 234902.
128. H. C. Hamaker, *Physica*, 1937, **4**, 1058.
129. V. A. Parsegian, *Van Der Waals Forces: A Handbook for Biologists, Chemists, Engineers, and Physicists*, Cambridge University Press, Cambridge, 2006.
130. R. J. Hunter, *Foundations of Colloid Science*, Oxford University Press, Oxford, 2nd edn, 2000.
131. D. A. Pink, B. Quinn, F. Peyronel and A. G. Marangoni, *J. Appl. Phys.*, 2013, **114**, 234901.
132. K. Suzuki and T. Ando, *J. Biochem.*, 1972, **72**, 1433.
133. L. Truelstrup Hansen and T. A. Gill, *J. Appl. Microbiol.*, 2000, **88**, 1049.
134. N. M. D. Islam, T. Itakura and T. Motohiro, *Bull. J. Soc. Sci. Fish.*, 1984, **50**, 1705.
135. M. Uyttendaele and J. Debevere, *Food Microbiol.*, 1994, **11**, 417.
136. H. K. Tolong, *Mechanisms of Interaction of Cationic Antimicrobial Peptides with the Cytoplasmic Membranes of E. coli and S. typhimurium*, MSc Thesis, Dalhousie University, Department of Food Science and Technology, Dalhousie, 2004.
137. S. Morein, A.-S. Andersson, L. Rilfors and G. Lindblom, *J. Biol. Chem.*, 1996, **271**, 6801.
138. J. N. Israelachvili, *Intermolecular and Surface Forces*, Oxford Academic Press, Oxford, 2006.
139. C. Vlachy, *Ann. Rev. Phys. Chem.*, 1999, **50**, 145.
140. M. Deserno, C. Holm and S. May, *Macromolecules*, 2000, **33**, 199.
141. R. R. Netz, *Phys. Rev. E.*, 1999, **60**, 3174.
142. R. R. Netz and D. Andelman, *Phys. Rep.*, 2003, **380**, 1.
143. C. Fleck, R. R. Netz and H. H. von Grünberg, *Biophys. J.*, 2002, **82**, 76.

144. I. Carmesin and K. Kremer, *Macromolecules*, 1988, **21**, 2819.
145. D. A. Pink, F. M. Hasan, B. E. Quinn, M. Winterhalter, M. Mohan and T. A. Gill, *J. Pept. Sci.*, 2014, **20**, 240.
146. L. Y. Yang, Z. H. Pei, S. Fujimoto and M. J. Blaser, *J. Bacteriol.*, 1992, **174**, 1258.
147. J. Dworkin, M. K. Tummuru and M. J. Blaser, *J. Bacteriol.*, 1995, **177**, 1734.
148. A. P. Moran, D. T. O'Malley, T. U. Kosunen and I. M. Helander, *Infect. Immun.*, 1994, **62**, 3922.
149. G. I. Perez-Perez, M. J. Blaser and J. H. Bryner, *Infect. Immun.*, 1986, **51**, 209.
150. B. W. Brooks, M. M. Garcia, R. H. Robertson and H. Lior, *Vet. Microbiol.*, 1996, **51**, 105.
151. S. Fujimoto, A. Takade, K. Amako and M. J. Blaser, *Infect. Immun.*, 1989, **59**, 2017.
152. J. M. D. Roberts, L. L. Graham, B. E. Quinn and D. A. Pink, *Biochim. Biophys. Acta*, 2013, **1828**, 1143.
153. R. Gomperts, E. Renner and M. Mehta, *Am. Lab.*, 2005, **37**, 12.
154. U. Landman, *Proc. Natl. Acad. USA*, 2005, **102**, 6671.
155. A. F. Voter, Introduction to the Kinetic Monte Carlo Method, in *Radiation Effects in Solids*, ed. K. E. Sickafus, E. A. Kotomin and B. P. Uberuaga, NATO Science Series, Springer, Dordrecht, 235, 2004.
156. P. Kratzer, arXiv,0904.2556 *[cond-mat.mtrl-sci]*.
157. E. Sanz and D. Marenduzzo, *J. Chem. Phys.*, 2010, **132**, 194102.
158. S. Jabbari-Farouji and E. Trizac, *J. Chem. Phys.*, 2012, **137**, 054107.
159. R. Mezzenga and P. Fischer, *Rep. Prog. Phys.*, 2013, **76**, 046601.
160. K. Binder and W. Paul, *Macromolecules*, 2008, **41**, 4537.
161. D. R. Mason and A. P. Sutton, *Phil. Trans. R. Soc. A*, 2005, **363**, 1961.
162. T. Murtola, A. Bunker, I. Vattulainen, M. Deserno and M. Karttunen, *Phys. Chem. Phys.*, 2009, **11**, 1869.
163. K. Binder, *Molec. Phys.*, 2010, **108**, 1797.
164. H.-P. Hsu and P. Grassberger, *J. Stat. Phys*, 2011, **144**, 597.
165. A. Milchev, D. I. Dimitrov and K. Binder, *Polymer*, 2008, **49**, 3611.
166. A. Cuetos, J. A. Anta and A. M. Puertas, *J. Chem. Phys.*, 2010, **133**, 154906.
167. S. Elli, L. Eusebio, P. Gronchi, F. Ganazzoli and M. Goisis, *Langmuir*, 2010, **26**, 15814.
168. K. Binder, J. Baschnagel and W. Paul, *Prog. Polym. Sci.*, 2003, **28**, 115.

169. L. F. van Heijkamp, I. M. de Schepper, M. Strobl, R. H. Tromp, J. R. Heringa and W. G. Bouwman, *J. Phys. Chem. A*, 2010, **114**, 2412.

170. R. Ettelaie, *Curr. Opin. Colloid Interface Sci.*, 2003, **8**, 415.

171. S. Auer and D. Frenkel, *Anu. Rev. Phys. Chem.*, 2004, **55**, 333.

172. R. de Vries and M. Cohen Stuart, *Curr. Opin. Colloid Interface Sci.*, 2006, **11**, 295.

173. Z. Adamczyk, *Curr. Opin. Colloid Interface Sci.*, 2012, **17**, 173.

174. K. Binder, *Anu. Rev. Mater. Res.*, 2008, **38**, 123.

175. S. Granick, S. K. Kumar, E. J. Amis, M. Antonietti, A. C. Balazs, S. K. Chakraborty, G. S. Grest, C. Hawker, P. Janmey, E. J. Kramer, R. Nuzzo, T. P. Russell and C. R. Safinya, *J. Polymer Sci.: B: Polymer Phys.*, 2003, **41**, 2755.

176. A. Corsi, A. Milchev, V. G. Rostiashvili and T. A. Vilgis, *Food Hydrocoll.*, 2007, **21**, 870.

177. S. Kim, K.-S. Lee, M. R. Zachariah and D. Lee, *J. Colloid Interface Sci.*, 2010, **344**, 353.

178. L. C.-K. Liau and C.-Y. Lin, *Colloids Surf. A: Physicochem. Eng. Aspects*, 2011, **388**, 70.

179. A. G. Marangoni, N. Acevedo, F. Maleky, E. Co, F. Peyronel, G. Mazzanti, B. Quinn and D. A. Pink, *Soft Matter*, 2012, **8**, 1275.

180. H. Shinto, *Adv. Powder Technol.*, 2012, **23**, 538.

181. R. L. McGreevy, *J. Phys.: Condens. Matter*, 2001, **13**, R877.

182. T. Arai and R. L. McGreevy, *J. Phys.: Condens. Matter*, 2005, **17**, S23.

183. K. Hagita, H. Okamoto, T. Arai, H. Kishimoto, N. Umesaki, Y. Shinohara and Y. Amemiya, Development of extended reverse Monte Carlo method for analysis of 2D-USAXS experimental data, in *CP832, Flow Dynamics, The Second International Conference on Flow Dynamics*, ed. M. Tokuyama and S. Maruyama, AIP, 2006.

184. E. D. den Aantrekker, R. R. Beumer, S. J. C. van Gerwen, M. H. Zwietering, M. van Schothorst and R. M. Boom, *Int. J. Food Microbiol.*, 2003, **87**, 1.

185. F. Poschet, A. H. Geeraerd, N. Scheerlinck, B. M. Nicolai and J. F. Van Impe, *Food Microbiol.*, 2003, **20**, 285.

186. B. de Lauzon, J. L. Volatier and A. Martin, *Publ. Hlth Nutr.*, 2004, **7**, 893.

187. R. Montville and D. Schaffner, *Appl. Envir. Microbiol.*, 2005, **71**, 746.

188. F. Poschet, A. H. Geeraerd, A. M. Van Loey, M. E. Hendrickx and J. F. Van Impe, *J. Food Control*, 2005, **16**, 873.

189. C. Ferrer, W. Tejedor, G. Klein, D. Rodrigo, M. Rodrigo and A. Martınez, *Eur. Food Res. Technol.*, 2006, **224**, 153.
190. C. Ferrer, D. Rodrigo, M. C. Pina, G. Klein, M. Rodrigo and A. Martinez, *Food Control*, 2007, **18**, 934.
191. M. Regier, E. H. Hardy, K. Knoerzer, C. V. Leeb and H. P. Schuchmann, *J. Food Eng.*, 2007, **81**, 485.
192. T. P. Oscar, *J. Food Protection*, 2009, **72**, 2078.
193. J. Qin and R. Lu, *Comput. Electron. Agric.*, 2009, **68**, 44.
194. T. Murakami, K. Takakura and T. Yamano, *J. Nutr. Sci. Vitaminol.*, 2010, **56**, 449.
195. M. C. Pina-Perez, M. M. Garcia-Fernandez, D. Rodrigo and A. Martinez-Lopez, *Foodborne Path. Dis.*, 2010, **7**, 459.
196. W. Slob, W. J. de Boer and H. van der Voet, *Food Chem. Toxicol.*, 2010, **48**, 178.
197. Y. Nakatani, T. Satoh, S. Saito, M. Watanabe, N. Yoshiike, S. Kumagai and Y. Sugita-Konishi, *Food Add. Contam.*, 2011, **28**, 471.
198. F. Sampedro, D. Rodrigo and A. Martínez, *Food Control*, 2011, **22**, 420.
199. T. Gibaud, N. Mahmoudi, J. Oberdisse, P. Lindner, J. S. Pedersen, C. L. P. Oliveira, A. Stradner and P. Schurtenberger, *Faraday Discuss.*, 2012, **158**, 267.
200. C. Marsh, *Theoretical Aspects of Dissipative Particle Dynamics*, D. Phil. Thesis, University of Oxford, Oxford, UK, 1998.
201. E. G. Flekkøy, P. V. Coveney and G. De Fabritiis, *Phys. Rev. E.*, 2000, **62**, 2140.
202. P. Español and M. Revenga, *Phys. Rev. E.*, 2003, **67**, 026705.
203. U. D. Schiller, *Dissipative Particle Dynamics: A Study of the Methodological Background*, Diploma Thesis, Condensed Matter Theory Group, Faculty of Physics, University of Bielefeld, Bielefeld, 2005.
204. I. V. Pivkin and G. E. Karniadakis, *J. Chem. Phys.*, 2006, **124**, 184101.
205. R. Kubo, *Rep. Prog. Phys.*, 1966, **29**, 255.
206. I. V. Pivkin and G. E. Karniadakis, *J. Computat. Phys.*, 2005, **207**, 114.
207. J. Smiatek, ESPResSo-Summer-School 2012, available at: http://espressomd.org/html/ess2012/Day4/T1-01-Hydrodynamics/talk2-DPD.pdf (accessed July 2014).
208. M. Revenga, M. Zúñiga and P. Español, *Comput. Phys. Commun.*, 1999, **121**, 122.
209. M. Whittle and E. Dickinson, *J. Colloid Interface Sci.*, 2001, **242**, 106.
210. R. D. Groot, *J. Chem. Phys.*, 2003, **118**, 11265.

211. N. Ter-Oganessian, *Active microrheology of semiflexible polymer solutions: computer simulations and scaling theory*, PhD Thesis, Technische Universität München, Physik Department, Lehrstuhl für Biophysik, E22, 2004.
212. S. Y. Trofimov, E. L. F. Nies and M. A. J. Michels, *J. Chem. Phys.*, 2005, **123**, 144102.
213. M. González-Melchor, E. Mayoral, M. E. Velázquez and J. Alejandre, *J. Chem. Phys.*, 2006, **125**, 224107.
214. A. F. Jakobsen, G. Besold and O. G. Mouritsen, *J. Chem. Phys.*, 2006, **124**, 094104.
215. V. Symeonidis, G. E. Karniadakis and B. Caswell, *J. Chem. Phys*, 2006, **125**, 184902.
216. W. Jiang, J. Huang, Y. Wang and M. Laradji, *J. Chem. Phys.*, 2007, **126**, 044901.
217. D. A. Fedosov, I. V. Pivkin and G. E. Karniadakis, *J. Computat. Phys.*, 2008, **227**, 2540.
218. A. Eriksson, M. N. Jacobi, J. Nystrom and K. Tunstrøm, *Europhys. Lett.*, 2009, **86**, 44001.
219. R. M. Füchslin, H. Fellermann, A. Eriksson and H.-J. Ziock, *J. Chem. Phys.*, 2009, **130**, 214102.
220. S. Litvinov, M. Ellero, X. Hu and N. A. Adams, *J. Chem. Phys.*, 2009, **130**, 021101.
221. K.-C. Huang, C.-M. Lin, H.-K. Tsao and Y.-J. Sheng, *J. Chem. Phys*, 2009, **130**, 24510.
222. A. Vázquez-Quesada, M. Ellero and P. Español, *J. Chem. Phys.*, 2009, **130**, 034901.
223. D. A. Fedosov, G. E. Karniadakis and B. Caswell, *J. Chem. Phys.*, 2010, **132**, 144103.
224. Z. Zhang and H. Guo, *J. Chem. Phys*, 2010, **133**, 144911.
225. K. Binder, T. Kreer and A. Milchev, *Soft Matter*, 2011, 7, 7159.
226. M. Lísal, J. K. Brennan and J. B. Avalos, *J. Chem. Phys.*, 2011, **135**, 204105.
227. M. Yamanoi, O. Pozo and J. M. Maia, *J. Chem. Phys.*, 2011, **135**, 044904.
228. S. Khani, M. Yamanoi and J. Maia, *J. Chem. Phys.*, 2013, **138**, 174903.
229. P. M. Kulkarni, C.-C. Fu, M. S. Shell and L. G. Leal, *J. Chem. Phys.*, 2013, **138**, 234105.
230. M. Langeloth, Y. Masubuchi, M. C. Böhm and F. Müller-Plathe, *J. Chem. Phys.*, 2013, **138**, 104907.
231. P. B. Warren, A. Vlasov, L. Anton and A. J. Masters, *J. Chem. Phys*, 2013, **138**, 204907.

232. N. Ter-Oganessian, B. Quinn, D. A. Pink and A. Boulbitch, *Phys. Rev. E*, 2005, **72**, 041510.
233. N. Ter-Oganessian, D. A. Pink and A. Boulbitch, *Phys. Rev. E*, 2005, **72**, 041511.
234. J. Uhde, N. Ter-Oganessian, D. A. Pink, E. Sackman and A. Boulbitch, *Phys. Rev, E*, 2005, **72**, 061916.
235. S. Succi, *The Lattice Boltzmann Equation: for Fluid Dynamics and Beyond*, Oxford University Press, Oxford, UK, 2001.
236. D. Raabc, *Model Simul. Mater. Sci. Eng.*, 2004, **12**, 13.
237. T. N. Phillips and G. W. Roberts, *J. Appl. Math.*, 2011, **76**, 790.
238. A. K. Datta, R. van der Sman, T. Gulati and A. Warning, *Faraday Discuss.*, 2012, **158**, 435.
239. D. B. Chirila, *Introduction to Lattice Boltzmann Methods*, 2010, available at: http://www.awi.de/fileadmin/user_upload/Research/Research_Divisions/Climate_Sciences/Paleoclimate_Dynamics/Modelling/Lessons/Einf_Ozeanographie/lecture_19_Jan_2010.pdf (accessed July 2014).
240. C. Peng, *The Lattice Boltzmann Method for Fluid Dynamics: Theory and Applications*, Department of Mathematics, Ecole Polytechnique Federale de Lausanne, 2010.
241. A. T. White and C. K. Chong, *J. Computat. Phys.*, 2011, **230**, 6367.
242. P. Lallemand and L.-S. Luo, *J. Computat. Phys.*, 2003, **184**, 406.
243. C. Pan, M. Hilpert and C. T. Miller, *Water Resour. Res.*, 2004, **40**, W01501.
244. J. Harting, H. J. Herrmann and E. Ben-Naim, *Europhys. Lett.*, 2008, **83**, 30001.
245. J. Kromkamp, D. van den Ende, D. Kandhai, R. van der Sman and R. Boom, *Chem. Eng. Sci.*, 2006, **61**, 858.
246. R. G. M. van der Sman, *Comput. Fluid.*, 2006, **35**, 849.
247. R. G. M. van der Sman and A. Graaf, *Rheol. Acta*, 2006, **46**, 3.
248. P. Yuan and L. Schaefer, *J. Fluid Eng.*, 2006, **128**, 142.
249. J. J. Derksen and H. E. A. Van Den Akker, *Chem. Eng. Res. Des.*, 2007, **85**, 697.
250. J. Harting, J. Chin, M. Venturoli and P. V. Coveney, *Phil. Trans. R. Soc. A*, 2005, **363**, 1895.
251. L. Velazquez-Ortega and S. Rodrıguez-Romo, *Can. J. Chem. Eng.*, 2008, **86**, 667.
252. Y. Y. Yan and Y. Q. Zu, *Int. J. Heat Mass Transfer*, 2008, **51**, 2519.
253. F. Jansen and J. Harting, *Phys. Rev. E*, 2011, **83**, 046707.
254. J. J. Derksen, *Phys. Fluid*, 2009, **21**, 083302.
255. J. J. Derksen and R. W. Prashant, *J. Non-Newtonian Fluid Mech.*, 2009, **160**, 65.

256. E. van der Zwan, R. van der Sman, K. Schroën and R. Boom, *J. Colloid Interface Sci.*, 2009, **335**, 112.
257. M. A. Hussein and T. Becker, *Food Biophys.*, 2010, **5**, 161.
258. J. Lee, V. Dünweg and J. Schumacher, *Comput. Math. Apps.*, 2010, **59**, 2374.
259. R. G. M. van der Sman, *Comput. Phys. Commun.*, 2010, **181**, 1562.
260. Z. Chai, B. Shi, Z. Guob and F. Rong, *J. Non-Newtonian Fluid Mech.*, 2011, **166**, 332.
261. S. Kondaraju, H. Farhat and J. S. Lee, *Soft Matter*, 2012, **8**, 1374.
262. H. Farhat, F. Celiker, T. Singh and J. S. Lee, *Soft Matter*, 2011, **7**, 1968.
263. H. M. Vollebregt, R. G. M. van der Sman and R. M. Boom, *Faraday Discuss.*, 2012, **158**, 80.
264. T. Kruger, S. Frijters, F. Gunther, B. Kaoui and J. Harting, *Eur. Phys. J. Special Topics*, 2013, **222**, 177.
265. F. Günther, F. Janoschek, S. Frijters and J. Harting, *Comput. Fluid*, 2013, **80**, 184.
266. E. Schlauch, M. Ernst, R. Seto, H. Briesen, M. Sommerfeld and M. Behr, *Comput. Fluis*, 2013, **86**, 199.
267. V. Stobiac, P. A. Tanguy and F. Bertrand, *Comput. Fluid*, 2013, **73**, 145.

Subject Index

Page numbers in *italics* indicate figures or tables